EXACT SOLUTIONS AND SCALAR FIELDS IN GRAVITY

EXACT SOLUTIONS AND SCALAR FIELDS IN GRAVITY

RECENT DEVELOPMENTS

Edited by

ALFREDO MACIAS
Universidad Autónoma Metropolitana–Itzapalapa
Mexico City, Mexico

JORGE L. CERVANTES-COTA
Institute Nacional de Investigaciones Nucleares
Mexico City, Mexico

CLAUS LÄMMERZAHL
University of Düsseldorf
Düsseldorf, Germany

Springer Science+Business Media, LLC

Library of Congress Cataloging-in-Publication Data

Exact solutions and scalar fields in gravity: recent developments/edited by Alfredo Macias, Jorge L. Cervantes-Cota, and Claus Lämmerzal.

p. cm.

Includes bibliographical references and index.

ISBN 978-1-4757-8200-4 ISBN 978-0-306-47115-5 (eBook)
DOI 10.1007/978-0-306-47115-5

1. Gravitation—Congresses. 2. Relativity (Physics)u—Congresses. 3. Scalar field theory—Congresses. I. Macías, A. (Alfredo) II. Cervantes Cota, Jorge Luis. III. Lämmerzahl, C. (Claus), 1956- IV. Mexican Meeting on Exact Solutions and Scalar Fields in Gravity (2000: Mexico, City)

QC178 .E94 2001
530.11—dc21

2001029948

Proceedings of the Mexican Meeting on Exact Solutions and Scalar Fields in Gravity, help October 2–6, 2000, in Mexico City, Mexico

ISBN 978-1-4757-8200-4

Originally published by Kluwer Academic / Plenum Publishers, New York in 2001
Softcover reprint of the hardcover 1st edition 2001

http://www.wkap.nl/

10 9 8 7 6 5 4 3 2 1

A C.I.P. record for this book is available from the Library of Congress

To the 65th birthday of Heinz Dehnen
and to the 60th birthday of Dietrich Kramer

CONTENTS

PREFACE

This book is dedicated to honour Professor Heinz Dehnen and Professor Dietrich Kramer on occasion of their 65th and 60th birthday, respectively. Their devotion to science and high level of energy and enthusiasm that they bring to their research on gravitation are much appreciated by their students and collaborators.

Friends, former students and colleagues of Heinz Dehnen and Dietrich Kramer came together in a meeting to celebrate and to express them their respect, admiration, and affection. The meeting took place in Mexico, since both Professors Dehnen and Kramer have contributed significantly to the development of different gravity groups there, as it is apparent in the brief biographies written below. Participants came to celebrate Heinz Dehnen and Dietrich Kramer's scientific and scholarly achievements, the inspirational quality of their teaching, their thoughtful guidance of graduated and ungraduated students, their graciousness as a colleague, their service to the gravity community.

Indeed, the papers in this volume are mainly based on the proceedings of the Mexican Meeting on Exact Solutions and Scalar Fields in Gravity that was held in honour of Heinz Dehnen and Dietrich Kramer. The conference took place on October 1st to 6th, 2000 at Centro de Investigación y de Estudios Avanzados (CINVESTAV) of the National Polytechnical Institute (IPN) in Mexico City. It was organized by Alfredo Macías, Jorge Cervantes–Cota, Claus Lämmerzahl, Tonatiuh Matos and Hernando Quevedo; their work contributed significantly to the success of the Meeting. We are specially grateful to Prof. Alberto García, Dean of the Physics Department, for the warm hospitality that was extended to all participants at CINVESTAV.

We wish to thank Universidad Autónoma Metropolitana–Iztapalapa and CINVESTAV–IPN for sponsoring this international endeavour.

The financial support of the Germany–Mexico exchange program of the DLR (Bonn)–CONACYT (Mexico City), ININ, and ICN–UNAM is gratefully acknowledged.

ALFREDO MACÍAS, JORGE L. CERVANTES-COTA, AND CLAUS LÄMMERZAHL

CONTRIBUTING AUTHORS

Eloy Ayón–Beato
Departamento de Física
Centro de Investigación y de Estudios Avanzados del IPN
Apdo. Postal 14–740, C.P. 07000, México, D.F., México.
E–mail: ayon@fis.cinvestav.mx

Alexander B. Balakin
Kazan State University, 420008 Kazan, Russia.

Abel Camacho
Departamento de Física,
Instituto Nacional de Investigaciones Nucleares.
Apartado Postal 18–1027, México, D. F., México.
E–mail: acamacho@nuclear.inin.mx

Jorge L. Cervantes–Cota
Departamento de Física,
Instituto Nacional de Investigaciones Nucleares (ININ),
P.O. Box 18–1027, México D.F. 11801, México.
E–mail:jorge@nuclear.inin.mx

Pablo Chauvet–Alducin
Departamento de Física , Universidad Autónoma Metropolitana–Iztapalapa,
P. O. Box. 55–534, México D. F., C.P. 09340 México.
E–mail: pcha@xanum.uam.mx

Francisco J. Chinea
Dept. de Física Teórica II, Ciencias Físicas,
Universidad Complutense de Madrid
E-28040 Madrid, Spain.
E–mail: chinea@eucmos.sim.ucm.es

Heinz Dehnen
Physics Department. University of Konstanz,
Box M 677. D–78457 Konstanz. Germany.
E–mail: Heinz.Dehnen@uni-konstanz.de

Frederick J. Ernst
FJE Enterprises. Potsdam, New York 13676, USA.
E–mail: gravity@slic.com, http://pages.slic.com/gravity

S. Formański
Institute of Physics, Technical University of Łódź,
Wólczańska 219. 93-005 Łódź, Poland.
E–mail: sforman@ck-sg.p.lodz.pl

Alberto García
Departamento de Física, CINVESTAV–IPN,
Apartado Postal 14–740, C.P. 07000, México, D.F., México.
E–mail: aagarcia@fis.cinvestav.mx

Hector García–Compeán
Departamento de Física, Centro de Investigación y Estudios Avanzados del IPN
P.O. Box 14-740, 07000, México D.F., México.
E–mail: compean@fis.cinvestav.mx

Francisco Siddhartha Guzmán
Departamento de Física, Centro de Investigación y de Estudios Avanzados del IPN,
AP 14–740, 07000 México D.F., México.
E–mail: siddh@fis.cinvestav.mx

V. D. Ivashchuk
Center for Gravitation and Fundamental Metrology, VNIIMS,
3-1 M. Ulyanovoy Str., Moscow, 117313, Russia.
Institute of Gravitation and Cosmology, Peoples' Friendship University,
6 Miklukho–Maklaya St., Moscow 117198, Russia.
E–mail: ivas@rgs.phys.msu.su

Markus King
Institute of Theoretical Physics
University of Tübingen, D–72076 Tübingen, Germany.
E–mail: markus.king@uni-tuebingen.de

Jaime Klapp
Instituto Nacional de Investigaciones Nucleares,
Ocoyoacac 52045, Estado de México, México.
E–mail: klapp@nuclear.inin.mx

Andreas Kleinwächter
Friedrich–Schiller–Universität Jena,
Theoretisch–Physikalisches–Institut,
Max–Wien–Platz 1, D–07743 Jena, Germany.
E-mail: kleinwaechter@tpi.uni-jena.de

Dietrich Kramer
Friedrich–Schiller–Universität Jena,
Theoretisch–Physikalisches–Institut,
Max–Wien–Platz 1, D–07743 Jena, Germany.
Dietrich.Kramer@tpi.uni-jena.de

Claus Lämmerzahl
Institut für Experimentalphysik
Heinrich–Heine–Universität Duesseldorf,
Universitaetsstrasse 1, Gebaeude 25.42, Raum O1.44 40225 Duesseldorf, Germany
E–mail: claus.laemmerzahl@uni-duesseldorf.de

Alfredo Macías
Departamento de Física, Universidad Autónoma Metropolitana–Iztapalapa,
Apartado Postal 55–534, C.P. 09340, México, D.F., México.
E–mail: amac@xanum.uam.mx

Vladimir S. Manko
Departamento de Física,
Centro de Investigación y de Estudios Avanzados del IPN,
A.P. 14-740, 07000 México D.F., México.
E–mail: vsmanko@fis.cinvestav.mx

Tonatiuh Matos
Departamento de Física, Centro de Investigación y de Estudios Avanzados del IPN,
AP 14–740, 07000 México D.F., México.
E–mail: tmatos@fis.cinvestav.mx

Reinhard Meinel
Friedrich-Schiller-Universität Jena,
Theoretisch-Physikalisches-Institut,
Max-Wien-Platz 1, D-07743 Jena, Germany.
E-mail: meinel@tpi.uni-jena.de

Vitaly N. Melnikov
Visiting Professor at Depto.de Fisica, CINVESTAV,
A.P.14-740, México 07000, D.F., México.
Center for Gravitation and Fundamental Metrology, VNIIMS,
3-1 M. Ulyanovoy Str., Moscow, 117313, Russia.
Institute of Gravitation and Cosmology, Peoples' Friendship University,
6 Miklukho-Maklaya St., Moscow 117198, Russia E-mail: melnikov@fis.cinvestav.mx

Eckehard W. Mielke
Departamento de Física, Universidad Autónoma Metropolitana-Iztapalapa,
Apartado Postal 55-534, C.P. 09340, México, D.F., México.
E-mail: ekke@xanum.uam.mx

Nikolai V. Mitskievich
Physics Department, CUCEI, University of Guadalajara, Guadalajara, Jalisco, México.
E-mail: nmitskie@udgserv.cencar.udg.mx

Cesar Mora
Departamento de Física, Universidad Autónoma Metropolitana, Iztapalapa
P.O. Box 55-534, CP 09340, México D.F., México
Departamento de Física, UPIBI-Instituto Politécnico Nacional,
Av. Acueducto s/n Col. Barrio La Laguna Ticomán,
CP 07340 México D.F., México.
E-mail: ceml@xanum.uam.mx

Marcos Nahmad
Departamento de Física,
Instituto Nacional de Investigaciones Nucleares (ININ),
P.O. Box 18-1027, México D.F. 11801, México.

Darío Núñez
Instituto de Ciencias Nucleares,
Universidad Nacional Autónoma de México,

A.P. 70-543, 04510 México D.F., México.
E–mail: nunez@fis.cinvestav.mx

Octavio Obregón
Instituto de Física de la Universidad de Guanajuato,
P.O. Box E–143, 37150, León Gto., México.
E–mail: octavio@ifug3.ugto.mx

Herbert Pfister
Institute of Theoretical Physics,
University of Tübingen, D–72076 Tübingen, Germany.
E-mail: herbert.pfister@uni-tuebingen.de

Luis O. Pimentel
Departamento de Física, Universidad Autónoma Metropolitana, Iztapalapa
P.O. Box 55–534, CP 09340, México D. F., México.
E-mail: lopr@xanum.uam.mx

Jerzy F. Plebański
Departamento de Fisica, CINVESTAV, 07000 Mexico D.F., México.
E–mail: pleban@fis.cinvestav.mx

Maciej Przanowski
Departamento de Fisica, CINVESTAV, 07000 Mexico D.F., Mexico.
Institute of Physics, Technical University of Lódź,
Wólczańska 219. 93-005 Lódź, Poland.
E–mail: przan@fis.cinvestav.mx

Dork Putzfeld
Insitute for Theoretical Physics, University of Cologne, D–50923 Köln, Germany.
E–mail: dp@thp.uni-koeln.de,

Hernando Quevedo
Instituto de Ciencias Nucleares,
Universidad Nacional Autónoma de México,
A.P. 70-543, 04510 México D.F., México.
E–mail: quevedo@nuclecu.unam.mx

Cupatitzio Ramírez
Facultad de Ciencias Físico Matemáticas,
Universidad Autónoma de Puebla,
P.O. Box 1364, 72000, Puebla, México.
E–mail: cramirez@fcfm.buap.mx

Mario A. Rodríguez–Meza
Departamento de Física,
Instituto Nacional de Investigaciones Nucleares (ININ),
P.O. Box 18-1027, México D.F. 11801, México.
E–mail: mar@nuclear.inin.mx

Eduardo Ruiz
Area de Física Teórica, Universidad de Salamanca,
37008 Salamanca, Spain

Miguel Sabido
Instituto de Física de la Universidad de Guanajuato,
P.O. Box E–143, 37150, León Gto., México.
E–mail: msabido@ifug3.ugto.mx

Franz E. Schunck
Institut für Theoretische Physik, Universität zu Köln,
50923 Köln, Germany.
E–mail: fs@thp.uni-koeln.de

Dominik J. Schwarz
Institut für Theoretische Physik, TU-Wien,
Wiedner Hauptstrass e 8–10, A–1040 Wien, Austria.

Leonardo Di G. Sigalotti
Instituto Venezolano de Investigaciones Científicas (IVIC)
Carretera Panamericana Km. 11, Altos de Pipe, Estado Miranda, Venezuela.

César A. Terrero–Escalante
Departamento de Física,
Centro de Investigación y de Estudios Avanzados del IPN,
Apdo. Postal 14–740, 07000, México, D.F., México.

L. Arturo Ureña–López
Departamento de Física, Centro de Investigación y de Estudios Avanzados del IPN, AP 14–740, 07000 México D.F., México.

Winfried Zimdahl
Universität Konstanz, PF M678 D-78457 Konstanz, Germany.
E–mail: zindahl@spock.physik.uni-konstanz.de

I

EXACT SOLUTIONS

SELF–GRAVITATING STATIONARY AXI-SYMMETRIC PERFECT FLUIDS: DIFFERENTIAL ROTATION AND SOME GEOMETRIC FEATURES

F.J. Chinea*
Dept. de Física Teórica II, Ciencias Físicas,
Universidad Complutense de Madrid
E-28040 Madrid, Spain

Keywords: Perfect fluids, differential rotations, Newtonian configurations.

1. NEWTONIAN CONFIGURATIONS

The equations for the interior of a self-gravitating barotropic perfect fluid in the classical (Newtonian) theory in an inertial frame are the following:

$$\frac{\partial \rho}{\partial t} + \nabla \cdot (\rho \mathbf{v}) = 0 \tag{1}$$

$$\frac{\partial \mathbf{v}}{\partial t} + (\mathbf{v} \cdot \nabla)\mathbf{v} = -\frac{1}{\rho}\nabla p - \nabla U \tag{2}$$

$$\Delta U = 4\pi G \rho \tag{3}$$

$$\mathcal{F}(\rho, p) = 0, \tag{4}$$

where ρ is the mass density, p the pressure, $\mathbf{v}(x, y, z, t)$ the velocity field of the fluid, and U the gravitational potential; all these functions may depend in principle on the space coordinates, $\mathbf{x} = (x, y, z)$, and on time t. G is the gravitational constant (in the following, $G = 1$). Equation (1) is the continuity equation, (2) is the Euler equation for the fluid, (3) the Poisson equation for the gravitational potential U, and (4) the barotropic equation of state. Some closed-form solutions of fluid configurations

*E–mail: chinea@eucmos.sim.ucm.es

Exact Solutions and Scalar Fields in Gravity: Recent Developments
Edited by Macias *et al.*, Kluwer Academic/Plenum Publishers, New York, 2001

with a compact boundary are the Maclaurin ellipsoids in the stationary axisymmetric case, and the Jacobi, Dedekind, and Riemann ellipsoids in the (in general) non-axisymmetric case [1, 2]. It is remarkable that the Jacobi ellipsoids, triaxial figures rotating around a principal axis as a whole (rigidly) with constant angular velocity, can be considered formally as stationary, due to the fact that the equations (1)-(3) can be transformed to the rotating frame anchored to the principal axes so that t does not appear explicitly; obviously, the analog of such non-axisymmetric configurations in general relativity (whatever they might be) would not be stationary, as they would emit gravitational radiation. Another interesting stationary axisymmetric configuration is a torus-like figure rotating around its symmetry axis. It was shown to exist (among other highly interesting results) in a very important paper of Poincaré [3], and was further analyzed in [4] by means of power series expansions.

In what follows, we shall concentrate on the stationary, axisymmetric case with azimuthal flow of equations (1)-(4); the z-axis is taken to be the symmetry axis. By imposing $\mathcal{L}_\xi \mathbf{v} = 0$ and $\mathcal{L}_\eta \mathbf{v} = 0$ (where $\mathcal{L}_\mathbf{w}$ denotes the Lie derivative in the direction of the vectorfield $\mathbf{w}$, $\xi = \partial_t$ and $\eta = \partial_\varphi$, φ being the azimuthal angle), one finds $\mathbf{v}(x, y, z, t) = \Omega(x^2 + y^2, z)\,(-y, x, 0)$, where $\Omega = \Omega(x^2 + y^2, z)$ is the angular velocity, depending in general on the position within the fluid (possible differential rotation). Taking this into account, one has $\frac{\partial \mathbf{v}}{\partial t} = 0$ and $(\mathbf{v} \cdot \nabla)\mathbf{v} = -\Omega^2\,(x, y, 0)$. From these and the integrability conditions for the set of three scalar equations arising from (2), one gets $\Omega_{,z} = 0$ (taking into account that ρ and p are functionally dependent, according to (4)), so that the angular velocity depends only on the distance to the rotation axis: $\Omega = \Omega(\sigma)$, where we have defined $\sigma \equiv x^2 + y^2$. It is easy to see that the continuity equation (1) is automatically satisfied under the assumed symmetries (which also require $\mathcal{L}_\xi \rho = \mathcal{L}_\eta \rho = 0$), and, in particular, the fluid flow has vanishing expansion. Equations (1)-(3) now reduce to the following ones:

$$\Omega^2(\sigma)\, x = \frac{1}{\rho} p_{,x} + U_{,x} \tag{5}$$

$$\Omega^2(\sigma)\, y = \frac{1}{\rho} p_{,y} + U_{,y} \tag{6}$$

$$0 = \frac{1}{\rho} p_{,z} + U_{,z}, \tag{7}$$

$$\Delta U = 4\pi\rho. \tag{8}$$

Note that equations (5)-(7) are equivalent to their following first integral:

$$U + \int_0^p \frac{dp'}{\rho(p')} - \frac{1}{2}\int_{\sigma_0}^{\sigma} \Omega^2(\sigma')\, d\sigma' + k = 0, \tag{9}$$

where σ_0 and k are constants.

The *rigidly rotating* case is characterized by vanishing shear and expansion. Such conditions reduce here to $v_{i,j} + v_{j,i} = 0$, which in the present case is equivalent to $\Omega' = 0$, where a prime denotes differentiation with respect to σ. In other words, we have $\Omega = \Omega_0 = \text{const.}$ for rigid rotation. Another interesting case is that of an *irrotational* rotation ($\nabla \times \mathbf{v} = 0$): $0 = v_{x,y} - v_{y,x} = -2\Omega - 2\sigma\Omega$; hence, $\Omega = \frac{\Omega_0}{x^2+y^2}$, where Ω_0 is a constant. In the irrotational case, $\| \mathbf{v} \| = \frac{|\Omega_0|}{\sqrt{x^2+y^2}}$, which blows up at the axis of rotation; however, one should bear in mind that regular toroidal-like irrotational configurations may be possible, as the rotation axis is outside the fluid in that case.

For positivity reasons, not all velocity profiles are allowed for a particular equation of state; for instance, in the case of *dust* (pressureless fluid), we have from (9): $\Delta U = 2\Omega^2 + 4(x^2+y^2)\Omega\Omega'$. By substituting this value of ΔU in (8), we obtain $2\pi\rho = \Omega^2 + 2(x^2+y^2)\Omega\Omega'$, which excludes differential rotations with $(\Omega^2)' < -\frac{\Omega^2}{\sigma}$, if the natural requirement $\rho > 0$ is made. In particular, an irrotational dust is not allowed.

Finally, let us say a word about the matching conditions. Outside the fluid, assuming a surrounding vacuum ($\rho = p = 0$, $\mathbf{v} = 0$), the equations to be solved reduce to

$$\Delta U = 0, \tag{10}$$

for the gravitational potential U. The matching conditions are derived from the physical requirement that the body forces (in the present case, the gravitational force) be continuous across the boundary:

$$(U_{\text{ext}} - U_{\text{int}})_{|\text{boundary}} = 0, \tag{11}$$

$$(\nabla U_{\text{ext}} - \nabla U_{\text{int}})_{|\text{boundary}} = 0. \tag{12}$$

The surface forces (let us call them $\mathbf{t}$) should also be continuous across the boundary. If $t_j = n^i T_{ij}$, where $\mathbf{n}$ is the normal to the boundary and T_{ij} the stress tensor, the following has to hold:

$$n^i (T_{ij(\text{ext})} - T_{ij(\text{int})})_{|\text{boundary}} = 0. \tag{13}$$

As $T_{ij(\text{ext})} = 0$ for vacuum, and $T_{ij(\text{int})} = p\delta_{ij}$ for a perfect fluid, the three conditions (13) reduce in the perfect fluid case to the single condition

$$p = 0. \tag{14}$$

The analytical expression for the boundary in the stationary axisymmetric case with azimuthal flow is thus $p(x^2 + y^2, z) = 0$, and on that surface (11) and (12) should hold. Obviously, the *dust* case is special, in that (14) does not characterize its boundary (14) is identically satisfied throughout space!); in that case, the boundary is given by the surface(s) where (11) and (12) are mutually consistent.

2. DIRICHLET BOUNDARY PROBLEM VS. FREE-BOUNDARY PROBLEM

Part of the technical difficulties associated with the problem of finding equilibrium configurations for self-gravitating fluids is due to the overdetermined character of conditions (11) and (12); in fact, the boundary is determined by the consistency of (11), (12), and (14). The classical equilibrium figures we have mentioned in the previous section were all obtained by means of Ansätze that turn out to be consistent (an important role is played by the fact that the Newtonian potential for an ellipsoidal body can be obtained in closed form in the case of constant density objects, and the resulting potential does not depend on the motion of the body, *i.e.* there are no Newtonian gravitomagnetic effects.) In that sense, the free-boundary problem is not really solved *ab initio* for those examples, but happy coincidences and insights are exploited.

In order to illustrate the fact that wild guesses are doomed to failure, let us consider the following naive attempt at getting a rigidly rotating configuration for a homogeneous (*i.e.* constant density) body. The integral (9) gives

$$U = \frac{1}{2}\Omega_0^2 (x^2 + y^2) - \frac{p}{\rho} + k, \tag{15}$$

and substituting (15) in (8), we get $\Delta p = 2\rho(\Omega_0^2 - 2\pi\rho)$. Assume now the (intuitively wrong!) Ansatz $p = p(r)$, with $r \equiv \sqrt{x^2 + y^2 + z^2}$. By integrating the previous equation for Δp, we find $p = \frac{1}{3}\rho(\Omega_0^2 - 2\pi\rho)r^2 - 2k_1 r^{-1} + k_2$, where k_1 and k_2 are constants; if we set $k_1 = 0$ for regularity at $r = 0$, the pressure can be expressed as $p = \frac{1}{3}\rho(\Omega_0^2 - 2\pi\rho)r^2 + p_c$, where p_c denotes the pressure at $r = 0$. By substituting p back in (15), we get $U = U_0 - p_c/\rho + (\Omega_0^2/6 + 2\pi\rho/3)r^2 - \Omega_0^2 z^2/2$, where U_0 is a constant. Let $r = r_0$ be the surface $p = 0$. Then, $r_0^2 = 3p_c\rho^{-1}(2\pi\rho - \Omega_0^2)^{-1}$, and the boundary value of U is

$$U_{|r=r_0} = U_0 + \frac{p_c\Omega_0^2}{\rho(2\pi\rho - \Omega_0^2)}(1 - P_2(\cos\theta)), \tag{16}$$

where P_2 is the appropriate Legendre polynomial, and θ the usual spherical coordinate. The exterior gravitational potential satisfying (10) and condition (11) for the boundary value (16) is easily found to be

$$U_{\rm ext} = [U_0 + \frac{p_c\Omega_0^2}{\rho(2\pi\rho - \Omega_0^2)}]\frac{r_0}{r} - \frac{p_c\Omega_0^2 r_0^3}{\rho(2\pi\rho - \Omega_0^2)}\frac{P_2(\cos\theta)}{r^3}; \tag{17}$$

but (as was to be expected from the obviously inadequate Ansatz), (12) is not satisfied:

$$\nabla U \neq \nabla U_{\rm ext} \tag{18}$$

on $r = r_0$, unless one of the following holds:

(a) $p_c = 0$ (giving $r_0 = 0$, *i.e.* no fluid), or

(b) $\Omega_0 = 0$ (static case).

Ignoring these two degenerate cases, there will be a discontinuity in the gravitational force, expressed by (18); it would be interpreted as the appearance of a surface layer at the fluid boundary, which artificially constrains the rotating body to keep the spherical shape we had assumed. By making virtue out of necessity, a similar behavior in the general relativistic case is sometimes "explained" *a posteriori* by an interpretation in terms of crusts, etc. But if we adopt a more conservative point of view (namely, if we stick to the perfect fluid case we started out with) such a situation is clearly undesirable and signals a failure in satisfying all the conditions inherent to the problem. All this is a consequence of the fact that we are not dealing with a Dirichlet-type problem (nor even with a Cauchy-type one!), but with a *free-boundary* problem. Unfortunately, the panoply of existing mathematical techniques for an exact treatment of such problems is yet quite limited.

Finally, let us mention that various interesting rigorous results for Newtonian configurations have been proven over the years. For lack of space, we refrain from discussing them here and giving an exhaustive list of references; please see [5] for further references, including those to deep mathematical results that have appeared since the classic monograph [6] was published.

3. INTERIOR SOLUTIONS WITH DIFFERENTIAL ROTATION IN GENERAL RELATIVITY

Clearly, all the difficulties associated with an exact treatment of rotating perfect fluid configurations in the Newtonian approach are tremendously increased in general relativity. As a matter of fact, it was only in 1968 that H.D. Wahlquist found for the first time an exact description of the interior of a particular general relativistic axisymmetric perfect fluid for the special case of stationary rigid rotation (vanishing shear of the fluid) [7]. Several other particular interior solutions for barotropic fluids rotating rigidly have been subsequently found. Unfortunately, none of them has been shown so far to possess a corresponding asymptotically flat exterior solution, in such a way that the overdetermined matching conditions (which in general relativity are the continuity of the metric and of its first derivatives across the boundary of the object) are satisfied. (Obviously, we are disregarding stationary axisymmetric configurations with extra Killing vectors, such as the case of cylindrical symmetry, which is irrelevant for astrophysical objects.)

The first stationary axisymmetric interior solution in general relativity for a barotropic perfect fluid with *differential* (*i.e.* non-rigid) rotation and only two Killing vectors appeared in 1990 [8]; the case considered there was one of vanishing vorticity (irrotational case). Let me now describe briefly the formalism we developed for finding that solution (the formalism itself was published later in detail in a separate, larger paper [9]).

An orthonormal tetrad of 1-forms, $\{\theta^0, \theta^1, \theta^2, \theta^3\}$ is introduced, with the choice $\theta^0 = u$ (where u is the velocity 1-form of the fluid). The presence of two commuting Killing fields $\xi^\dagger$ and $\eta^\dagger$ such that they can be identified with $\xi^\dagger = \partial_t$ and $\eta^\dagger = \partial_\varphi$ is assumed (where t is a time coordinate and φ a cyclic one, and $\xi^\dagger$ denotes the contravariant expression corresponding to the 1-form ξ, etc.). The flow is assumed to be azimuthal: $u \wedge \xi \wedge \eta = 0$. If we are interested in non-trivial situations, we have to make sure (perhaps *a posteriori*, after some solution is found) that we are not dealing with a static case, by checking that there does *not* exist a timelike Killing vector $\zeta^\dagger$ such that $\zeta \wedge d\zeta = 0$. For the reasons given above, we find it preferable that there be no additional independent Killing vectors (especially if the extra Killing field commutes with $\xi^\dagger$ and $\eta^\dagger$, which could be interpreted as giving rise to a cylindrical or some such physically undesirable symmetry). By a standard reduction, $\theta^0 = u$ and θ^1 are chosen as lying in the $\{\xi, \eta\}$ subspace. Due to the symmetries, the coordinates t and φ are ignorable, and the coefficients of the tetrad 1-forms (and of all the differential forms that appear in the formulation) may only depend on two additional coordinates. The energy-momentum tensor is taken to be $T_{\alpha\beta} = (\mu + p)u_\alpha u_\beta + p g_{\alpha\beta}$, where $g_{\alpha\beta}$ is the spacetime metric, u_α the velocity 1-form, p the pressure, and μ the energy density of the fluid. Hodge duality in the $\{\theta^2, \theta^3\}$ subspace is defined by $*\theta^2 = \theta^3$ and $*\theta^3 = -\theta^2$. It is an important feature of the system under consideration that one can write $\sigma = \theta^1 \otimes_s s$ and $\omega_{\text{tensor}} = \theta^1 \wedge w$, where σ is the shear tensor and ω_{tensor} the vorticity 2-tensor; s ("deformity") and w ("roticity") are 1-forms which carry the information for the shear and the vorticity, respectively. With a being the acceleration of the fluid, ν a connection 1-form in the $\{\theta^2, \theta^3\}$

subspace, b a 1-form which reduces in vacuum to $b = \rho^{-1}\, d\rho$ (ρ being the Weyl canonical coordinate), and the definition $\tilde{\theta^2} = \theta^2$, $\tilde{\theta^3} = -\theta^3$ (extended by linearity to any 1-form), the full set of equations to be considered is the following:

$$d u = a \wedge u + w \wedge \theta^1 , \tag{19}$$
$$d \theta^1 = (b - a) \wedge \theta^1 + s \wedge u , \tag{20}$$
$$d \theta^2 = -\nu \wedge \theta^3 , \tag{21}$$
$$d \theta^3 = \nu \wedge \theta^2 , \tag{22}$$
$$d a = w \wedge s , \tag{23}$$
$$d b = 0 , \tag{24}$$
$$d w = -(b - 2a) \wedge w , \tag{25}$$
$$d s = (b - 2a) \wedge s , \tag{26}$$
$$d * (w - s) + 2a \wedge *w + 2(a - b) \wedge *s = 0 , \tag{27}$$
$$d * a + b \wedge *a + \frac{1}{2} w \wedge *w - \frac{1}{2} s \wedge *s = \frac{1}{2}(\mu + 3p)\theta^2 \wedge \theta^3 , \tag{28}$$
$$d * b + b \wedge *b = 2p\theta^2 \wedge \theta^3 , \tag{29}$$
$$d \nu + a \wedge *b - a \wedge *a + \frac{1}{4}(s - w) \wedge *(s - w) = \frac{1}{2}(\mu + p)\theta^2 \wedge \theta^3 , \tag{30}$$
$$d \tilde{b} + b \wedge \tilde{b} + 2(a - b) \wedge \tilde{a} - \frac{1}{2}(s - w) \wedge (\tilde{s} - \tilde{w}) + 2\nu \wedge *\tilde{b} = 0 , \tag{31}$$
$$d * \tilde{b} + b \wedge *\tilde{b} + 2(a - b) \wedge *\tilde{a} - \frac{1}{2}(s - w) \wedge *(\tilde{s} - \tilde{w}) - 2\nu \wedge \tilde{b} = 0 . \tag{32}$$

This equation corresponds to (2.62) of reference [9]. Please note that there is a typographical error in the last sign of that equation.

$$d p + (\mu + p) a = 0. \tag{33}$$

Equations (19)-(22) are the first structural equations of Cartan, (23)-(26) their integrability conditions (first Bianchi identities), (27)-(32) the field equations (in particular, (28) is the Raychaudhuri equation), and (33) the contracted second Bianchi identity (Euler equation for the fluid). While u and θ^1 are fixed, there is a remaining gauge freedom of spatial rotations in the $\{\theta^2, \theta^3\}$ subspace. Under such rotations, however, a, b, s, and w remain invariant. The $\tilde{}$ operation is not intrinsically defined, and gauge rotations change it, but the equations remain invariant in form as a set. Fluids with a barotropic equation of state are obviously characterized by the condition $d\mu \wedge dp = 0$; by using (23) and the exterior derivative of (33), it is easy to see that (disregarding the degenerate case $\mu + p = 0$) a is closed, and w and s are collinear: $w \wedge s = 0$.

The irrotational solution ($w = 0$) presented in [8] was found by imposing the Ansatz

$$b \wedge s = 0. \tag{34}$$

A word is in order concerning that choice. It was inspired by the fact that, for a Newtonian fluid rotating rigidly, one has $b \wedge w = 0$ (with the appropriate definitions of b and w in the underlying Euclidean metric); it seemed then natural to impose (34) as a simplifying assumption in the irrotational case, with the role of w taken over by s. Although this Ansatz did work as an aid in finding the solution, we should bear in mind, however, that $b \wedge w = 0$ is satisfied due to the fact that in the Newtonian case the vorticity lines are straight lines, whereas in a relativistic fluid one would expect a dynamical (multipole-like) distribution for w or s.

In order to arrive at the solution in [8], a further Ansatz of a more mathematical nature was imposed; the end result was that all the quantities of the solution, including the metric, can be written in terms of a single function $T(\rho)$ (ρ means here a coordinate!), which satisfies the second-order ordinary differential equation

$$T_{\rho\rho} + (\frac{1}{\rho} + \frac{k\rho}{T^2})T_\rho = 0, \tag{35}$$

where k is a constant, and subindices denote ordinary derivatives. Equation (35) can be transformed to a first-order one, which turns out to be an Abel equation of the first kind. The (non-constant) angular velocity distribution for the rotating fluid is given by

$$\Omega(\rho) = l \int_{\rho_1}^{\rho} \frac{\rho'}{T(\rho')} d\rho', \tag{36}$$

where l and ρ_1 are constants. The equation of state is that of stiff matter, $\mu = p$, and the solution is of Petrov type I. The surface area of the body ($p = 0$, $t =$ const.) is finite. The solution admits no further Killing vectors.

A few other differentially rotating interior solutions have subsequently been found [10]. Needless to say, none of the differentially rotating interior solutions with no extra Killing fields has yet been shown to admit a matching asymptotically flat exterior vacuum field, a fact similar to that mentioned above for rigidly rotating solutions.

4. VACUUM AS A "PERFECT FLUID". THE "RIGID ROTATION" AND THE "IRROTATIONAL" GAUGES

Equations (19)-(32) still hold when $\mu = p = 0$ (and (33) becomes a trivial identity.) One important difference, however, is the following.

Now $\theta^0 = u$ is no longer related to the velocity of a fluid, so one is free to redefine θ^0 and θ^1, allowing for a hyperbolic rotation. It is an interesting technical point that, under this additional gauge, it is always possible to make either s or w vanish (the integrability conditions for doing so are satisfied) [9]. Let us briefly examine the case $s = 0$ (for obvious reasons, we shall call it the rigid-rotation gauge.) We can choose functions ρ, z, u, ψ, and χ such that $b = \rho^{-1}\, d\rho$, $*b = \rho^{-1}\, dz$, $a = du$, $w = \rho^{-1}e^{2u}\, d\psi$, $*w = e^{-2u}\, d\chi$ (ρ and z can be interpreted as quasi-cylindrical coordinates, while χ is the twist potential). The Ernst potential $\mathcal{E}$ [11] is simply $\mathcal{E} \equiv e^{2u} + i\chi$, which satisfies the Ernst equation $d * d\mathcal{E} + b \wedge *d\mathcal{E} = 2(\mathcal{E} + \bar{\mathcal{E}})^{-1} d\mathcal{E} \wedge *d\mathcal{E}$. The relation among the Ernst formulation and the 1-forms discussed in the previous section is given by

$$\frac{d\mathcal{E}}{\mathcal{E} + \bar{\mathcal{E}}} = a + \frac{i}{2} * w. \tag{37}$$

Let us note in passing that the symmetry inherent in the expressions given above for the 1-forms and their duals can be implemented by means of the Kramer-Neugebauer transformation [12]:

$$u \longleftrightarrow -u + \frac{1}{2}\ln\rho, \quad \psi \longleftrightarrow i\chi. \tag{38}$$

A more unconventional formulation for the vacuum case may be obtained by means of the alternative "irrotational" gauge [9]: Now we have $w = 0$, and we find $b = \rho^{-1}\, d\rho$, $*b = \rho^{-1}\, dz$, $a = du$, $s = \rho e^{-2u}\, d\psi$, $*s = \rho^{-2}e^{2u}\, d\chi$, for appropriate ρ, z, u, ψ, and χ. In analogy with the Ernst formulation, we can introduce $\mathcal{F} \equiv \rho^2 e^{-2u} + i\chi$, which satisfies $d * d\mathcal{F} + b \wedge *d\mathcal{F} = 2(\mathcal{F} + \bar{\mathcal{F}})^{-1} d\mathcal{F} \wedge *d\mathcal{F}$. The analog of (37) is

$$\frac{d\mathcal{F}}{\mathcal{F} + \bar{\mathcal{F}}} = b - a + \frac{i}{2} * s, \tag{39}$$

while a discrete symmetry transformation, analogous to the Kramer-Neugebauer one, is now [9]

$$u \longleftrightarrow -u + \frac{3}{2}\ln\rho, \quad \psi \longleftrightarrow i\chi. \tag{40}$$

By examining the equations for the rigid-rotation and irrotational gauges, it is seen that both pictures can be related by the transformation [9]

$$a \longleftrightarrow -a + b, \quad w \longleftrightarrow s \tag{41}$$

or, in terms of functions rather than 1-forms, $u \longleftrightarrow -u + \ln\rho$, $\chi_r \longleftrightarrow \chi_d$, where χ_r denotes the function associated with the roticity, while χ_d is the corresponding one for the deformity.

It is an interesting question which one of the two discussed gauges (or perhaps an intermediate one?) is more convenient for problems such as the matching between interior and exterior solutions.

5. SOME GEOMETRIC FEATURES OF A RIGIDLY ROTATING PERFECT FLUID INTERIOR SOLUTION

It is well known that the analog of some standard Newtonian results for rotating perfect fluids, such as the existence of an equatorial symmetry plane, the spherical shape of a static configuration, or the fact that straight lines parallel to the rotation axis intersect the two-dimensional surface of the body on at most two points, are not yet fully proven in general relativity. In order to get some intuition in this new regime, it is worthwhile to analize some of the salient features of particular relativistic solutions.

We are fortunate to have one such (rigidly rotating) interior solution, due to D. Kramer [13] , whose main quantities can be expressed in comparatively simple explicit analytical terms. Recently, we carried out an analysis of some geometric properties of that solution [14]. It had been already pointed out [13] that the body is oblate. By computing the Gauss-Bonnet integral on the bounding 2-surface, we can conclude that the object is topologically a sphere.

One interesting fact is that the Gaussian curvature of the 2-surface becomes negative at the equator (and near it) for high enough rotation rates, and in fact tends to an infinite negative value at the equator for rotation rates approaching the maximum one allowed by the solution. The length of parallels increases monotonically from pole to equator for a certain range of the rotation parameter, but it presents a maximum away from the equator when sufficiently high rotation rates are considered; when the terminal rotation speed is reached, the length along the equator vanishes. Our intuition with Euclidean geometry would lead us to believe that a fission of the body takes place for the maximal speed of rotation; but a more detailed investigation of the three-dimensional geometry *within* the body (for constant time) shows that the geodesic distance between a point in the boundary and the axis of rotation increases as the point moves from the pole to the equator along a meridian (for *any* rotation rate). Geodesic distances *from the center* to a point in a meridian exhibit a similar monotonically increasing behavior, when the point varies from one pole to the equator (all these conclusions are made technically possible by the fact that in the Kramer solution the geodesics can be expressed in terms of quadratures) [14]. Finally, it can be shown that the analog of the Newtonian result that straight lines parallel to the axis of rotation have at most two points of intersection with the bounding 2-surface (we may call this "vertical convexity" for short) also holds for the Kramer solution. Obviously, we have to generalize appropriately the concept of "straight line parallel to the axis"; we do it in the obvious way, by considering geodesics which intersect the equatorial plane orthogonally and possess a constant azimuthal coordinate. Once this is done, vertical convexity can be proven, by looking at how the pressure varies along such geodesics, and making use of the remarkable fact that the pressure is harmonic (for a 2-dimensional flat-space Laplacian) in the Kramer solution.

Acknowledgments

It is for me an honor and a pleasure to participate in the celebration for Prof. Dehnen and Prof. Kramer. I thank Dr. T. Matos for his invitation to do so. The work described here has been or is currently supported by Projects PB89-0142, PB92-0183, PB95-0371, and PB98-0772 of Dirección General de Enseñanza Superior e Investigación Científica, Spain. I am indebted to other members of some of those Projects (L. Fernández-Jambrina, L.M. González-Romero, C. Hoenselaers, F. Navarro-Lérida, and M.J. Pareja) for discussions.

References

[1] H. Lamb, *Hydrodynamics*, Dover, New York, 1945.

[2] S. Chandrasekhar, *Ellipsoidal Figures of Equilibrium*, Dover, New York, 1987.

[3] H. Poincaré, *Acta. Math.* **7** (1885) 259.

[4] A.B. Basset, *Amer. J. Math.* **11** (1889) 172.

[5] F.J. Chinea, *Ann. Phys. (Leipzig)* **9** (2000) SI–38.

[6] L. Lichtenstein, *Gleichgewichtsfiguren rotierender Flüssigkeiten*, Springer, Berlin, 1933.

[7] H.D. Wahlquist, *Phys. Rev.* **172** (1968) 1291.

[8] F.J. Chinea and L.M. González–Romero, *Class. Quantum Grav.* **7** (1990) L99.

[9] F.J. Chinea and L.M. González–Romero, *Class. Quantum Grav.* **9** (1992) 1271.

[10] J.M.M. Senovilla, *Class. Quantum Grav.* **9** (1992) L167; F.J. Chinea, *Class. Quantum Grav.* **10**, 2539 (1993); A. García, *Class. Quantum Grav.* **11** (1994) L45.

[11] F.J. Ernst, *Phys. Rev.* **167** (1968) 1175.

[12] D. Kramer and G. Neugebauer, *Commun. Math. Phys.* **10** (1968) 132.

[13] D. Kramer, *Class. Quantum Grav.* **1** (1984) L3; *Astron. Nach.* **307** (1986) 309.

[14] F.J. Chinea and M.J. Pareja, *Class. Quantum Grav.* **16** (1999) 3823.

NEW DIRECTIONS FOR THE NEW MILLENNIUM

Frederick J. Ernst *
FJE Enterprises. Potsdam, New York 13676, USA

Keywords: Exact solutions, potentials formaliosm, Geroch group.

1. INTRODUCTION

As we stand at the threshold of a new millennium, perhaps it would not be amiss to reflect upon what has been accomplished since Einstein presented general relativity to the world, to ponder where we should attempt to go from here, and to speculate about what will actually take place in the future.

Almost immediately the immense mathematical complexity of the Einstein field equations was recognized. Nevertheless, because of the minuscule spacetime curvature induced by a typical star, one was able to employ simple approximation methods to make predictions about the astronomical phenomena that were then accessible. Encouraged by relativity's ability to account for perihelion precession, bending of light rays and gravitational red shift, some people began to wonder what else this theory might predict when gravity was really strong.

The early musings about the Schwarzschild solution were flawed, but from the mid–century work of Martin Kruskal a much richer picture emerged. Then, while involved in a purely mathematical investigation, Roy Kerr happened upon the single most important discovery, an exact solution representing a rotating black hole.

My humble contribution, in the complex potential formalism [1], was to focus people's attention upon a seemingly impossible problem that actually turned out to be solvable, the study of spacetimes with two commuting Killing vectors.

*http://pages.slic.com/gravity

2. THE QUEST FOR THE GEROCH GROUP AND ITS GENERALIZATIONS

In 1972 Robert Geroch [2] speculated that there might exist a transformation group acting upon an infinite hierarchy of complex potentials, using which one might be able to transform any one stationary axisymmetric spacetime into any other such spacetime. William Kinnersley and D. M. Chitre [3] painstakingly worked out the detailed structure of the infinitesimal elements of Geroch's group (actually an electrovac generalization of Geroch's group). By summing series they were able to deduce the Kerr metric in several ways. Similarly they were able to generalize certain solutions of Tomimatsu and Sato. Their most important contribution, however, was that they demonstrated convincingly that Geroch was on the right track.

A number of people, among whom were Isidore Hauser and the author [4], observed that the so-called "Ernst equations" were associated with a "linear system." With this observation the mathematical machinery of Riemann–Hilbert problems could be brought to bear upon vacuum and electrovac spacetimes with two commuting Killing vectors. This had ramifications both for practical calculation and for development of proofs that such Geroch groups actually existed in a well-defined mathematical sense.

In the early 1980's, Hauser and Ernst succeeded in using the homogeneous Hilbert problem (HHP) formalism to prove Geroch's conjecture and its electrovac generalization [5]. This involved showing that, under appropriately chosen premises, solving the HHP was equivalent to solving a certain Fredholm integral equation of the second kind. Development of the proof was facilitated by the fact that solutions of the elliptic Ernst equation are analytic if they are $\mathbf{C}^3$.

¿From the Hauser–Ernst homogeneous Hilbert problem formalism for stationary axisymmetric fields was derived the integral equation formalism of Nail Sibgatullin [6], which has been used very effectively by Vladimir Manko here at CINVESTAV to generate a large number of exact stationary axisymmetric solutions.

Georgii Alekseev [7] observed that if one admits solutions that exist nowhere on the symmetry axis, the Kinnersley–Chitre transformations do not tell the whole story. He found that a certain nonsingular integral equation describes the larger picture. Much later, Isidore Hauser and I found in our study of spacetimes admitting two commuting spacelike Killing vectors that the Alekseev equation was again germane. In this case, one can have spacetimes that are even $\mathbf{C}^\infty$ without being analytic, so it is truly amazing that an Alekseev type equation remains applicable.

Our discovery of the generalized Alekseev equation permitted us to identify a transformation group that is significantly larger than that associated with Kinnersley–Chitre transformations. This involved, among other things, showing that, under appropriately chosen premises, solving the generalized Alekseev equation was equivalent to solving a certain Fredholm equation of the second kind. The complete proof [8] is rather lengthy.

This proof is divided into six sections:

1 Statement of our generalized Geroch conjecture

2 Alekseev–type singular integral equation equivalent to our HHP

3 Fredholm integral equation equivalent to Alekseev-type equation

4 Existence and properties of the HHP solution

5 Derivatives of $\mathcal{F}$ and H

6 Proof of our generalized Geroch conjecture

It should be emphasized that, in order to formulate properly a generalized Geroch conjecture, years of work were required before we even arrived at the formulation that we spelled out in Sec. 1. This started with a complete analysis of the Weyl case, where all objects we used could be worked out explicitly.

In developing this proof, we had to be very careful to specify the domains of all the potentials we were employing, whereas in the proofs that we developed for the Geroch conjecture itself in the early 1980s, the assumption of an $\mathbf{C}^3$ $\mathcal{E}$–potential sufficed to guarantee analyticity. Initially we employed in this more recent work quite general domains for our $\mathcal{E}$-potentials, but it turned out that nothing was gained thereby, and we eventually elected to use certain diamond shaped regions bounded by null lines. In certain cases these diamond shaped regions were truncated when one reached $\rho = 0$, the analog of the "axis" in the case of the elliptic Ernst equation. In the case of the hyperbolic Ernst equation, $\rho = 0$ is where colliding wave solutions typically exhibit either horizons or singular behavior.

We employed one of many linear systems for the Ernst equation. It involved a 2×2 matrix field $\mathcal{F}$ dependent upon the spectral parameter τ. The domain of this field was specified with equal care. Then we identified a set $\mathcal{S}_{\mathcal{F}}$ of permissible $\mathcal{F}$–potentials and stated a homogeneous Hilbert problem on arcs, analogous to the HHP that we had formulated in 1980 on a closed contour to handle the far simpler problem we were considering back then. Finally, we stated our new generalized Geroch conjecture, which involved an unspecified subset of the set $\mathcal{S}_{\mathcal{F}}$ of permissible $\mathcal{F}$–potentials.

Identifying this subset only came in Sec. 6 of our paper, where we showed that the members of the subset could be determined in terms of a simple differentiabilty criterion imposed on the *Abel transforms* of the initial data (the $\mathcal{E}$–potential) on two intersecting null lines.

Before we ever got to this stage in the proof, it was necessary, in Sec. 2, to derive an Alekseev–type singular integral equation, and then show that, making appropriate differentiability assumptions, the HHP and this Alekseev–type were equivalent. Then, in Sec. 3, we derived from the Alekseev–type equation a Fredholm integral equation of the second kind, and showed that, making appropriate differentiability assumptions, the Alekseev–type equation and this Fredholm equation were equivalent. In Sec. 4 we employed the *Fredholm alternative theorem* to prove the existence and uniqueness of a solution of the Fredholm equation, the Alekseev–type equation and the HHP. In Sec. 5 we studied the partial derivatives of the various potentials that are constructed from the HHP solution.

In Sec. 6, we found that we could identify the subset of permissible $\mathcal{F}$-potentials $\mathcal{S}_{\mathcal{F}}$ by demanding at least $\mathbf{C}^3$ differentiability for the generalized Abel transforms of the initial data, We then showed that the HHP solution $\mathcal{F}$ was a member of the selected subset of $\mathcal{S}_{\mathcal{F}}$. The generalized Geroch group was thus identified.

Finally, we see that the optimism born of Geroch's speculation of 1972 was well-founded, not only for the elliptic Ernst equation but for the hyperbolic Ernst equation as well. I am pleased that this work could be completed during the lifetime of my 77 year old colleague, Isidore Hauser, who has contributed so much to proving mathematically that which everyone else seems to think needs no proof other than a little hand–waving.

3. THE PERFECT FLUID JOINING PROBLEM

During the last two years he and I have developed, using our homogeneous Hilbert problem formalism, a straightforward procedure [9] for determining the axis values of the complex $\mathcal{E}$–potential that describes a *not necessarily flat* vacuum metric joined smoothly across a zero pressure surface to a given perfect fluid (or other type of) interior solution. Of course, once the axis values are known, one can use Sibgatullin's formalism to generate the complete exterior field, the study of which may throw light upon possible physical interpretation of the given interior solution. We hope that our procedure will be useful to those who are skilled at getting exact perfect fluid solutions but are not familiar with Riemann–Hilbert technology.

4. FUTURE DIRECTIONS

I, for one, am ready now to move on to other endeavors. Two months ago I gave up funding by the National Science Foundation in order to regain that spirit of complete freedom that marked the earliest steps in my mathematical odyssey, which commenced when I was a mere ten years old and was exposed to a book concerning the general theory of relativity.

A number of possible endeavors, e.g., writing a history of the two Killing vector problem or developing an electronic archive of actual stationary axisymmetric solutions, including all their salient properties, have already occurred to me, and more possibilities appear each month. At this time I would like to share with you some of my initial thoughts concerning the future.

4.1. GENERAL

We scientists must look beyond our own profession to the broader context in which life is evolving. One does not have to look at much television to see where most people will be going in the early part of the next millennium. An incredible number of the most popular TV shows glorify miracles and the martial arts, with witches, vampires, gods, angels, demons and other supernatural beings outnumbering ordinary humans. I am thinking of programs such as Buffy the Vampire Slayer, Angel, Xena, Hercules, Charmed, Cleopatra 2525, Highlander, The Raven, The Crow, Touched by an Angel and many others.

On the other hand, one is hard put to enumerate TV shows that glorify logic, science or mathematics. Except for a few public broadcasting selections such as Nature, one tends to see rather negative depictions of scientists, unless they practice medicine. Well, medicine is really more of an art than a science, isn't it? In the Star Trek series, even today, they don't get the concept of spacetime or quantum theory right, although someone has obviously provided the writers with a lot of the scientific jargon, and the actors sometimes actually know how to pronounce the terms they employ.

I think we scientists could do a lot more to make elusive ideas accessible to the general public, and so give science a chance of surviving more than a few decades into the new millennium. Of course, we would have to start with an effort to get across the novel ideas that underlay the physics of the seventeenth through nineteenth centuries. Then, at least, the starship will no longer stop when the engines are turned off.

Surely the internet could be exploited more effectively to bring such education to a wide audience. Best of all, it would be fun to try, knowing that the basic tools are improving dramatically even as we speak.

4.2. RADIATING AXISYMMETRIC SYSTEMS

For me, one of the most interesting areas of current research is that of axisymmetric radiating systems. Here, the existence of one spacelike Killing vector provides one complex potential that is independent of the azimuthal angle. That potential is real in the special case considered by Herman Bondi et al. [10] but is complex in the more general case considered by Rayner Sachs [11]. Much work has been done, especially by the relativity research group at Pittsburgh, writing computer code [12] to solve the associated field equations in an effective manner. It is not, however, outside the realm of possibilities that additional insight into this problem might be gained by perturbing in some unspecified manner the linear system associated with the Ernst equation for a stationary axisymmetric field. For this reason, I have begun a study of the existing literature on radiating axisymmetric systems, and would recommend such a course to anyone who is intent upon remaining at the cutting edge of the Einstein equation solution business. Even if it turns out that no additional insights or no exact solutions emerge from such an investigation, it will still be fun to play with the computer codes.

5. TIME FOR A NEW WORLD ORDER?

I am not sure that science has a chance of surviving very long, as long as it remains aloof from everyday life. The fault lies not in the practice of science alone. The pressures of unbridled capitalism, unrestrained by competition with communism, are likely to decimate the natural world, while ethnic and religious factions are likely to be at one another's throats, spurred on by those who would profit from such schisms. One can only hope that labor unions will finally see that they, like major corporations, must adopt an international perspective, and one can only hope that the next person to create a major religion will finally get it right, creating not more schism but rather a more complete unification, designed to ease humanity (and other living creatures) into a more stable and more productive society based upon a more enlightened, less aloof, science.

Acknowledgments

The research that is reported here was supported by grants PHYS–93–07762, PHYS–96–01043 and PHYS–98–00091 from the National Science Foundation to FJE Enterprises.

References

[1] F.J. Ernst, *Phys. Rev.* **167** (1968) 1175; **168** (1968) 1415.

[2] G. Geroch, *J. Math. Phys.* **13** (1972) 394;

[3] W. Kinnersley and D.M. Chitre, *J. Math. Phys.* **18** (1977) 1538; **19** (1978) 1926; **19** (1978) 2037.

[4] I. Hauser and F.J. Ernst, *J. Math. Phys.* **21** (1980) 1126; **21** (1980) 1418.

[5] I. Hauser and F.J. Ernst, *J. Math. Phys.* **22** (1981) 1051; in *Galaxies, Axisymmetric Systems and Relativity*, ed. M.A.H. MacCallum (Cambridge University Press, 1985); in *Gravitation and Geometry*, eds. W. Rindler and A. Trautman (Bibliopolis, Naples, 1987).

[6] N.R. Sibgatullin, *Oscillations and Waves in Strong Gravitational and Electromagnetic Fields* (Springer-Verlag, Berlin, 1991).

[7] G.A. Alekseev, *Sov. Phys. Dokl. (USA)* **30** (1985) 565; *Trudy Matem. Inst. Steklova* **176** (1987) 215.

[8] I. Hauser and F.J. Ernst, to be published in *J. Gen. Rel. Grav.*. Preprint gr-qc/0002049.

[9] I. Hauser and F.J. Ernst, preprint gr-qc/0010058.

[10] H. Bondi, M.G.J. van der Burg and A.W.K. Metzner, *Proc. Roy. Soc.* (London) **A269** (1962) 21.

[11] R.K. Sachs, *Proc. Roy. Soc.* (London) **A270** (1962) 103.

[12] N.T. Bishop, R. Gomez, L. Lehner, M. Maharaj and J. Winicour, Phys. Rev. **D56** (1997) 6298. Preprint gr-qc/9705033.

ON MAXIMALLY SYMMETRIC AND TOTALLY GEODESIC SPACES

Alberto Garcia*
Departamento de Física,
Centro de Investigación y de Estudios Avanzados del IPN
Apdo. Postal 14-740, 07000 México DF, México

Abstract The purpose of this note is the demonstration of the following theorem: the metric of an (n+1)–dimensional space V_{n+1} admitting a totally geodesic maximally symmetric hypersurface V_n is reducible to $ds^2 = g(v)dv^2 + g_{ij}(x^k)dx^i dx^j$, $(i, j = 1, ..., n)$.

Keywords: Geodesic spaces, maximally symmetric spaces, exact solutions.

1. INTRODUCTION

Nowadays it is common to find in the Physics literature old geometrical concepts coming from the end of the nineteen and beginning of the twenty centuries used quite freely and sometime far from their original understanding. On this respect, in this work in honor to Professors Heinz Dehnen and Dietrich Kramer, I shall consider the concept of maximally symmetric totally geodesic sub–spaces, arriving at the theorem stated in the abstract.

Most of the basic material on the subject is taken with slight modifications from the Weinberg [1] and Einsenhart [2] textbooks. Since in these books different conventions in the definition of the Riemann–Christoffel tensor and related quantities are used, to facilitate the translation, the correspondence is given: $R^{\lambda}_{\mu\nu\kappa}(W) = \Gamma^{\lambda}_{\mu\nu,\kappa} - \Gamma^{\lambda}_{\mu\kappa,\nu} + \Gamma^{\eta}_{\mu\nu}\Gamma^{\lambda}_{\kappa\eta} - \Gamma^{\eta}_{\mu\kappa}\Gamma^{\lambda}_{\nu\eta} = -R^{\lambda}_{\mu\nu\kappa}(E), R_{\mu\nu}(W) = R^{\lambda}{}_{\mu\lambda\nu}(W), R_{\mu\nu}(E) = R^{\lambda}{}_{\mu\nu\lambda}(E), R_{\mu\nu}(W) = R_{\mu\nu}(E), R(W) = R(E)$, where in parenthesis stands W and E to denote correspondingly Weinberg and Eisenhart definitions. These letters W and E, followed by page numbers, will refer to statements or equations stated in the corresponding pages of the quoted books.

*E–mail: aagarcia@fis.cinvestav.mx

In the forthcoming sections we briefly review the main concepts yielding to the definition of maximally symmetric totally geodesic spaces. To start with, Sec.2 is devoted to Killing vectors and their defining equations. In Sec.3 we deal with the properties of maximally symmetric spaces. Sec.4 is addressed to coordinate representations of maximally symmetric spaces. The definition of maximally symmetric sub–spaces is given in Sec. 5, with emphasis to the case of a hypersurface. In Sec.6 the characterization of totally geodesic hypersurfaces is given and the explicit metric allowing for a maximally symmetric totally geodesic hypersurface is reported. In Sec.7 maximally symmetric totally geodesic sub–spaces are considered. In Sec.8 we established that the BTZ black hole solution is not a locally maximally symmetric totally geodesic hypersurface of the (3+1) conformally flat stationary axisymmetric PC metric.

2. KILLING EQUATIONS AND INTEGRABILITY CONDITIONS

Under a given coordinate transformation $x \longrightarrow x'$, a metric $g_{\mu\nu}(x)$ is said to be form–invariant if the transformed metric $g'_{\mu\nu}(x')$ is the same function of its argument x'^{μ} as the original metric $g_{\mu\nu}(x)$ was of its argument x^{μ}, i.e. $g'_{\mu\nu}(x) = g_{\mu\nu}(x)$ for all x. In general, at any given point the law of transformation is $g_{\mu\nu}(x) = \partial x'^{\alpha}/\partial x^{\mu}\ \partial x'^{\beta}/\partial x^{\nu}\ g'_{\alpha\beta}(x')$ thus, for a form–invariant metric under transformation one replaces $g'_{\mu\nu}(x')$ by $g_{\mu\nu}(x')$, and consequently

$$g_{\mu\nu}(x) = \frac{\partial x'^{\alpha}}{\partial x^{\mu}} \frac{\partial x'^{\beta}}{\partial x^{\nu}} g_{\alpha\beta}(x') \tag{1}$$

which can be thought of as system of equations to determine x'^{μ} as function of x^{ν}, $x'^{\mu} = x'^{\mu}(x^{\nu})$. Any transformation fulfilling (1) is called an isometry. One way to look for solutions of (1) is to consider an infinitesimal coordinate transformation

$$x'^{\mu} = x^{\mu} + \epsilon\ \xi^{\mu}(x), \ |\epsilon| \ll 1, \tag{2}$$

thus to the first order in ϵ, Eq. (1) gives rise to the Killing equation

$$\frac{\partial \xi^{\mu}(x)}{\partial x^{\rho}} g_{\mu\sigma}(x) + \frac{\partial \xi^{\mu}(x)}{\partial x^{\sigma}} g_{\rho\mu}(x) + \xi^{\mu}(x) \frac{\partial g_{\rho\sigma}(x)}{\partial x^{\mu}} = 0, \tag{3}$$

or, introducing covariant derivatives,

$$\xi_{\sigma;\rho} + \xi_{\rho;\sigma} = 0, \tag{4}$$

for the Killing vector $\xi_\mu(x)$ of the metric $g_{\rho\sigma}(x)$. Because of the general decomposition $\xi_{\mu;\nu} = (\xi_{\mu;\nu} + \xi_{\nu;\mu})/2 + (\xi_{\mu;\nu} - \xi_{\nu;\mu})/2$, the Killing equation, first parenthesis equated to zero, impose $n(n+1)/2$ conditions on $\xi_\mu(x)$.

Integrability conditions of the Killing equation.
Using the above Killing equations in the Ricci identity, commutators of covariant derivatives,

$$\xi_{\sigma;\rho;\mu} - \xi_{\sigma;\mu;\rho} = -\xi_\lambda R^\lambda{}_{\sigma\rho\mu}, \tag{5}$$

by adding to this equation its two cyclic permutations, one arrives at

$$\begin{aligned} \xi_{\sigma;\rho;\mu} &- \xi_{\sigma;\mu;\rho} + \xi_{\mu;\sigma;\rho} - \xi_{\mu;\rho\sigma} + \xi_{\rho;\mu;\sigma} - \xi_{\rho;\sigma;\mu} \\ &= -\xi_\lambda \left(R^\lambda_{\sigma\rho\mu} + R^\lambda_{\mu\sigma\rho} + R^\lambda_{\rho\mu\sigma} \right) = 0, \end{aligned} \tag{6}$$

the last equality to zero arising from the properties of the Riemann tensor. The right hand side containing the derivatives of ξ , using the derivatives of the Killing equation (4), amounts to twice the expression $\xi_{\sigma;\rho;\mu} - \xi_{\sigma;\mu;\rho} - \xi_{\mu;\rho\sigma} = 0$, which substituted in (5) yields

$$\xi_{\mu;\rho;\sigma} = -\xi_\lambda R^\lambda{}_{\sigma\rho\mu}. \tag{7}$$

From the Ricci identity

$$\xi_{\mu;\rho;\sigma;\nu} - \xi_{\mu;\rho;\nu;\sigma} = -\xi_{\lambda;\rho} R^\lambda{}_{\mu\sigma\nu} - \xi_{\mu;\lambda} R^\lambda{}_{\rho\sigma\nu}, \tag{8}$$

after the substitution of the third derivative of ξ from (7) on the left hand side of (8) one obtains

$$\xi_{\mu;\rho;\sigma;\nu} = -\xi_{\lambda;\nu} R^\lambda{}_{\sigma\rho\mu} - \xi_\lambda R^\lambda{}_{\sigma\rho\mu;\nu}. \tag{9}$$

Therefore, second and higher derivatives of the $\xi's$ are expressible linearly and homogeneously in terms of the $\xi's$ and their first derivatives.

On the other hand, equations (8), or equivalently (9), are the integrability conditions of (7), which by using (4) amounts to the integrability conditions

$$\begin{aligned} &\xi_\lambda \left(R^\lambda{}_{\nu\rho\mu;\sigma} - R^\lambda{}_{\sigma\rho\mu;\nu} \right) + \xi_{\lambda;\kappa} \left(\delta^\kappa_\sigma R^\lambda{}_{\nu\rho\mu} - \delta^\kappa_\nu R^\lambda{}_{\sigma\rho\mu} + \delta^\kappa_\rho R^\lambda{}_{\mu\sigma\nu} \right. \\ &- \left. \delta^\kappa_\mu R^\lambda{}_{\rho\sigma\nu} \right) = 0, \end{aligned} \tag{10}$$

imposing linear relations among ξ_λ, and $\xi_{\lambda;\rho}$ at any given point.

Since higher derivatives of ξ_μ, starting from the second one, are determined at any point X as linear combinations of $\xi_\mu(X)$ and $\xi_{\mu;\nu}(X)$, then the function $\xi_\mu(x)$, if it exits, can be constructed as a Taylor series in $x^\mu - X^\mu$ within a finite neighborhood of X, thus any Killing vector of the metric $g_{\mu\nu}(x)$ can be expressed as

$$\xi_\rho(x) = A^\nu_\rho(x, X)\xi_\nu(X) + B^{\mu\nu}_\rho(x, X)\xi_{\mu;\nu}(X), \tag{11}$$

where $A^\nu_\rho(x, X)$ and $B^{\mu\nu}_\rho(x, X)$, the same for all Killing vectors at the neighborhood of X, are functions of the metric $g_{\mu\nu}(x)$ and X. Nevertheless, whether (4) is solvable for a given set of initial data $\xi_\lambda(X)$, and $\xi_{\lambda;\rho}(X)$ depends on the above integrability conditions (10), imposing linear relations among ξ_λ, and $\xi_{\lambda;\rho}$ at any given point X. The maximal dimension of the space of initial data constructed on $(\xi_\lambda(X), \xi_{\lambda;\rho}(X))$, being $n + n(n-1)/2 = n(n+1)/2$, is reduced by the number of relations (10). Thus , if there were no relations (10) on the data ($\xi_\lambda(X)$,and $\xi_{\lambda;\rho}(X)$), then each Killing vector $\xi_\mu(x)$ of a given $g_{\mu\nu}(x)$ was uniquely determined by the independent quantities $\xi_\mu(X)$ and $\xi_{\mu;\nu}(X)$, in number equal respectively to n and $n(n-1)/2$. Hence, a V_n allows for at most an $n(n+1)/2$ parameter group of Killing vectors $\xi_\mu(x)$.

The so–called maximally symmetric space V_n , or space of constant curvature, admits the maximal number of $n(n+1)/2$ Killing vectors. Hence, the equations (10) do not impose conditions on $\xi_\lambda(X)$, and $\xi_{\lambda;\rho}(X)$ at any point X, and therefore they acquire respectively n and $n(n-1)/2$ arbitrary values at X. The vanishing of the factor multiplying $\xi_{\lambda;\rho}$ in (10) yields to the constancy of the scalar curvature R.

3. MAXIMALLY SYMMETRIC SPACES

A maximally symmetric space is, by definition, a space which admits a continuous group of motions of $n(n+1)/2$ parameters. On the other hand, a maximally symmetric space is characterized, according to Weinberg, by the following property:W379, a maximally symmetric space is necessarily homogeneous and isotropic about all points. Weinberg worked out in details the structure of the Killing vectors, which arises from the the homogeneous and isotropic properties of a considered space:

- W378:*A metric space is said to be homogeneous if there exit infinitesimal isometries (2) that carry any point into any other point in its immediate neighborhood. That is, the metric must admit Killing vectors that at any point take all possible values. In particular, in a V_n one can choose a set of n Killing vectors ${\xi^{(\mu)}}_\nu(x; X)$ with*

$$ {\xi^{(\mu)}}_\nu(X; X) = \delta^\mu_\nu, \tag{12}$$

- W378: *A metric space is said to be isotropic about a given point if there exit infinitesimal isometries (2) that leave the point X fixed,* *so that $\xi_\mu(X) = 0$, and for which the first derivatives $\xi_{\mu;\nu}(X)$ at X takes all possible values, subject only to the antisymmetry condition (4).In particular, in a V_n one can choose a set of $n(n-1)/2$ Killing vectors $\xi^{(\mu\nu)}{}_\lambda(x;X)$ such that*

$$\begin{aligned}\xi^{(\mu\nu)}{}_\lambda(x;X) &= -\xi^{(\nu\mu)}{}_\lambda(x;X),\\ \xi^{(\mu\nu)}{}_\lambda(X;X) &= 0,\\ \xi^{(\mu\nu)}{}_{\lambda;\rho}(X;X) &= \left[\frac{\partial}{\partial x^\rho}\xi^{(\mu\nu)}{}_\lambda(x;X)\right]_{x=X} = \delta^\mu_\lambda\delta^\nu_\rho - \delta^\mu_\rho\delta^\nu_\lambda \quad (13)\end{aligned}$$

,

where upper indices in parenthesis denote the particular Killing vector we are dealing with.

Moreover, for metric spaces that are isotropic about *every* point there exist Killing vectors $\xi^{(\mu\nu)}_\lambda(x;X)$ and $\xi^{(\mu\nu)}_\lambda(x;X+dX)$ that satisfy the above initial conditions at X and at $X+dX$, respectively. As a corollary one has: W379, *any space that is isotropic about every point is also homogeneous.*

Constancy of the scalar curvature.
Since at any given point, for maximally symmetric spaces, one can assign to $\xi_{\lambda;\kappa}$ arbitrary values, the antisymmetric part of its coefficient

$$\frac{1}{2}(\xi_{\lambda;\kappa} - \xi_{\kappa;\lambda})\left(\delta^\kappa_\mu R^\lambda{}_{\rho\sigma\nu} - \delta^\kappa_\rho R^\lambda{}_{\mu\sigma\nu} + \delta^\kappa_\nu R^\lambda{}_{\sigma\rho\mu} - \delta^\kappa_\sigma R^\lambda{}_{\nu\sigma\mu}\right) = 0, \quad (14)$$

must vanish

$$\begin{aligned}&\delta^\kappa_\mu R^\lambda{}_{\rho\sigma\nu} - \delta^\kappa_\rho R^\lambda{}_{\mu\sigma\nu} + \delta^\kappa_\nu R^\lambda{}_{\sigma\rho\mu} - \delta^\kappa_\sigma R^\lambda{}_{\nu\rho\mu}\\ =\ &\delta^\lambda_\mu R^\kappa{}_{\rho\sigma\nu} - \delta^\lambda_\rho R^\kappa{}_{\mu\sigma\nu} + \delta^\lambda_\nu R^\kappa{}_{\sigma\rho\mu} - \delta^\lambda_\sigma R^\kappa{}_{\nu\rho\mu} \quad (15)\end{aligned}$$

Contracting κ and μ, using the cyclic rule fulfilled by $R^\kappa{}_{\sigma\rho\mu}$ and the definition of $R_{\sigma\rho}$, one obtains

$$(n-1)R_{\lambda\rho\sigma\nu} = g_{\lambda\sigma}R_{\nu\rho} - g_{\lambda\nu}R_{\sigma\rho} = -g_{\rho\sigma}R_{\nu\lambda} + g_{\rho\nu}R_{\sigma\lambda}. \quad (16)$$

The last equality has been added because of the antisymmetric character in the first two indices λ and ρ of the tensor $R_{\lambda\rho\sigma\nu}$. From that part of the above equation involving the Ricci tensor, contracting in λ and ρ, one arrives at

$$R_{\rho\sigma} = \frac{R}{n}g_{\rho\sigma},\ R = R^\lambda_\lambda. \quad (17)$$

Thus, substituting (17) in (16) one obtains the expression

$$R_{\lambda\rho\sigma\nu} = \frac{R}{n(n-1)}\left(g_{\nu\rho}g_{\lambda\sigma} - g_{\rho\sigma}g_{\lambda\nu}\right) \quad (18)$$

The constancy of R results from the Bianchi identity

$$R^{\lambda}{}_{\mu\nu\sigma;\epsilon} + R^{\lambda}{}_{\mu\sigma\epsilon;\nu} + R^{\lambda}{}_{\mu\epsilon\nu;\sigma} = 0. \tag{19}$$

Contracting in λ and ν one obtains

$$R_{\mu\sigma;\epsilon} + R^{\lambda}{}_{\mu\sigma\epsilon;\lambda} - R_{\mu\epsilon;\sigma} = 0 \tag{20}$$

contracting again in σ and μ one gets

$$\left[R^{\mu}{}_{\epsilon} - \frac{1}{2}\delta^{\mu}{}_{\epsilon}R\right]_{;\mu} = 0 \tag{21}$$

substituting (17) in the above equation, one arrives at

$$\left[\frac{1}{n} - \frac{1}{2}\right]\frac{\partial}{\partial x^{\epsilon}}R = 0. \tag{22}$$

Hence any space of three or more dimensions in which (17) holds *everywhere* will have R constant. Moreover, for maximally symmetric spaces of two dimensions the scalar curvature R occurs to be also constant. A necessary and sufficient condition that the curvature at every point of space be independent of the orientation is that the Riemann tensor is expressible as (18).

Introducing the curvature constant K, one has

$$\begin{aligned} R =: -n(n-1)K,\ R_{\sigma\rho} &= -(n-1)Kg_{\sigma\rho},\ R_{\lambda\rho\sigma\nu} \\ &= K\left(g_{\sigma\rho}g_{\lambda\nu} - g_{\nu\rho}g_{\lambda\sigma}\right). \end{aligned} \tag{23}$$

On the other hand, via the homogeneous property at a given point x in a maximally symmetric space, there exist independent Killing vectors for which $\xi_{\mu}(x)$ takes any values one likes, thus

$$R^{\lambda}{}_{\sigma\rho\mu;\nu} = R^{\lambda}{}_{\nu\rho\mu;\sigma} \tag{24}$$

which by cyclic permutation of ρ, μ and σ gives

$$\left[R^{\lambda}{}_{\sigma\rho\mu} + R^{\lambda}{}_{\rho\mu\sigma} + R^{\lambda}{}_{\mu\sigma\rho}\right]_{;\nu}$$

$$= R^{\lambda}{}_{\nu\rho\mu;\sigma} + R^{\lambda}{}_{\nu\mu\sigma;\rho} + R^{\lambda}{}_{\nu\sigma\rho;\mu}, \tag{25}$$

which, by virtue of the properties of the Riemann tensor, is identically zero, and therefore it gives no information.

It becomes then clear that maximally symmetric spaces are unique in the sense that any two maximally symmetric spaces of the same signature and the same constant curvature are isometric and have the same intrinsic properties, and that the equations of the isometric correspondence involve $n(n+1)/2$ arbitrary constants.

4. REPRESENTATION OF MAXIMALLY SYMMETRIC SPACES

In general, one may pose the question if there exists a system of coordinates in a space of constant curvature such that the metric is of the form

$$\begin{aligned} ds^2 &= \Phi^{-2}C_{\mu\nu}dx^\mu dx^\nu = \Phi^{-2}\left[C_{11}(dx^1)^2 + \dots + C_{nn}(dx^n)^2\right] \\ &= \Phi^{-2}\left[e_1(dx^1)^2 + \dots + e_n(dx^n)^2\right], \end{aligned} \tag{26}$$

where C is a diagonal matrix with non–zero diagonal elements $C_{\mu\nu} = \pm\delta_{\mu\nu}, e_\mu := C_{\mu\mu} = \pm 1$, for $\mu = 1, ..., n$, depending upon the signature.

The substitution in the Riemann tensor yields

$$\begin{aligned} \frac{\partial^2\Phi}{\partial x^\mu \partial x^\nu} &= 0, \\ \Phi\left(e_\mu \frac{\partial^2\Phi}{\partial x^{\mu 2}} + e_\nu \frac{\partial^2\Phi}{\partial x^{\nu 2}}\right) &= e_\mu e_\nu \left[K + \sum e_\lambda \left(\frac{\partial\Phi}{\partial x^\lambda}\right)^2\right], \ \mu \neq \nu \end{aligned} \tag{27}$$

in the right hand side there is no summation neither in μ nor in ν.

¿From the first equations one arrives at

$$\Phi = X_1(x^1) + \dots + X_n(x^n), \tag{28}$$

where $X_\mu(x^\mu)$ depend only on their arguments. Substituting Φ in the second equations one obtains

$$X_\mu(x^\mu) = e_\mu(ax^{\mu 2} + 2b_\mu x^\mu + c_\mu), \tag{29}$$

where is no summation in μ. The constants a, b_ν, and c_ν are arbitrary, and the constant curvature K occurs to be related with them through

$$K = 4\sum e_\mu(ac_\mu - b_\mu^2). \tag{30}$$

Hence, one can give various alternative representations of the metric of the studied space. In particular, the Riemman form arises when a is different from zero; accomplishing first a shifting of the $x's$, $x'^\mu = x^\mu + b_\mu/a$, followed by scaling transformations $x''^\nu = x'^\nu/(\sum e_\mu C_\mu)^2$, one arrives at the Riemann form

$$ds^2 = \frac{\left[e_1(dx^1)^2 + \dots + e_n(dx^n)^2\right]}{\left[1 + \frac{K}{4}(e_1(x^1)^2 + \dots + e_n(x^n)^2)\right]^2}. \tag{31}$$

Representation as embedding in a S_{n+1}.

Let us consider the metric (26) given as

$$ds^2 = \Phi^{-2}C_{\mu\nu}dx^\mu dx^\nu, \ \Phi = 1 + \frac{K}{4}C_{\mu\nu}x^\mu x^\nu, \tag{32}$$

this metric can be written as

$$ds^2 = C_{\mu\nu}d(\Phi^{-1}x^\mu)d(\Phi^{-1}x^\nu) - d(\Phi^{-1})d(\Phi^{-1}C_{\mu\nu}x^\mu x^\nu). \tag{33}$$

Because of the relation

$$\Phi^{-1}C_{\mu\nu}x^\mu x^\nu = \frac{4}{K}\left(1 - \frac{1}{\Phi}\right), \tag{34}$$

one has

$$\begin{aligned} -d(\Phi^{-1})d(\Phi^{-1}C_{\mu\nu}x^\mu x^\nu) &= \frac{1}{K}\left(d\frac{2}{\Phi}\right)^2 = \epsilon\left(d\frac{2k}{\Phi}\right)^2, \\ k^2 : &= \frac{\epsilon}{K}, \ \epsilon = \pm 1. \end{aligned} \tag{35}$$

Defining the new coordinates

$$z^\mu = \frac{x^\mu}{\Phi}, \ z^{n+1} = \frac{2k}{\Phi} - k, \ \mu = 1, ..., n, \tag{36}$$

one obtains the relation

$$C_{\mu\nu}z^\mu z^\nu + \epsilon(z^{n+1})^2 = \epsilon k^2 = \frac{1}{K}, \ (\mu, \nu = 1, ..., n), \tag{37}$$

thus, one embeds the metric (32) in a S_{n+1}

$$ds^2 = C_{\mu\nu}dz^\mu dz^\nu + \epsilon(dz^{n+1})^2, \ (\mu, \nu = 1, ..., n), \tag{38}$$

by restricting the variables $z's$ to the embedding condition (37).

The inverse transformations are given by

$$x^\mu = \frac{2z^\mu}{1 + z^{n+1}/k} = \frac{2z^\mu}{1 \pm \sqrt{1 - \epsilon C_{\mu\nu}z^\mu z^\nu / k^2}}, \ (\mu, \nu = 1, ..., n), \tag{39}$$

One can easily check that the metric (32) is endowed with $n(n+1)/2$ Killing vectors:

$$k^\mu = \Omega^\mu{}_\nu x^\nu + \frac{\epsilon}{2k}x^\mu \epsilon_\nu x^\nu + k\epsilon\left(1 - \frac{\epsilon}{k^2}C_{\mu\nu}x^\mu x^\nu\right), \ \Omega_{\mu\nu} = -\Omega_{\nu\mu}, \tag{40}$$

where the $n(n-1)/2$ parameters of $\Omega_{\mu\nu}$ and the n constants ϵ_μ determine the total of $n(n+1)/2$ parameters. The indices are raised and lowered by the constant metric $C_{\mu\nu}$.

Representation a la Weinberg

¿From the above metric (38), by substituting the differential of z^{n+1}

$$dz^{n+1} = -\frac{C_{\mu\nu}z^\mu dz^\nu}{z^{n+1}}, \ \epsilon(dz^{n+1})^2 = K\frac{(C_{\mu\nu}z^\mu dz^\nu)^2}{1 - KC_{\mu\nu}z^\mu z^\nu}, \tag{41}$$

one obtains

$$ds^2 = C_{\mu\nu}dz^\mu dz^\nu + K\frac{(C_{\mu\nu}z^\mu dz^\nu)^2}{1 - KC_{\mu\nu}z^\mu z^\nu}. \tag{42}$$

where $C_{\mu\nu}$ is a $n \times n$ matrix diagonal matrix with elements $e_\mu = \pm 1, \mu = 1, ..., n$. Restoring the $x's$ typing, one has

$$ds^2 = e_1(dx^1)^2 + ... + e_n(dx^n)^2 + K\frac{(e_1x^1dx^1 + ... + e_nx^ndx^n)^2}{1 - K(e_1(x^1)^2 + ... + e_n(x^n)^2)}. \tag{43}$$

In general, one may consider $C_{\mu\nu}$ in (42) as a constant $n \times n$ matrix, with K being the curvature constant.

In particular, for a maximally symmetric space–time metric with three positive and one negative eingenvalue one can set $C_{\mu\nu} = \eta_{\mu\nu}$, hence

$$ds^2 = d\mathbf{x}^2 - dt^2 + \frac{K(\mathbf{x}d\mathbf{x} - t\,dt)^2}{1 - K(\mathbf{x}^2 - t^2)}, \tag{44}$$

where the vector notation is used; this metric is an alternative representation of the well known de–Sitter space,

5. MAXIMALLY SYMMETRIC SUB–SPACES

As Weinberg points out in W395: *In many cases of physical importance, the whole space is not maximally symmetric, but it can be decomposed into maximally symmetric subspaces... The maximal symmetry of a family of subspaces imposes very strong constraints on the metric of the whole space.*

The main theorem on maximally symmetric spaces, see W396, can be formulated as follows: *It is always possible to choose the x^i, $i = 1, ..., n$, coordinates of the maximally symmetric n–dimensional subspaces V_n and the coordinates y^a, $a = n+1, ..., m$, of the m–n–dimensional subspaces V_{m-n} so that the metric of the whole m–dimensional space V_m is given by*

$$ds^2 = g_{\mu\nu}dx^\mu dx^\nu = g_{ab}(y^c)dy^a dy^b + f(y^c)\tilde{g}_{ij}(x^k)dx^i\, dx^j, \tag{45}$$

where $g_{ab}(y^c)$ and $f(y^c)$ are functions of the y^a–coordinates alone, and $\tilde{g}_{ij}(x^k)$ are functions of the x^i–coordinates alone, thus $\tilde{g}_{ij}(x^k)$ is by itself the metric of an n–dimensional maximally symmetric space. (The indices $a, b, c, ... = n+1, ..., m$ run over the $m-n$ labels of the y–coordinates, $i, j, k, ... = 1, ..., n$ take values over the n labels of the x–coordinates, and Greek indices run 1,...,m).

Maximally symmetric subspaces V_n of n–dimensions (with $n(n+1)/2$ independent Killing vectors $\xi^i(x, y)$) are defined as those subspaces with constants $y^a = \text{const.}$, for which the metric of the whole space is invariant under a group of infinitesimal transformations

$$\begin{aligned} x^i &\longrightarrow x'^i = x^i + \epsilon\xi^i(x,y), \\ y^a &\longrightarrow y'^a = y^a, \end{aligned} \tag{46}$$

thus, with respect to these transformations the y^a–components of $\xi(x,y)$ vanish, i.e. $\xi^a(x,y) = 0$. This means that (46) is an isometry of the whole metric $g_{\mu\nu}$, and consequently the vector $\xi^\mu(x,y) = (\xi^i(x,y), 0)$ has to be a solution of the Killing equation for the whole space.

Specializing this theorem for a maximally symmetric hypersurface V_n immersed in a $n+1$-dimensional space V_{n+1}, the corresponding V_{n+1} metric decomposes as follows:

$$ds^2 = g_{\mu\nu}dx^\mu dx^\nu = g(y)dy^2 + f(y)\tilde{g}_{ij}(x^k)dx^i dx^j, \tag{47}$$

where Greek indices run $1, ..., n+1$, and Latin indices run over $1, ..., n$.

We shall demonstrate that for a maximally symmetric hypersurface V_n which, at the same time, is a totally geodesic hypersurface, the function $f(v)$ entering in (47) has to be a constant (different from zero), which by scaling transformations of the x–coordinates can be equated to 1, $f(v) \to 1$.

6. TOTALLY GEODESIC HYPERSURFACES

An n-dimensional hypersurface V_n is called a *totally geodesic hypersurface* of an enveloping V_{n+1}, if all the geodesics of V_n are geodesics of V_{n+1}. For totally geodesic hypersurfaces there exists the following theorem, stated in the Eisenhart's textbook [2], E183:

A necessary and sufficient condition that a V_{n+1} admits a family of totally geodesic hypersurfaces is that its fundamental form be reducible to

$$ds^2 = a_{ij}(x^k)dx^i dx^j + B(x^k, y)(dy)^2,\ i, ..., k = 1, ..., n, \tag{48}$$

where a_{ij} are independent of y and $B(x^k, x^{n+1} =: y)$ is any function of x^k and y. In comparison with the Eisenhart's coordinates, we have replaced $y^1, ...y^n$ correspondingly by $x^1, ..., x^n$, and y^{n+1} by y. Notice that in (48), in front of $a_{ij}(x^k)$ stands the unit as a factor.

In the Eisenhart's Chapter IV, the theory of n–dimensional subspaces V_n immersed in a space V_m is developed. Let us consider sub–spaces V_n immersed in a space V_m, with $m > n+1$. The metrics of V_m and V_n are given correspondingly by

$$V_m : ds_m^2 = a_{\alpha\beta}(y^\sigma)dy^\alpha dy^\beta, (\alpha, \beta = 1, ..., m), \tag{49}$$

and

$$V_n : ds_n^2 = g_{ij}(x^k)dx^i dx^j, (i,j = 1,...,n), \tag{50}$$

where the immersion of V_n is defined through the functions

$$y^\alpha = F^\alpha(x^1,...,x^n), \tag{51}$$

with the Jacobian matrix $\frac{\partial F^\alpha}{\partial x^i}$ of rank n.

Let Greek indices $\alpha,...,\nu$ take the values $1,...,m$, while indices $i,...,k$ run $1,...,n$ and $a,...,c$ run $n+1,...,m$. In V_m, one can chose a system of real unit vectors $\xi^\alpha_{|a}$ mutually orthogonal to one another,

$$a_{\alpha\beta}\xi^\alpha{}_{a|}\xi^\beta{}_{b|} = \pm\delta_{a|b|}, \tag{52}$$

and normal to V_n,

$$a_{\alpha\beta}y^\alpha{}_{,i}\xi^\beta{}_{a|} = 0, \tag{53}$$

where $y^\alpha_{,i} := \frac{\partial y^\alpha}{\partial x^i}$. Differentiating covariantly these equations with respect to the metric of V_n, i.e. with respect to x^i and the metric $g_{ij}(x^k)$ one arrives at

$$\Omega_{a|ij} = a_{\alpha\beta}\, y^\alpha_{,i;j}\, \xi^\beta_{a|} + [\mu\nu,\beta]_a y^\mu_{,i} y^\nu_{,j} \xi^\beta_{a|}, \tag{54}$$

which can be rewritten as

$$-\Omega_{a|ij} = a_{\alpha\beta}\, y^\alpha_{,i}\, \xi^\beta_{a|,j} + [\mu\beta,\nu]_a y^\nu_{,i} y^\mu_{,j} \xi^\beta_{a|}, \tag{55}$$

where $[\mu\nu,\beta]_a$ is the Christoffel symbol with respect to $a_{\alpha\beta}$ and evaluated at points of V_n,

$$2[\mu\nu,\beta]_a = a_{\mu\beta,\nu} + a_{\beta\nu,\mu} - a_{\mu\nu,\beta}. \tag{56}$$

Specializing the theory for a space of $n+1$–dimensions V_{n+1} and its hypersurface V_n, the metric of the space V_{n+1} is given by

$$V_{n+1} : ds^2_{n+1} = a_{\alpha\beta}(y^\sigma)dy^\alpha dy^\beta, (\alpha,\beta = 1,...,n+1), \tag{57}$$

while the metric of the hypersurface V_n is given as

$$V_n : ds_n^2 = g_{ij}(x^k)dx^i dx^j, (i,j = 1,...,n). \tag{58}$$

The immersed hypersurface V_n is defined by equations of the form

$$y^\alpha = F^\alpha(x^1,...,x^n). \tag{59}$$

A unit normal to V_n vector ξ^α has to fulfill the orthogonality condition

$$a_{\alpha\beta}y^\alpha_{,i}\xi^\beta = 0, \tag{60}$$

where $y^\alpha_{,i} = \frac{\partial y^\alpha}{\partial x^i}$, and the normalization condition:

$$a_{\alpha\beta}\xi^\alpha\xi^\beta = \pm 1. \tag{61}$$

These quantities determine the symmetric covariant tensor Ω_{ij}:

$$\Omega_{ij} = a_{\alpha\beta}\,\xi^\beta\, y^\alpha_{,i;j} + [\mu\nu,\beta] y^\mu_{,i} y^\nu_{,j}\xi^\beta, \tag{62}$$

where, as before, $[\mu\nu,\beta]_a$ is the Christoffel symbol with respect to $a_{\alpha\beta}$.

Moreover, in Eisenhart's book, E183, one reads:
...a necessary and sufficient condition that a V_n be a totally geodesic hypersurface is that $\frac{1}{R} = 0$, and from $\frac{1}{R} = \Omega_{ij}\frac{dx^i}{ds}\frac{dx^j}{ds}$, that $\Omega_{ij} = 0$.

6.1. MAXIMALLY SYMMETRIC TOTALLY GEODESIC HYPERSURFACE

Applying this theory to the maximally symmetric space (47), with the identification

$$y^1 = x^1, ..., y^n = x^n,\ y^{n+1} = y, a_{ij} = f(y)\tilde{g}_{ij}, \tag{63}$$

the normal to V_n unit vector ξ^μ and $y^\alpha_{,i}$ are given by

$$y^\alpha_{,i} = \delta^\alpha_i, \xi_\mu = \sqrt{g(y)}\delta^\mu_y, \xi^\mu = \frac{1}{\sqrt{g(y)}}\delta^\mu_y, \tag{64}$$

which, substituted in (60), imply $a_{iy} = 0$, these conditions are identically fulfilled because of the structure of the considered metric (47). The substitution in Ω_{ij} yields

$$\Omega_{ij} = -a_{ij,y}\frac{1}{\sqrt{g(y)}} = -f(y)_{,y}\frac{1}{\sqrt{g(y)}}\tilde{g}_{ij}(x^k). \tag{65}$$

Therefore, a space V_{n+1} admits a totally geodesic maximally symmetric hypersurface V_n iff :

$$\Omega_{ij} = 0 \rightarrow f(y)_{,y} = 0, f(y) = const. \rightarrow 1. \tag{66}$$

Consequently, the metric of V_{n+1}, admitting a totally geodesic maximally symmetric hypersurface V_n, is reducible to

$$ds^2 = g(y)dy^2 + g_{ij}(x^k)dx^i dx^j. \tag{67}$$

Some properties of a totally geodesic hypersurface are: the lines of curvature of a totally geodesic hypersurface are indeterminate, and the normals to a totally geodesic hypersurface are parallel in the enveloping space.

7. TOTALLY GEODESIC SUB–SPACES

A V_n immersed in a V_m, $m > n+1$, is called totally geodesic subspace of V_m if all the geodesics of V_n are geodesics of V_m.

A necessary and sufficient condition for a V_n immersed in a V_m, $m > n+1$, be totally geodesic is that

$$\Omega_{a|ij} = 0, \ (a = n+1, ..., m), \ (i, j = 1, ..., n), \tag{68}$$

where $\Omega_{a|ij}$, see (54), are given by

$$\Omega_{a|ij} = a_{\alpha\beta}\, y^{\alpha}_{,i;j}\, \xi^{\beta}_{|a} + [\mu\nu, \beta]_a y^{\mu}_{,i} y^{\nu}_{,j} \xi^{\beta}_{|a}, \tag{69}$$

where $\xi^{\beta}_{|a}$, $a = n+1, ..., m$, denote the components of a particular choice of mutually orthogonal one to another and normal to V_n vectors. The functions $\xi^{\beta}_{|a}$ and $[\mu\nu, \beta]_a$ are invariants for transformations of coordinates x^i in V_n, $y^{\alpha}_{,ij}$ are the components of a symmetric tensor in the $x's$ and $y^{\alpha}_{,i}$ are components of a vector. Hence it follows from (69), that for each value of a the $\Omega_{a|ij}$ are the components of a symmetric tensor in V_n.

One can write $\Omega_{a|ij}$, as $-\Omega_{a|ij} = a_{\alpha\beta}\, y^{\alpha}_{,i}\, \xi^{\beta}_{|a,j} + [\mu\nu, \beta]_a y^{\mu}_{,i} y^{\nu}_{,j} \xi^{\beta}_{|a}$.

In E186 one encounters the following statement: the sub–spaces $y^a =$ const., $a = n+1, ..., m$, are totally geodesic in V_m if the metric of V_m can be decomposed as

$$ds^2 = a_{ij}(y^k)\, dy^i\, dy^j + a_{ab}(y^k, y^c) dy^a dy^b, \ (a, c = n+1, ..., m), \tag{70}$$

$i, k = 1, ..., n$, where the functions $a_{ij}(y^k)$ are independent of $y^{n+1}, ..., y^m$. In general, a_{bc} are functions of y^i and y^c.

7.1. MAXIMALLY SYMMETRIC TOTALLY GEODESIC SUB–SPACES

If metric components a_{ab} of the above metric are independent of y^k, that is $a_{ab} = a_{ab}(y^c)$, the sub-spaces $y^c =$ const. are maximally symmetric totally geodesic in V_m if the metric of V_m can be decomposed as

$$ds^2 = a_{ab}(y^c)\, dy^a dy^b + a_{ij}(y^k)\, dy^i\, dy^j, \ (a, b, c = m+1, ..., n), \tag{71}$$

$i, j, k = 1, ..., m$.

8. THE 2+1 BTZ SOLUTION AS HYPERSURFACE OF THE 3+1 PC METRIC

Let us now consider the Weyl zero Plebański-Carter [A] metric [3, 4]. We shall show that this metric (being locally maximally symmetric by itself) admits locally as a (maximally symmetric) hypersurface the BTZ space-time [5], which occurs to be not totally geodesic.

In order to establish this assertion, we start with the conformally flat Plebański-Carter [A] metric, written as

$$\begin{aligned} ds^2_{W0PC} &= \frac{\Delta}{P}dp^2 + \frac{P}{\Delta}(d\tau + q^2 d\sigma)^2 \\ &\quad + \frac{\Delta}{Q}dq^2 - \frac{Q}{\Delta}(d\tau - p^2 d\sigma)^2, \end{aligned} \tag{72}$$

with structural functions

$$\begin{aligned} P &= \frac{J^2}{4} + Mp^2 + \frac{1}{l^2}p^4, \\ Q &= \frac{J^2}{4} - Mq^2 + \frac{1}{l^2}q^4, \\ \Delta &= p^2 + q^2. \end{aligned} \tag{73}$$

Accomplishing the coordinate transformations

$$\begin{aligned} \tau &= l\phi, \qquad \sigma = -\frac{2}{Jl}t, \\ p^2 &= \frac{1}{2}\left(r^2v^2 - l^2M + \sqrt{(r^2v^2 - l^2M)^2 + J^2l^2(v^2-1)}\right), \\ q^2 &= \frac{1}{2}\left(-r^2v^2 + l^2M + \sqrt{(r^2v^2 - l^2M)^2 + J^2l^2(v^2-1)}\right), \end{aligned} \tag{74}$$

the metric (72) is reducible to

$$ds^2_{W0PC} = (v^2-1)^{-1}\, l^2\, dv^2 + v^2 ds^2_{BTZ}, \tag{75}$$

where ds^2_{BTZ} corresponds to the metric of the well known BTZ black hole of the (2+1)-gravity:

$$\begin{aligned} ds^2_{BTZ} &= -\left(\frac{J^2}{4r^2} - M + \frac{1}{l^2}r^2\right)dt^2 + \frac{dr^2}{\frac{J^2}{4r^2} - M + \frac{1}{l^2}r^2} \\ &\quad + r^2\left(\frac{-J}{2r^2}dt + d\phi\right)^2. \end{aligned} \tag{76}$$

Consequently, the metric (75), ds^2_{W0PC}, allows locally for the locally maximally symmetric subspace BTZ, ds^2_{BTZ}, but because of the presence of the factor v^2 in equation (75) in front of this subspace metric, the space BTZ is not a totally geodesic hypersurface.

On may consider the BTZ metric as a spatial slice of the W0PC metric given as

$$ds^2_{W0PC} = l^2 d\theta^2 + \cosh^2\theta ds^2_{BTZ}, \tag{77}$$

in the sense of limiting transition, namely by setting θ tends to zero in (77), in the slice $\theta = 0$ the metric ds^2_{W0PC} becomes the ds^2_{BTZ} metric.

8.1. THE STATIC BTZ SOLUTION AS ANTI–DE–SITTER METRIC

Consider the BTZ metric (76), equate in it J to zero, and replace t by τ, $t = \tau$, the resulting metric is the so called static BTZ:

$$ds^2 = -(-M + \frac{r^2}{l^2})d\tau^2 + \frac{1}{-M + \frac{r^2}{l^2}}\, dr^2 + r^2 d\phi^2. \tag{78}$$

Accomplishing in the above expression the coordinate transformations

$$\begin{aligned} \tau &= \frac{l}{\sqrt{M}}\operatorname{arctanh}\left(\frac{t}{z}\right), \\ r &= \sqrt{M}\sqrt{z^2 - t^2 + l^2}, \\ \phi &= \frac{1}{\sqrt{M}}\operatorname{arcsinh}\left(\frac{y}{\sqrt{z^2 - t^2 + l^2}}\right), \end{aligned} \tag{79}$$

one arrives at

$$ds^2 = dy^2 + dz^2 - dt^2 - \frac{(ydy + zdz - tdt)^2}{y^2 + z^2 - t^2 + l^2}. \tag{80}$$

The inverse transformations are given by

$$\begin{aligned} t &= \frac{l}{\sqrt{M}}\sqrt{-M + \frac{r^2}{l^2}}\sinh\left(\frac{\sqrt{M}}{l}\tau\right), \\ z &= \frac{l}{\sqrt{M}}\sqrt{-M + \frac{r^2}{l^2}}\cosh\left(\frac{\sqrt{M}}{l}\tau\right), \\ y &= \frac{l}{\sqrt{M}}\sinh\left(\phi\sqrt{M}\right), \end{aligned} \tag{81}$$

where z and t are restricted to the values

$$z^2 - t^2 = \frac{l^2}{\sqrt{M}}\left(-M + \frac{r^2}{l^2}\right) \geq 0. \tag{82}$$

The jacobian $J = \frac{\partial(t,z,y)}{\partial(\tau,\phi,q)}$ occurs to be $J = -\frac{r^2}{l\sqrt{M}} \cosh\left(\sqrt{M}\phi\right)$.

On the other hand, if one were considering the embedding into four dimensions:

$$\begin{aligned} ds^2 &= -dx^2 + dy^2 + dz^2 - dt^2 \\ & y^2 - x^2 + z^2 - t^2 = l^2, \end{aligned} \tag{83}$$

the transformations to the metric (78) amount to

$$\begin{aligned} x \pm y &= \frac{r}{\sqrt{M}} \exp\left(\pm\sqrt{M}\,\phi\right), \\ z \pm t &= \frac{l}{\sqrt{M}} \sqrt{-M + \frac{r^2}{l^2}} \exp\left(\pm\frac{\sqrt{M}}{l}\tau\right). \end{aligned} \tag{84}$$

Thus, the static BTZ is just a maximally symmetric 3–dimensional space, which can be identified with the three dimensional anti–de–Sitter space.

Acknowledgments

I dedicate this work to Heinz Dehnen and Dietrich Kramer. This research was partially supported by CONACYT: Grant 32138.

References

[1] S. Weinberg, Gravitation and Cosmology: principles and applications of the general theory of relativity. (John Wiley and Sons, Inc. New York, 1972).

[2] L.P. Eisenhart, Riemannian Geometry. (Princeton Univ. Press, USA, sixth printing, 1966).

[3] M. Cataldo, S. del Campo, A. García, "BTZ black hole from (3+1) gravity", gr-qc/ 0004023.

[4] B. Carter, Commun. Math. Phys. 10, 280 (1968), idem, Phys. Lett. A 26, 399 (1968); J. Plebański, Ann Phys. (USA) 90, 196 (1975).

[5] M. Bañados, C. Teitelboim and J. Zanelli, Phys. Rev. Lett. 69, 1849 (1992).

DISCUSSION OF THE THETA FORMULA FOR THE ERNST POTENTIAL OF THE RIGIDLY ROTATING DISK OF DUST

Andreas Kleinwächter *
Friedrich-Schiller-Universität Jena
Theoretisch-Physikalisches-Institut
Max-Wien-Platz 1, D-07743 Jena, Germany

Abstract The exact global solution of the rigidly rotating disk of dust[1] is given in terms of ultraelliptic functions. Here we discuss the "theta formula" for the Ernst potential[2]. The space-time coordinates of the problem enter the arguments of these functions via ultraelliptic line integrals which are related to a Riemann surface.

The solution is reformulated so as to make it easier to handle and all integrals are transformed into definite real integrals. For the axis of symmetry and the plane of the disk these general formulae can be reduced to standard elliptic functions and elliptic integrals.

Keywords: Ernst potentials, rotating dust disk, exact solutions.

1. THE THETA FORMULA FOR THE ERNST POTENTIAL

The problem of the relativistically rotating disk of dust was first formulated and approximately solved by Bardeen and Wagoner[3, 4]. It was analytically solved by Neugebauer and Meinel[5, 6, 1]. In[1] the global solution is given in terms of ultraelliptic functions. Here we start from the solution for the Ernst potential in terms of hyperelliptic theta functions[2]. The space-time of the rigidly rotating disk of dust is described by Weyl–Lewis–Papapetrou coordinates

$$ds^2 = e^{-2U}[e^{2k}(d\rho^2 + d\zeta^2) + \rho^2 d\phi^2] - e^{2U}(dt + a d\phi)^2 \,. \qquad (1)$$

Due to the axisymmetry of the problem, the metric functions U, a and k are functions of ρ and ζ alone. In these coordinates, the vacuum Einstein

* E-mail: kleinwaechter@tpi.uni-jena.de

equations are equivalent to the Ernst equation

$$(\Re f)(f_{,\rho\rho} + f_{,\zeta\zeta} + \frac{1}{\rho} f_{,\rho}) = f_{,\rho}^2 + f_{,\zeta}^2 \tag{2}$$

for the complex Ernst potential $f(\rho, \zeta) = e^{2U}(\rho, \zeta) + i\, b(\rho, \zeta)$ where the imaginary part is related to the metric coefficient a via $a_{,\rho} = \rho\, e^{-4U} b_{,\zeta}$ and $a_{,\zeta} = -\rho\, e^{-4U} b_{,\rho}$.
The boundary value problem for the Ernst potential is described in detail in[7]. These theta functions were first introduced by Rosenhain[8] and are a generalization of the Jacobian theta functions. They depend on five arguments: two main arguments and three moduli. The definition of the four Jacobi theta functions $\vartheta_1, \ldots, \vartheta_4$ and of the 16 Rosenhain functions $\vartheta_{n,k}$ $(n, k = 1, 2, 3, 4)$ can be found in the appendix. The solution in terms of ultraelliptic theta functions. is:

$$f = e^{-(\gamma_0 u + \gamma_1 v + \mu w)} \frac{\vartheta_{4,4}(\alpha_0 u + \alpha_1 v - C_1, \beta_0 u + \beta_1 v - C_2;\, B_{11}, B_{22}, B_{12})}{\vartheta_{4,4}(\alpha_0 u + \alpha_1 v + C_1, \beta_0 u + \beta_1 v + C_2;\, B_{11}, B_{22}, B_{12})}. \tag{3}$$

The solution exists and is regular everywhere outside the disk in the range $0 < \mu \leq \mu_0 = 4.62966\ldots$, where $\mu = 2\Omega^2 \rho_0^2 e^{-2V_0}$ (Ω : constant angular velocity, $\exp(2V_0) = \exp(2U(0,0))$, ρ_0 : radius of the disk). The arguments $\alpha_0, \alpha_1, \beta_0, \beta_1, \gamma_0, \gamma_1, C_1, C_2, B_{11}, B_{22}, B_{12}$ of the "theta formula" for f depend via ultraelliptic line integrals on the normalized space-time coordinates $x = \rho/\rho_0$, $y = \zeta/\rho_0$ and on the parameter μ. The functions u, v and w (also depending on x, y, μ) are explicitly known. In the following the arguments (using normalized coordinates $x = \rho/\rho_0, y = \zeta/\rho_0$) are explained. The convention for the root W_1 is $\Re W_1 < 0$ for $(x, y) \notin \{(r, 0) : 0 \leq r \leq 1\}$. The functions u, v and w are given by

$$u = \int_{-i}^{i} \frac{h\, \mathrm{d}X}{W_1}, \quad v = \int_{-i}^{i} \frac{h\, X \mathrm{d}X}{W_1}, \quad w = \int_{-i}^{i} \frac{h\, X^2 \mathrm{d}X}{W_1}. \tag{4}$$

$$h = \frac{\ln(\sqrt{1 + \mu^2(1 + X^2)^2} + \mu(1 + X^2))}{\pi i \sqrt{1 + \mu^2(1 + X^2)^2}}, \quad W_1 = \sqrt{(X - y)^2 + x^2} \tag{5}$$

The normalization parameters $\alpha_0, \alpha_1, \beta_0, \beta_1, \gamma_0, \gamma_1$, the moduli B_{mn} of the theta functions and the quantities C_1, C_2 are defined via integrals on the two sheets of the hyperelliptic Riemann surface (see Figure 1) related to

$$W = \mu \sqrt{(X - X_1)(X - \overline{X_1})(X - X_2)(X - \overline{X_2})(X - (y + ix))(X - \overline{(y + ix)})} \tag{6}$$

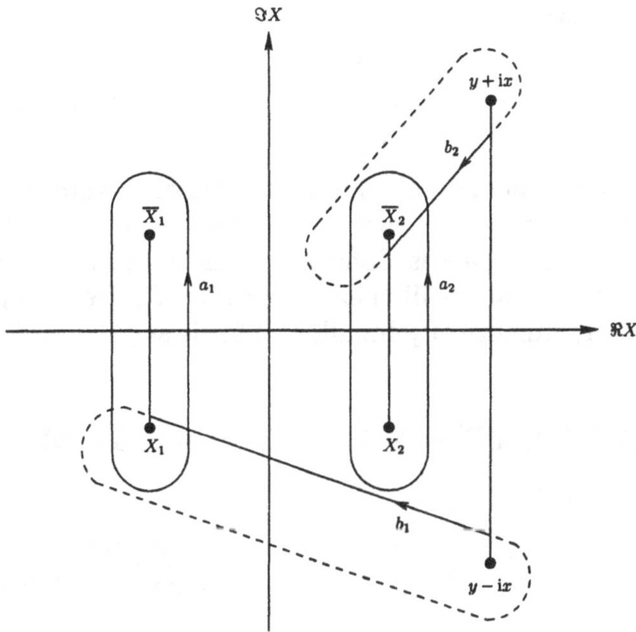

Figure 1. Riemann surface with cuts between the branch points X_1 and $\bar{X}_1$, X_2 and $\bar{X}_2$, $y - ix$ and $y + ix$. Also shown are the four periods a_m and b_m $(m = 1, 2)$. Unbroken lines belong to the upper sheet, dashed lines to the lower one.

with $X_1 = -\frac{1}{\sqrt{2\mu}}(\sqrt{\sqrt{1+\mu^2} - \mu} + i\sqrt{\sqrt{1+\mu^2} + \mu})$ and $X_2 = -\overline{X_1}$.

The value of (6) in the upper sheet of the surface is defined by $W \to +\mu X^3$ for $X \to \infty$.

The two normalized Abelian differentials of the first kind are defined by

$$\mathrm{d}\omega_1 := \alpha_0 \frac{\mathrm{d}X}{W} + \alpha_1 \frac{X\mathrm{d}X}{W}, \quad \mathrm{d}\omega_2 := \beta_0 \frac{\mathrm{d}X}{W} + \beta_1 \frac{X\mathrm{d}X}{W} \tag{7}$$

$$\text{with} \quad \oint_{a_m} \mathrm{d}\omega_n \stackrel{!}{=} i\pi\delta_{mn} \quad (m = 1, 2;\, n = 1, 2)\,. \tag{8}$$

The values of γ_0 and γ_1 have to be determined by integration over an Abelian differential of the third kind

$$\mathrm{d}\omega := \gamma_0 \frac{\mathrm{d}X}{W} + \gamma_1 \frac{X\mathrm{d}X}{W} + \mu \frac{X^2\mathrm{d}X}{W} \tag{9}$$

$$\text{with the requirement:} \quad \oint_{a_m} d\omega \stackrel{!}{=} 0 \quad (m = 1, 2)\,. \tag{10}$$

The remaining quantities are defined to be

$$B_{mn} := \oint_{b_m} d\omega_n \quad \text{and} \quad C_m := \int_{\infty^+}^{y-ix} d\omega_m \quad (m = 1,2\,;\, n = 1,2)\,. \tag{11}$$

The goal is now to calculate all the arguments once and for all so that for applications it is not necessary to consider the Riemann surface and Abelian differentials. This is done in the next section. Using these results, the theta formula will then be recast. The resulting formulae can be very easily numerically handled. This is summarized in the last section.

2. CALCULATION OF THE ARGUMENTS

Using the notation

$$\xi_1 \;:=\; -\frac{1}{\sqrt{2\mu}}\sqrt{\sqrt{1+\mu^2}-\mu}\,, \eta_1 \;:=\; \frac{1}{\sqrt{2\mu}}\sqrt{\sqrt{1+\mu^2}+\mu}\,, \tag{12}$$

$$\xi_2 \;:=\; \frac{1}{\sqrt{2\mu}}\sqrt{\sqrt{1+\mu^2}-\mu}\,, \quad \eta_2 \;:=\; \frac{1}{\sqrt{2\mu}}\sqrt{\sqrt{1+\mu^2}+\mu}\,, \tag{13}$$

$$\xi_3 \;:=\; y\,, \qquad \eta_3 \;:=\; x\,. \tag{14}$$

the cuts are than denoted by $V(\xi_1, \eta_1), V(\xi_2, \eta_2)$ and $V(\xi_3, \eta_3)$ with:

$$V(\xi_i, \eta_i) := \{\, \xi_i + s\eta_i \;:\; s \in (-1,1)\,\}\,. \tag{15}$$

Now the behavior of the "fundamental root" (6) is carefully investigated. Therefore the "elementary root" for the complex quantity $X = r + i\,s$ jumping at the cut $V(a,b)$ is introduced as:

$$\mathcal{W}(r,s;a,b) \;:=\; (X-a)\sqrt{1+\frac{b^2}{(X-a)^2}} \qquad (r, s \text{ and } a, b \text{ real}) \tag{16}$$

where the choice $\Re\sqrt{\ldots} \geq 0$ is made. The fundamental root (6) can be built up from (16) using the definitions (12)–(14) by:

$$W(r,s;\mu,x,y) = \mu\,\mathcal{W}(r,s;\xi_1,\eta_1)\,\mathcal{W}(r,s;\xi_2,\eta_2)\,\mathcal{W}(r,s;\xi_3,\eta_3)\,. \tag{17}$$

The radicand $\mathcal{R}(r,s;a,b)$ of (16) is given by:

$$1+\frac{b^2}{(X-a)^2} \;=\; \mathcal{R}_R(r,s;a,b) \,+\, i\,\mathcal{R}_I(r,s;a,b)\,, \tag{18}$$

$$\mathcal{R}_R(r,s;a,b) \;:=\; 1+\frac{b^2[(r-a)^2-s^2]}{[(r-a)^2+s^2]^2}\,, \tag{19}$$

$$\mathcal{R}_I(r,s;a,b) \;:=\; -2s\,\frac{(r-a)b^2}{[(r-a)^2+s^2]^2}\,. \tag{20}$$

With the notation $\mathcal{W} = (X-a)(\mathcal{S}_R + i\,\mathcal{S}_I)$ one finds further:

$$\mathcal{S}_R(r,s;a,b) = \frac{1}{\sqrt{2}}\sqrt{\sqrt{\mathcal{R}_R^2+\mathcal{R}_I^2}+\mathcal{R}_R}\,, \tag{21}$$

$$\mathcal{S}_I(r,s;a,b) = \operatorname{sign}(\mathcal{R}_I)\frac{1}{\sqrt{2}}\sqrt{\sqrt{\mathcal{R}_R^2+\mathcal{R}_I^2}-\mathcal{R}_R}\,, \tag{22}$$

$$\operatorname{sign}(\mathcal{R}_I) = -\operatorname{sign}(s)\operatorname{sign}(r-a)\,. \tag{23}$$

The roots that appear here are roots of positive, real quantities and must always be taken to have positive sign. The definition

$$g_\pm(r,s;a,b) := \frac{1}{\sqrt{2}}\sqrt{\sqrt{\mathcal{R}_R^2+\mathcal{R}_I^2}\pm\mathcal{R}_R} \tag{24}$$

$$= \frac{1}{\sqrt{2}}\Big\{\frac{\sqrt{[\,(r-a)^2+s^2+b^2\,]^2-4b^2s^2}}{(r-a)^2+s^2} \pm \{1+\frac{b^2[(r-a)^2-s^2]}{[(r-a)^2+s^2]^2}\}\Big\}^{1/2} \tag{25}$$

leads to the following representation of the elementary root

$$\mathcal{W}(r,s;a,b) = \mathcal{W}_R(r,s;a,b) + i\,\mathcal{W}_I(r,s;a,b)\,, \tag{26}$$

$$\mathcal{W}_R(r,s;a,b) = \operatorname{sign}(r-a)\,[\,|r-a|g_+(r,s;a,b) + |s|\,g_-(r,s;a,b)\,], \tag{27}$$

$$\mathcal{W}_I(r,s;a,b) = \operatorname{sign}(s)\,[\,|s|\,g_+(r,s;a,b) - |r-a|\,g_-(r,s;a,b)\,]\,. \tag{28}$$

From these equations one can observe the following properties for $\mathcal{W}$ and W which will be used later to calculate the integrals on the Riemann surface.

$$\lim_{\varepsilon\to 0}\mathcal{W}(a\pm\varepsilon,s;a,b) = \pm\sqrt{b^2-s^2} \qquad (\,s\le b\,)\,, \tag{29}$$

$$\mathcal{W}(r,s=0;a,b) = \operatorname{sign}(r-a)\sqrt{(r-a)^2+b^2}\,, \tag{30}$$

$$\mathcal{W}(r,-s;a,b) = \overline{\mathcal{W}(r,s;a,b)}\,. \tag{31}$$

For the fundamental root it leads finally to

$$W(r,-s;\mu,x,y) = \overline{W(r,s;\mu,x,y)}\,, \tag{32}$$

$$W(r=\xi_i-\varepsilon,s;\mu,x,y) = -W(r=\xi_i+\varepsilon,s;\mu,x,y) \tag{33}$$

$$(i\in\{1,2,3\}\,,\ |s|<\xi_i\,,\ \varepsilon \text{ sufficiently small})\,,$$

$$W(r,s=0;\mu,x,y) = \operatorname{sign}[(r-\xi_1)(r-\xi_2)(r-\xi_3)] \times\mu\sqrt{(r-\xi_1)^2+\eta_1^2}\sqrt{(r-\xi_2)^2+\eta_2^2}\sqrt{(r-\xi_3)^2+\eta_3^2}\,. \tag{34}$$

In the following formulae (i,j,k) is a permutation of $(1,2,3)$. With

$$G_R(s;a_i) := \mathcal{W}_R(\xi_i,s;\xi_j,\eta_j)\,\mathcal{W}_R(\xi_i,s;\xi_k,\eta_k) - \mathcal{W}_I(\xi_i,s;\xi_j,\eta_j)\,\mathcal{W}_I(\xi_i,s;\xi_k,\eta_k)\,,$$

$$G_I(s;a_i) := \mathcal{W}_R(\xi_i,s;\xi_j,\eta_j)\,\mathcal{W}_I(\xi_i,s;\xi_k,\eta_k) + \mathcal{W}_I(\xi_i,s;\xi_j,\eta_j)\,\mathcal{W}_R(\xi_i,s;\xi_k,\eta_k)$$

one gets for the fundamental root along the cut $V(\xi_i,\eta_i)$:

$$\begin{aligned} W_\pm^{a_i}(s;\mu,x,y) &:= \lim_{\varepsilon\to 0} W(\xi_i\pm\varepsilon,s;\mu,x,y) \\ &= \pm\mu\sqrt{\eta_i^2-s^2}\,\mathcal{W}(\xi_i,s;\xi_j,\eta_j)\,\mathcal{W}(\xi_i,s;\xi_k,\eta_k) \\ &= \pm\mu\sqrt{\eta_i^2-s^2}\,[G_R(s;a_i)+i\,G_I(s;a_i)]\,. \end{aligned}$$

For the calculation of the normalizing coefficients $\alpha_0,\alpha_1,\beta_0,\beta_1,\gamma_0,\gamma_1$ and the quantities B_{11},B_{12},B_{22} the following notations are introduced:

$$\mathcal{A}_i^n(\mu,x,y) := \oint_{a_i}\frac{X^n\,\mathrm{d}X}{W(X)} = 2i\int_{-\eta_i}^{\eta_i}\frac{(\xi_i+i\,s)^n\,\mathrm{d}s}{W_+^{a_i}(s;\mu,x,y)}\,. \qquad (35)$$

As an example $\mathcal{A}_i^0$ is investigated:

$$\begin{aligned} \mathcal{A}_i^0(\mu,x,y) &= 2i\int_{-\eta_i}^{\eta_i}\frac{G_R(s;a_i)-i\,G_I(s;a_i)}{G_R^2(s;a_i)+G_I^2(s;a_i)}\frac{\mu^{-1}\,ds}{\sqrt{\eta_i^2-s^2}} \\ &= 2\int_{-\eta_i}^{\eta_i}\frac{G_I(s;a_i)}{G_R^2(s;a_i)+G_I^2(s;a_i)}\frac{\mu^{-1}\,ds}{\sqrt{\eta_i^2-s^2}} \\ &+ 2i\int_{-\eta_i}^{\eta_i}\frac{G_R(s;a_i)}{G_R^2(s;a_i)+G_I^2(s;a_i)}\frac{\mu^{-1}\,ds}{\sqrt{\eta_i^2-s^2}}. \end{aligned}$$

Due to the symmetry properties of $G_I(s;a_i)$ the real part of the equation above vanishes. Nevertheless for the calculation of B_{11},B_{12},B_{22} it is suitable to use the corresponding integrals over the interval $(0,\eta_i)$. Therefore the integrals

$$\mathcal{L}_i^0(\mu,x,y) := \int_0^{\eta_i}\frac{G_R(s;a_i)}{G_R^2(s;a_i)+G_I^2(s;a_i)}\frac{\mu^{-1}\,ds}{\sqrt{\eta_i^2-s^2}}, \qquad (36)$$

$$\mathcal{L}_i^1(\mu,x,y) := \int_0^{\eta_i}\frac{\xi_iG_R(s;a_i)+sG_I(s;a_i)}{G_R^2(s;a_i)+G_I^2(s;a_i)}\frac{\mu^{-1}\,ds}{\sqrt{\eta_i^2-s^2}}, \qquad (37)$$

$$\mathcal{L}_i^2(\mu,x,y) := \int_0^{\eta_i}\frac{(\xi_i^2-s^2)G_R(s;a_i)+2\xi_isG_I(s;a_i)}{G_R^2(s;a_i)+G_I^2(s;a_i)}\frac{\mu^{-1}\,ds}{\sqrt{\eta_i^2-s^2}}, \qquad (38)$$

$$\mathcal{I}_i^0(\mu,x,y) := \int_0^{\eta_i}\frac{G_I(s;a_i)}{G_R^2(s;a_i)+G_I^2(s;a_i)}\frac{\mu^{-1}\,ds}{\sqrt{\eta_i^2-s^2}}, \qquad (39)$$

$$\mathcal{I}_i^1(\mu,x,y) \;:=\; \int\limits_0^{\eta_i} \frac{\xi_i G_I(s;a_i) - sG_R(s;a_i)}{G_R^2(s;a_i)+G_I^2(s;a_i)} \frac{\mu^{-1}\,ds}{\sqrt{\eta_i^2 - s^2}}, \tag{40}$$

$$\mathcal{I}_i^2(\mu,x,y) \;:=\; \int\limits_0^{\eta_i} \frac{(\xi_i^2 - s^2)G_I(s;a_i) - 2\xi_i s G_R(s;a_i)}{G_R^2(s;a_i)+G_I^2(s;a_i)} \frac{\mu^{-1}\,ds}{\sqrt{\eta_i^2 - s^2}} \tag{41}$$

are defined and hence

$$\mathcal{A}_i^n(\mu,x,y) = 4\,i\;\mathcal{L}_i^n(\mu,x,y)\,. \tag{42}$$

For the calculation of the moduli B_{mn} and of C_1, C_2 integrals of the type, using (34),

$$\mathcal{K}_{c,d}^n(\mu,x,y) = \int\limits_c^d \frac{r^n\,\mathrm{d}r}{W(r,s=0;\mu,x,y)} \qquad (n=0,1)\,, \tag{43}$$

and the integrals $\mathcal{I}_i^0$ and $\mathcal{I}_i^1$ $(i = 1,2)$ are used. The only imaginary contribution in this context is given by

$$i\,(\alpha_0\mathcal{L}_1^0 + \alpha_1\mathcal{L}_1^1) = \frac{i\,\pi}{4}, \qquad i\,(\beta_0\mathcal{L}_1^0 + \beta_1\mathcal{L}_1^1) = 0\,, \tag{44}$$

$$i\,(\alpha_0\mathcal{L}_2^0 + \alpha_1\mathcal{L}_2^1) = 0\,, \qquad i\,(\beta_0\mathcal{L}_2^0 + \beta_1\mathcal{L}_2^1) = \frac{i\,\pi}{4}, \tag{45}$$

$$i\,(\alpha_0\mathcal{L}_3^0 + \alpha_1\mathcal{L}_3^1) = -\frac{i\,\pi}{4}, \qquad i\,(\beta_0\mathcal{L}_3^0 + \beta_1\mathcal{L}_3^1) = -\frac{i\,\pi}{4}\,. \tag{46}$$

The equations (44) and (45) follow immediately from what precedes and (46) is a consequence of them.

Calculation of $\alpha_0, \alpha_1, \beta_0, \beta_1$:

Due to the equations (7),(8) and (42) the following linear systems are to be solved:

$$\begin{pmatrix} \mathcal{L}_1^0 & \mathcal{L}_1^1 \\ \mathcal{L}_2^0 & \mathcal{L}_2^1 \end{pmatrix} \begin{pmatrix} \alpha_0 \\ \alpha_1 \end{pmatrix} = \begin{pmatrix} \pi/4 \\ 0 \end{pmatrix},$$

$$\begin{pmatrix} \mathcal{L}_1^0 & \mathcal{L}_1^1 \\ \mathcal{L}_2^0 & \mathcal{L}_2^1 \end{pmatrix} \begin{pmatrix} \beta_0 \\ \beta_1 \end{pmatrix} = \begin{pmatrix} 0 \\ \pi/4 \end{pmatrix}.$$

With $D_a := \mathcal{L}_1^0\mathcal{L}_2^1 - \mathcal{L}_2^0\mathcal{L}_1^1$ the solution is

$$\alpha_0 = \frac{\pi}{4D_a}\mathcal{L}_2^1, \quad \alpha_1 = -\frac{\pi}{4D_a}\mathcal{L}_2^0, \quad \beta_0 = -\frac{\pi}{4D_a}\mathcal{L}_1^1, \quad \beta_1 = \frac{\pi}{4D_a}\mathcal{L}_1^0\,. \tag{47}$$

Calculation of γ_1 and γ_2:
The definition (9) and the condition (13) lead to the system

$$\begin{pmatrix} \mathcal{L}_1^0 & \mathcal{L}_1^1 \\ \mathcal{L}_2^0 & \mathcal{L}_2^1 \end{pmatrix} \begin{pmatrix} \gamma_0 \\ \gamma_1 \end{pmatrix} = -\mu \begin{pmatrix} \mathcal{L}_1^2 \\ \mathcal{L}_2^2 \end{pmatrix}$$

with the solution

$$\gamma_0 = -\mu \frac{\mathcal{L}_1^2 \mathcal{L}_2^1 - \mathcal{L}_2^2 \mathcal{L}_1^1}{D_a}, \qquad \gamma_1 = -\mu \frac{\mathcal{L}_1^0 \mathcal{L}_2^2 - \mathcal{L}_2^0 \mathcal{L}_1^2}{D_a}. \tag{48}$$

Calculation of the moduli B_{11}, B_{22}, B_{12} and of C_1, C_2:
It turns out that for the calculation of these quantities two cases have to be discussed separately, namely $y > \Re X_2$ (that means the cut $V(\xi_3, \eta_3)$ is on the right of the cut $V(\xi_2, \eta_2)$; see Figure 2) and $y < \Re X_2$ (that means the cut $V(\xi_3, \eta_3)$ is on the left of $V(\xi_2, \eta_2)$; see Figure 3). In both cases the path of integration is divided in suitable parts.
The case $y > \Re X_2$:

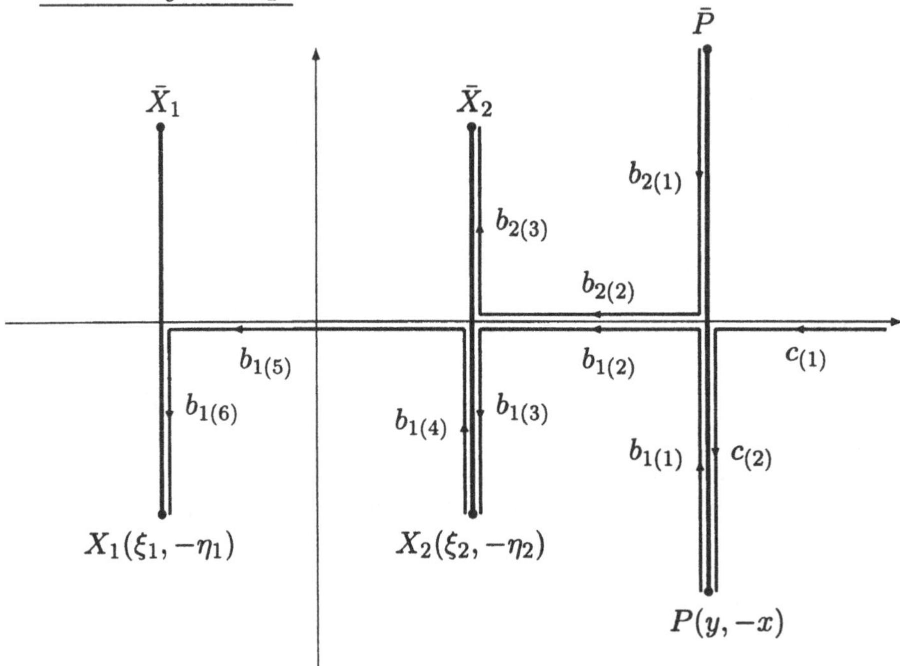

FIG. 2 — Complex plane with cuts $V(\xi_1, \eta_1)$, $V(\xi_2, \eta_2)$ and $V(\xi_3 = y, \eta_3 = x)$ for the case $y > \Re X_2$ (This case corresponds to the Riemann surface shown in Figure 1);
$B_{mn} = \int_{b_m} d\omega_n = \sum_l \int_{b_{m(l)}} d\omega_n$, $\quad C_n = \int_{+\infty}^{P} d\omega_n = \sum_l \int_{c_{(l)}} d\omega_n$.

From Figure 2 follows:

$$B_{11} = 2\{\alpha_0(\mathcal{I}_3^0 + \mathcal{K}_{y,\xi_2}^0 + 2\mathcal{I}_2^0 + \mathcal{K}_{\xi_2,\xi_1}^0 + \mathcal{I}_1^0) + \tag{49}$$

$$\alpha_1(\mathcal{I}_3^1 + \mathcal{K}_{y,\xi_2}^1 + 2\mathcal{I}_2^1 + \mathcal{K}_{\xi_2,\xi_1}^1 + \mathcal{I}_1^1)\}, \tag{50}$$

$$B_{22} = 2\{\beta_0(\mathcal{I}_3^0 + \mathcal{K}_{y,\xi_2}^0 + \mathcal{I}_2^0) + \beta_1(\mathcal{I}_3^1 + \mathcal{K}_{y,\xi_2}^1 + \mathcal{I}_2^1)\}, \tag{51}$$

$$B_{12} = 2\{\beta_0(\mathcal{I}_3^0 + \mathcal{K}_{y,\xi_2}^0 + 2\mathcal{I}_2^0 + \mathcal{K}_{\xi_2,\xi_1}^0 + \mathcal{I}_1^0) + \tag{52}$$

$$\beta_1(\mathcal{I}_3^1 + \mathcal{K}_{y,\xi_2}^1 + 2\mathcal{I}_2^1 + \mathcal{K}_{\xi_2,\xi_1}^1 + \mathcal{I}_1^1) - \frac{i\pi}{4}\}, \tag{53}$$

$$B_{21} = 2\{\alpha_0(\mathcal{I}_3^0 + \mathcal{K}_{y,\xi_2}^0 + \mathcal{I}_2^0) + \alpha_1(\mathcal{I}_3^1 + \mathcal{K}_{y,\xi_2}^1 + \mathcal{I}_2^1) - \frac{i\pi}{4}\}, \tag{54}$$

$$C_1 = \alpha_0(\mathcal{K}_{\infty,y}^0 + \mathcal{I}_3^0) + \alpha_1(\mathcal{K}_{\infty,y}^1 + \mathcal{I}_3^1) + \frac{i\pi}{4}, \tag{55}$$

$$C_2 = \beta_0(\mathcal{K}_{\infty,y}^0 + \mathcal{I}_3^0) + \beta_1(\mathcal{K}_{\infty,y}^1 + \mathcal{I}_3^1) + \frac{i\pi}{4}. \tag{56}$$

The case $y < \Re X_2$:

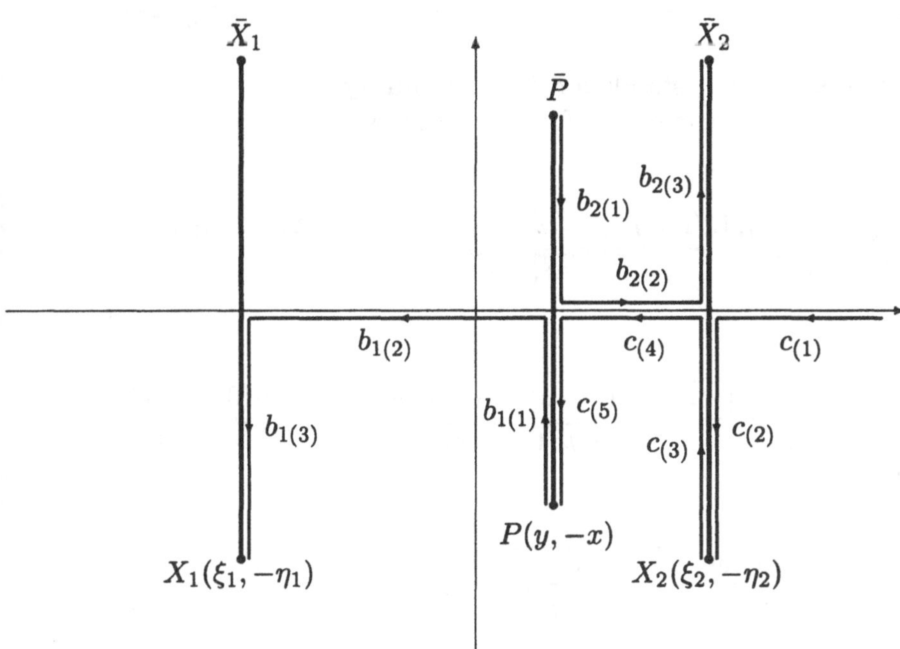

FIG. 3 — Complex plane with cuts $V(\xi_1,\eta_1)$,$V(\xi_2,\eta_2)$ and $V(\xi_3 = y, \eta_3 = x)$ for the case $y < \Re X_2$; $B_{mn} = \int_{b_m} d\omega_n = \sum_l \int_{b_{m(l)}} d\omega_n$, $C_n = \int_{+\infty}^{P} d\omega_n = \sum_l \int_{c_{(l)}} d\omega_n$.

From Figure 3 follows:

$$B_{11} = 2\{\alpha_0(\mathcal{I}_3^0 + \mathcal{K}_{y,\xi_1}^0 + \mathcal{I}_1^0) + \alpha_1(\mathcal{I}_3^1 + \mathcal{K}_{y,\xi_1}^1 + \mathcal{I}_1^1)\}, \tag{57}$$

$$B_{22} = 2\{\beta_0(-\mathcal{I}_3^0 + \mathcal{K}_{y,\xi_2}^0 - \mathcal{I}_2^0) + \beta_1(-\mathcal{I}_3^1 + \mathcal{K}_{y,\xi_2}^1 - \mathcal{I}_2^1)\}, \tag{58}$$

$$B_{12} = 2\{\beta_0(\mathcal{I}_3^0 + \mathcal{K}_{y,\xi_1}^0 + \mathcal{I}_1^0) + \beta_1(\mathcal{I}_3^1 + \mathcal{K}_{y,\xi_1}^1 + \mathcal{I}_1^1) + \frac{i\pi}{4}\}, \tag{59}$$

$$B_{21} = 2\{\alpha_0(-\mathcal{I}_3^0 + \mathcal{K}_{y,\xi_2}^0 - \mathcal{I}_2^0) + \alpha_1(-\mathcal{I}_3^1 + \mathcal{K}_{y,\xi_2}^1 - \mathcal{I}_2^1) + \frac{i\,\pi}{4}\} \quad (60)$$

$$\begin{aligned} C_1 &= \alpha_0(\mathcal{K}_{\infty,\xi_2}^0 + 2\mathcal{I}_2^0 + \mathcal{K}_{\xi_2,y}^0 + \mathcal{I}_3^0) \\ &+ \alpha_1(\mathcal{K}_{\infty,\xi_2}^1 + 2\mathcal{I}_2^1 + \mathcal{K}_{\xi_2,y}^1 + \mathcal{I}_3^1) + \frac{i\,\pi}{4}, \end{aligned} \quad (61)$$

$$\begin{aligned} C_2 &= \beta_0(\mathcal{K}_{\infty,\xi_2}^0 + 2\mathcal{I}_2^0 + \mathcal{K}_{\xi_2,y}^0 + \mathcal{I}_3^0) \\ &+ \beta_1(\mathcal{K}_{\infty,\xi_2}^1 + 2\mathcal{I}_2^1 + \mathcal{K}_{\xi_2,y}^1 + \mathcal{I}_3^1) - \frac{i\,\pi}{4}. \end{aligned} \quad (62)$$

The factor 2 is due to the fact that on the Riemann surface (see Figure 1) one also has to integrate the corresponding return path in the lower sheet for calculating the moduli. With the general result $B_{12} = B_{21}$ a further simplification of the formulae is possible.

The imaginary parts of the equations follow directly from (44),(45) and (46).

The symbol ∞ in $\mathcal{K}_{\infty,y}^n$ stands for "+ real infinity".

Using symmetry properties the functions u, v and w (4) can be calculated:

$$u(\mu,x,y) = \frac{2}{\pi}\int_0^1 \frac{\ln[\sqrt{1+\mu^2(1-s^2)^2} + \mu(1-s^2)]}{\sqrt{1+\mu^2(1-s^2)^2}} \, \frac{\mathcal{W}_R(0,s;y,x)}{\mathcal{W}_R^2 + \mathcal{W}_I^2} \, \mathrm{d}s\,, \quad (63)$$

$$v(\mu,x,y) = \frac{2}{\pi}\int_0^1 \frac{\ln[\sqrt{1+\mu^2(1-s^2)^2} + \mu(1-s^2)]}{\sqrt{1+\mu^2(1-s^2)^2}} \, \frac{s\mathcal{W}_I(0,s;y,x)}{\mathcal{W}_R^2 + \mathcal{W}_I^2} \, \mathrm{d}s\,, \quad (64)$$

$$w(\mu,x,y) = -\frac{2}{\pi}\int_0^1 \frac{\ln[\sqrt{1+\mu^2(1-s^2)^2} + \mu(1-s^2)]}{\sqrt{1+\mu^2(1-s^2)^2}} \, \frac{s^2\mathcal{W}_R(0,s;y,x)}{\mathcal{W}_R^2 + \mathcal{W}_I^2} \, \mathrm{d}s\,. \quad (65)$$

3. TRANSFORMATION OF THE THETA FORMULA

Now, using relations between Rosenhain's theta functions, the solution for the Ernst potential is transformed in such a way that all arguments become real.

$$\vartheta_{4,4}(x_1,x_2;B_{11},B_{22},B_{12}) =$$

$$\vartheta_{3,3}(x_1+x_2,x_1-x_2;\; B_{11}+2B_{12}+B_{22}, B_{11}-2B_{12}+B_{22}, B_{11}-B_{22})$$

$$-\vartheta_{2,2}(x_1+x_2,x_1-x_2;\,B_{11}+2B_{12}+B_{22},B_{11}-2B_{12}+B_{22},B_{11}-B_{22}) \tag{66}$$

It is convenient to introduce the following combinations of the arguments.

$$\begin{aligned}
L &:= \exp\{-(\gamma_0 u+\gamma_1 v+\mu w)\}\,, & S &:= (\alpha_0+\beta_0)u + (\alpha_1+\beta_1)v, && (67)\\
T &:= B_{11} + B_{22} + 2\,\Re B_{12}\,, & A &:= (\alpha_0-\beta_0)u + (\alpha_1-\beta_1)v, && (68)\\
B &:= B_{11} + B_{22} - 2\,\Re B_{12}\,, & C &:= \Re(C_1+C_2), && (69)\\
R &:= B_{11} - B_{22}\,, & D &:= \Re(C_1-C_2). && (70)
\end{aligned}$$

The results of the last section lead immediately to the equations (upper sign corresponds to $y>\Re X_2$ and lower sign to $y<\Re X_2$):

$$\begin{gathered}
B_{12}=\Re B_{12}\mp i\frac{\pi}{2}, \quad C_1=\Re C_1 \stackrel{+}{+} i\frac{\pi}{4}, \quad C_2=\Re C_2 \pm i\frac{\pi}{4},\\
B_{11}+B_{22}+2B_{12}=T\mp i\pi, \quad B_{11}+B_{22}-2B_{12}=B\pm i\pi,\\
y>\Re X_2: \quad C_1+C_2=\Re(C_1+C_2)+i\frac{\pi}{2}, \quad C_1-C_2=\Re(C_1-C_2),\\
y<\Re X_2: \quad C_1+C_2=\Re(C_1+C_2), \quad C_1-C_2=\Re(C_1-C_2)+i\frac{\pi}{2}.
\end{gathered}$$

From the definition of Rosenhain's theta functions one can find the relations:

$$\begin{aligned}
\vartheta_{3,3}(x_1,x_2;\,T\pm i\pi,B\mp i\pi,R) &= \vartheta_{4,4}(x_1,x_2;\,T,B,R)\,,\\
\vartheta_{2,2}(x_1,x_2;\,T\pm i\pi,B\mp i\pi,R) &= \vartheta_{2,2}(x_1,x_2;\,T,B,R)\,,\\
\vartheta_{4,4}(x_1\pm i\frac{\pi}{2},x_2;\,T,B,R) &= \vartheta_{3,4}(x_1,x_2;\,T,B,R)\,,\\
\vartheta_{4,4}(x_1,x_2\pm i\frac{\pi}{2};\,T,B,R) &= \vartheta_{4,3}(x_1,x_2;\,T,B,R)\,,\\
\vartheta_{2,2}(x_1\pm i\frac{\pi}{2},x_2;\,T,B,R) &= \pm i\,\vartheta_{1,2}(x_1,x_2;\,T,B,R)\,,\\
\vartheta_{2,2}(x_1,x_2\pm i\frac{\pi}{2};\,T,B,R) &= \pm i\,\vartheta_{2,1}(x_1,x_2;\,T,B,R)\,.
\end{aligned}$$

Using these properties equation (66) leads to the result:

For $y>\Re X_2$:

$$f = L\,\frac{\vartheta_{3,4}(S-C,A-D;\,T,B,R)+i\,\vartheta_{1,2}(S-C,A-D;\,T,B,R)}{\vartheta_{3,4}(S+C,A+D;\,T,B,R)-i\,\vartheta_{1,2}(S+C,A+D;\,T,B,R)}\,. \tag{71}$$

For $y<\Re X_2$:

$$f = L\,\frac{\vartheta_{4,3}(S-C,A-D;\,T,B,R)+i\,\vartheta_{2,1}(S-C,A-D;\,T,B,R)}{\vartheta_{4,3}(S+C,A+D;\,T,B,R)-i\,\vartheta_{2,1}(S+C,A+D;\,T,B,R)}\,. \tag{72}$$

With $\vartheta^{\pm}_{n,k} := \vartheta_{n,k}(S \pm C, A \pm D; T, B, R)$ one has

$$e^{2U} = L\,\frac{\vartheta^{-}_{3,4}\vartheta^{+}_{3,4} - \vartheta^{-}_{1,2}\vartheta^{+}_{1,2}}{[\vartheta^{+}_{3,4}]^2 + [\vartheta^{+}_{1,2}]^2}\,, \quad b = L\,\frac{\vartheta^{-}_{3,4}\vartheta^{+}_{1,2} + \vartheta^{+}_{3,4}\vartheta^{-}_{1,2}}{[\vartheta^{+}_{3,4}]^2 + [\vartheta^{+}_{1,2}]^2} \quad (y > \Re X_2)\,, \tag{73}$$

$$e^{2U} = L\,\frac{\vartheta^{-}_{4,3}\vartheta^{+}_{4,3} - \vartheta^{-}_{2,1}\vartheta^{+}_{2,1}}{[\vartheta^{+}_{4,3}]^2 + [\vartheta^{+}_{2,1}]^2}\,, \quad b = L\,\frac{\vartheta^{-}_{4,3}\vartheta^{+}_{2,1} + \vartheta^{+}_{4,3}\vartheta^{-}_{2,1}}{[\vartheta^{+}_{4,3}]^2 + [\vartheta^{+}_{2,1}]^2} \quad (y < \Re X_2)\,. \tag{74}$$

4. APPLICATIONS

The arguments (67)–(70) of the equations (74),(73) are given by (47), (48), [(49)–(56) for $y > \Re X_2$] or [(57)–(61) for $y < \Re X_2$] and (63)–(65). These quantities can be calculated as shown above with the three types of integrals (43), (36)–(38), (39)–(40).
These integrals themselves can be numerically evaluated very easily. If the y-coordinate reaches $\Re X_2$ from the left or right the formulae (74) and (73) converge to the same result. The limiting procedure $(\xi_3 = y, \eta_3 = x) \to (\xi_2, \eta_2)$ can be done in an analytical manner. In this region a little more attention must be paid to the numerical evaluation of the integrals. The formulae (74) and (73) are also very well suited to derive formulae for the special cases when one is in the disk $(y = 0, x \leq 1)$ or on the axis $(x = 0)$. This can be found in [9].

Acknowledgments

The author would like to thank Gernot Neugebauer, Reinhard Meinel and Marcus Ansorg for many discussions. The support of the DFG is gratefully acknowledged.

5. APPENDIX

Jacobian theta functions $\quad \vartheta_n = \vartheta_n(v; B)$:

$$\begin{aligned}
\vartheta_1 : &= \sum_{n=-\infty}^{\infty} (-1)^n \exp\{[\tfrac{1}{2}(2n+1)]^2 B + (2n+1)v\}\,, \\
\vartheta_2 : &= \sum_{n=-\infty}^{\infty} \exp\{[\tfrac{1}{2}(2n+1)]^2 B + (2n+1)v\}\,, \\
\vartheta_3 : &= \sum_{n=-\infty}^{\infty} \exp\{n^2 B + 2nv\}\,,
\end{aligned}$$

$$\vartheta_4 : = \sum_{n=-\infty}^{\infty} (-1)^n \exp\{n^2 B + 2nv\}\,.$$

Rosenhain's theta functions: $\vartheta_{n,k} = \vartheta_{n,k}(v, w;\, B_{11}, B_{22}, B_{12})$:

$$\begin{aligned}
\vartheta_{1,k} : &= \sum_{m=-\infty}^{\infty} (-1)^m \exp\{[\tfrac{1}{2}(2m+1)]^2 B_{11} + 2(m+\tfrac{1}{2})v\} \\
&\times\ \vartheta_k(w + (m+\tfrac{1}{2})B_{12};\, B_{22})\,, \qquad (75) \\
\vartheta_{2,k} : &= \sum_{m=-\infty}^{\infty} \exp\{[\tfrac{1}{2}(2m+1)]^2 B_{11} + 2(m+\tfrac{1}{2})v\} \\
&\times\ \vartheta_k(w + (m+\tfrac{1}{2})B_{12};\, B_{22})\,, \qquad (76) \\
\vartheta_{3,k} : &= \sum_{m=-\infty}^{\infty} \exp\{m^2 B_{11} + 2mv\}\, \vartheta_k(w + mB_{12};\, B_{22})\,, \qquad (77) \\
\vartheta_{4,k} : &= \sum_{m=-\infty}^{\infty} (-1)^m \exp\{m^2 B_{11} + 2mv\}\, \vartheta_k(w + mB_{12};\, B_{22}). \qquad (78)
\end{aligned}$$

References

[1] G. Neugebauer and R. Meinel, *Phys. Rev. Lett.* **75** (1995) 3046.

[2] G. Neugebauer, A. Kleinwächter and R. Meinel, *Helv. Phys. Acta* **69** (1996) 472.

[3] J.M. Bardeen and R.V. Wagoner, *Astrophys. J.* **158** (1969) L65.

[4] J.M. Bardeen and R.V. Wagoner, *Astrophys. J.* **167** (1971) 359.

[5] G. Neugebauer and R. Meinel, *Astrophys. J.* **414** (1993) L97.

[6] G. Neugebauer and R. Meinel, *Phys. Rev. Lett.* **73** (1994) 2166.

[7] G. Neugebauer, *Ann. Phys.* (Leipzig) **9** (2000) 342.

[8] G. Rosenhain, *Crelle's J. für Math.* **40** (1850) 319.

[9] A. Kleinwächter, *Ann. Phys.* (Leipzig) **9** (2000) 99.

THE SUPERPOSITION OF NULL DUST BEAMS IN GENERAL RELATIVITY

Dietrich Kramer *
FSU Jena

Abstract In this contribution we consider the gravitational fields arising from the superposition of two null dust beams which propagate in opposite spatial directions. Exact time–independent solutions are given for the interaction region of outgoing and incoming Vaidya solutions and for the non- linear superposition of two cylindrically symmetric pencils of light propagating in the positive and negative axial directions.

Keywords: Null dust beams, exact solutions in GR.

1. INTRODUCTION

Null dust (or pure radiation, or incoherent radiation) represents a simple matter source in classical gravity and describes the flux of massless particles (photons) with a fixed propagation direction given by the geodesic null vector field **k**. We assume that **k** is affinely parametrized,

$$k_m k^m = 0, \quad k^m k^n{}_{;m} = 0. \tag{1}$$

The energy–momentum tensor of null dust,

$$T_{mn} = \mu k_m k_n, \quad \mu > 0, \tag{2}$$

where μ denotes the energy density with respect to an observer characterized by its four–velocity **u**, $u^m u_m = -c^2$. The energy–momentum tensor (2) has a similar structure as that of dust (pressure–free perfect fluid); this explains the notation "null dust".
The most popular null dust solutions describe

- Vaidya's shining star [1]

$$ds^2 = r^2 d\Omega^2 - 2dudr - \left(1 - \frac{2m(u)}{r}\right) du^2, \tag{3}$$

* Dietrich.Kramer@tpi.uni-jena.de

where $m(u)$ is any monotonically decreasing function of the retarded time.

- Kinnersley's photon rocket [2]

$$\begin{aligned} ds^2 &= \frac{r^2}{P^2} d\Omega^2 - 2dudr - 2H du^2, \\ P &= \cosh\alpha(u) + \sinh\alpha(u)\cos\vartheta, \\ H &= -r(\log P)_{,u} + \frac{K}{2} - \frac{m(u)}{r}. \end{aligned} \tag{4}$$

This solution is a member of the Robinson–Trautman class and describes the gravitational field of an accelerated point singularity.

- Bonnor's pencil of light [3]

$$ds^2 = dx^2 + dy^2 - 2dudv - 2H du^2, \; H_{,xx} + H_{,yy} = \eta, \; H = H(x, y, u). \tag{5}$$

These three solutions are algebraically special (at least two of the null eigendirections of the Weyl tensor coincide) and of Kerr–Schild type,

$$g_{mn} = \eta_{mn} - 2H k_m k_n, \tag{6}$$

and the eigendirections are non–twisting.

We will construct exact solutions with the energy–momentum tensor

$$T_{mn} = \mu_1 k_m k_n + \mu_2 l_m l_n, \quad k_m k^m = 0 = l_m l^m, \tag{7}$$

which is composed of two null dust terms with the propagation vectors **k** and **l**. In order to get static (or stationary) solutions we can assume, without loss of generality, that the two beams have equal energy density, $\mu_1 = \mu_2 = \mu$.

The *cylindrically* symmetric case with *radial* components of the two propagation vectors (the axial and azimuthal components being zero) leads to the general solution (in terms of double null coordinates u and v)

$$ds^2 = -2\exp(f(u) + g(v))dudv + r^2 d\varphi^2 + dz^2, \quad r = u + v, \tag{8}$$

with arbitrary real functions $f(u)$ and $g(v)$. This solution becomes *static* for the special choice $f(u) = u$, $g(v) = v$. It has been assumed that no gravitational Einstein-Rosen waves survive when the null dust is switched off.

The *spherically* symmetric case and *cylindrically* symmetric configurations with *axial* components of the propagation vectors will be treated in the following two sections.

2. THE SUPERPOSITION OF SPHERICALLY SYMMETRIC BEAMS OF NULL DUST

The field equations can be cast into the simultaneous system

$$Y_{,uv} = Z_{,\xi}, \quad \xi_{,uv} = Z_{,Y}, \quad Z := -Y^{-1/2}\exp\xi \tag{9}$$

of two partial differential equations of second order for the metric functions $Y = Y(u,v)$ and $\xi = \xi(u,v)$. Numerical studies of the differential equations equivalent to (9) are reported in [4].

We did not succeed in finding the general solution to this system. In [6] the derivation of a *static* solution has been given in detail. Introducing ξ as one of the two relevant space–time coordinates one obtains the metric in *closed* form:

$$\begin{aligned} \mathrm{d}s^2 &= \frac{\mathrm{e}^\xi}{r}(-\mathrm{d}t^2 + J^2\mathrm{d}\xi^2) + r^2\mathrm{d}\Omega^2, \quad r = \sqrt{1+2\xi}\,J, \\ J &= -\frac{\mathrm{e}^\xi}{\sqrt{1+2\xi}} - \mathrm{i}\sqrt{\frac{\pi}{2\mathrm{e}}}\mathrm{erf}\Big(\mathrm{i}\sqrt{(1+2\xi)/2}\Big). \end{aligned} \tag{10}$$

The expression for J contains the error function of an imaginary argument. The metric (10) corresponds to the solution given by [5], but not in closed form and with quite another interpretation.

The solution (10) can be matched across null hypersurfaces to the incoming and outgoing Vaidya solutions. The boundary conditions (at $v=$ const, say) imply the parametric representations of m and u in terms of ξ,

$$\begin{aligned} r &= r(\xi) = \sqrt{1+2\xi}\,J, \\ m &= m(\xi) = \frac{r}{2} - \frac{(r+\mathrm{e}^\xi)^2\mathrm{e}^{-\xi}}{2(1+2\xi)}, \\ u &= u(\xi) = -2\int \frac{r\mathrm{e}^\xi}{r+\mathrm{e}^\xi}\,d\xi. \end{aligned} \tag{11}$$

These equations fix the evolution of the boundary, $r = r(u)$, and determine the mass as a (monotonically decreasing) function of retarded time, $m = m(u)$.

The exact solution (10) is regular outside the curvature singularity at $r = 0$ and gives the spherically symmetric gravitational field in the space–time region where incoming and outgoing null dust beams of equal intensity overlap. For instance, the reflection of radiation emitted from a spherically symmetric source, with the aid of a perfectly reflecting concentric shell, can lead to a static interaction region in which the obtained solution holds.

3. THE GRAVITATIONAL FIELD OF TWO COUNTER–MOVING BEAMS OF LIGHT

3.1. THE PROBLEM TO BE SOLVED. BASIC ASSUMPTIONS

In a pioneering paper, Bonnor [3] published the exact solution (5) for a cylindrically symmetric pencil of light (i.e. for null dust) propagating along the axis. The particular solution with constant mass density was smoothly matched to an exterior plane–wave vacuum field. Bonnor pointed out that the Einstein equations allow a linear superposition of parallel beams if they shine in the *same* direction. The question arises: What about the "superposition" of two pencils propagating in *opposite* spatial directions, one in the positive z–direction, the other in the negative one?

The calculation of null geodesics (trajectories of test photons) in Bonnor's solution for a pencil of light shows that test photons with a velocity component opposite to the propagation of the pencil are in general focussed to the axis. This behaviour has to be avoided in the wanted gravitational field with two source terms due to the pencils of opposite propagation directions. The exact solution given below does satisfy the regularity condition at the axis. The photons move along null geodesics of their *total* field.

To attack this problem and to construct an exact solution we first of all formulate the simplifying assumptions which have a clear physical or geometrical meaning. We confine ourselves to the (nonlinear) superposition of two stright steady beams of circular cross sections and equal intensities. To have cylindrical symmetry we consider only the situation when the counter–moving beams of light penetrate each other and have a common symmetry axis and the photons forming the sources move along curves with constant spatial distance from the axis. The two beams propagate in the positive and negative z–directions. The gravitational field produced by this configuration should be time–independent and cylindrically symmetric. These remarks lead us to the crucial *assumption:*

The solution we are interested in admits three commuting Killing vectors.

For static cylindrically symmetric metrics one usually assumes that the Killing vectors ∂_t (staticity), ∂_φ (axial symmetry) and ∂_z (translational symmetry) are hypersurface–orthogonal; the metric with the coordinates (r, z, φ, t) has diagonal form and the metric components depend only on the radial coordinate r. It turns out that there is no solution of the Einstein equations with an energy–momentum tensor of the form

(7) if the three commuting Killing vectors are hypersurface–orthogonal. Hence we are led to the conclusion:

At least two of the three Killing vectors are twisting.

3.2. THE STATIC SOLUTION

Let us first consider the *static* case when the two spacelike Killing vectors ∂_z and ∂_φ are not hypersurface–orthogonal. The corresponding space–time metric reads

$$ds^2 = \exp(2K)(dx^2 - c^2dt^2) + \exp(-2U)W^2d\varphi^2 + \exp(2U)(dz + Ad\varphi)^2, \tag{12}$$

where K, U, W and A are functions of the radial coordinate x. Numbering the coordinates according to $x^m = (x, z, \varphi, ct)$ we have to demand the relations

$$k^1 = k^3 = 0, \quad l^1 = l^3 = 0, \quad k^2 = k^4, \quad -l^2 = l^4 = k^4 \tag{13}$$

between the contravariant components of the null vectors $\mathbf{k}$ and $\mathbf{l}$. The field equations reduce to the three conditions

$$R = 0, \quad R_{11} = 0, \quad S := R_{22}R_{33} - R_{23}^2 = 0. \tag{14}$$

The first two of these equations imply

$$K = -\ln(W) + \int W^{-1}\,dx, \tag{15}$$

$$A = 2\int \exp(-2U)\sqrt{W^2U_x^2 - W_x - WW_xU_x + W_x^2}\,dx \tag{16}$$

(subscripts denoting derivatives with respect to x). These relations enable us to express the metric functions K and A in terms of W and U (and their derivatives). Inserting the expressions (15) and (16) into the remaining field equation $S = 0$ one obtains a very complicated ordinary differential equation for W and U. Fortunately, it factorizes in the particular case $K = U$. The quantity S defined in (14) then splits into the product of two expressions. Putting one of them equal to zero one gets the second–order differential equation

$$WW_{xx} - 1 + 4W_x - 3W_x^2 = 0 \tag{17}$$

which can be solved by introducing $p = W_x$ as a new radial coordinate. It results

$$\ln W = \int \frac{p\,dp}{1 - 4p + 3p^2}, \qquad p = W_x, \tag{18}$$

and in terms of the radial coordinate r defined by $r^2 = p - 1$ the resulting static metric takes the simple explicit form

$$\begin{aligned} ds^2 &= Y^{-3}dr^2 + Y^{-2/3}\left[-c^2dt^2 + r^2Y d\varphi^2 + \left(dz + \frac{2}{\sqrt{3}}\lambda r^2 d\varphi\right)^2\right], \\ Y &= 1 - \lambda^2 r^2, \qquad \lambda = \text{constant}. \end{aligned} \tag{19}$$

This solution contains the arbitrary real parameter λ, for $\lambda = 0$ the space–time becomes flat. At the axis $r = 0$ the metric is regular; the first derivatives of all metric functions vanish at the axis and the square of the Killing vector $\eta = \partial_\varphi$,

$$\eta^m \eta_m = r^2 Y^{-2/3}(1 + \lambda^2 r^2/3) \tag{20}$$

is zero at $r = 0$. This behaviour is important for the interpretation of the two–component null dust solution (18) as an axisymmetric gravitational field. Notice that constants of integration have been chosen such that the regularity condition at the axis is satisfied.

The null vectors $\mathbf{k}$ and $\mathbf{l}$ given by (13) together with

$$k^4 = l^4 = \sqrt{\frac{2}{3\kappa}}\lambda Y^{4/3} \tag{21}$$

(κ – –Einstein's gravitational constant) are *geodesic*, shearfree, non-expanding, but *twisting*. They have zero radial and azimuthal contravariant components.

An observer at rest in the metric (18), i.e. an observer with the 4–velocity

$$\mathbf{u} = (\mathbf{k} + \mathbf{l})(-2k_m l^m)^{-1/2}, \tag{22}$$

measures the energy density

$$\mu = T_{mn}u^m u^n = -k_m l^m = \frac{4}{3}\lambda^2(1 - \lambda^2 r^2)^2 \tag{23}$$

which is positive everywhere, has its maximal value at the axis $r = 0$, decreases monotonically with increasing distance from the symmetry axis and goes to zero at $r = \lambda^{-1}$ which corresponds to an infinite physical distance $L = \int Y^{-3/2}\, dr$ from the axis. The energy density cannot be prescribed, it results from our ansatz $K = U$ which makes the field equations tractable. It is much more difficult to start the integration procedure with a given $\mu = \mu(r)$.

After a determined linear transformation (with constant coefficients) of the ignorable coordinates our interior solution (18) can be smoothly matched to the exterior Levi–Civita metric

$$ds^2 = \rho^{(n^2-1)/2}(d\rho^2 - c^2 dT^2) + \rho^{n+1} d\Phi^2 + \rho^{-n+1} dZ^2. \tag{24}$$

For matching it is convenient to use the physical distance L from the symmetry axis as the radial coordinate in the interior and exterior metrics instead of the coordinates r and ρ which are related to L by the relations

$$r = L(1+\lambda^2 L^2)^{-1/2}, \qquad \rho = L^{4/(n^2+3)} \tag{25}$$

The boundary conditions at the surface $L = L_0$ imply the relation

$$n^2 = \frac{1+3\lambda^2 L_0^2}{1+\frac{1}{3}\lambda^2 L_0^2} \tag{26}$$

between the real parameters λ in the interior, n in the exterior, and the radius L_0 of the boundary surface.

3.3. THE STATIONARY SOLUTION

Now we will treat the *stationary* case when the Killing vectors ∂_t (timelike) and ∂_φ (spacelike) are not hypersurface–orthogonal. In this case we can start with the non–diagonal line element

$$\mathrm{d}s^2 = \exp(2K)(\mathrm{d}x^2+\mathrm{d}z^2) + \exp(-2U)W^2\mathrm{d}\varphi^2 - \exp(2U)(c\mathrm{d}t + A\mathrm{d}\varphi)^2. \tag{27}$$

The derivation of the stationary solution is very similar to the procedure applied in the static case. The resulting stationary metric

$$\begin{aligned} \mathrm{d}s^2 &= Y^{-3}\mathrm{d}r^2 + Y^{-2/3}\left[\mathrm{d}z^2 + r^2 Y\mathrm{d}\varphi^2 - \left(c\mathrm{d}t + \frac{2}{\sqrt{3}}\lambda r^2\mathrm{d}\varphi\right)^2\right] \\ Y &= 1+\lambda^2 r^2, \qquad \lambda = \text{constant} \end{aligned} \tag{28}$$

can be obtained from its static counterpart (18) with the aid of the *complex* substitution

$$ct \to \mathrm{i}z, \qquad z \to \mathrm{i}ct, \qquad \lambda \to \mathrm{i}\lambda. \tag{29}$$

The interior solution (27) can be matched to an exterior vacuum solution which is also related to (24) by a complex substitution.

The two solutions (18) and (27) describe physically different situations. In both cases the photons which form the source of the gravitational field move along screwed lines on the cylinder mantle $r =$ const (**k** and **l** have no r–components!) and surround the axis $r = 0$. However, in the *static* case the photons of the beam with propagation vector **k** and those with **l** move on the *same* trajectories in opposite directions whereas in the *stationary* case they move on *different* screwed curves with the same sense of revolution. Thus the total rotation of the source is zero because of compensation in the static solution (18) and it does not vanish in the stationary solution (27). In any case rotational effects are responsible for balancing the gravitational attraction such that time–independent configurations can exist at all and mutual focusing does not occur.

The derivation of the exact null fluid solutions (18) and (27) was first given in our paper [7]. For a discussion and illustration of the trajectories (screwed motion), see [9]. Therein Fermi coordinates along the world line of a fixed point on the cylinder mantle are introduced to confirm the interpretation of the photon trajectories.

3.4. OTHER SOLUTIONS

We have mentioned that the quantity S defined in (14) factorizes for $K = U$. This holds also for the special choice $K = -2U$. In both cases one can put either of the two factors in S equal to zero. In this way one obtains four different static solutions which are listed in [8]. The resulting mass density profiles plotted in [8] are well-behaved only in the solution treated here.

The attempt to start with a metric admitting an Abelian group G_3 where none of the Killing vectors is hypersurface–orthogonal did not lead to more general solutions. The existence of three twisting Killing vectors is not compatible with the structure (13) of the propagation vectors. Hence there are only two possibilities: the non–twisting Killing vector is either timelike or spacelike, i.e. the gravitational field is either static or stationary and the corresponding metrics are (12) and (27), respectively.

One can interchange the coordinates z and φ in the static solution (18) to obtain a metric for counter–rotating null dust components, but then the axis is not regular.

4. DISCUSSION

Under certain conditions the superposition of two null dust components propagating in opposite spatial directions gives rise to static (or stationary) gravitational fields in the interaction region. Examples considered in this article are the static spherically symmetric solution (10) derived in [6], the static cylindrically symmetric solution (18) derived in [7], and its stationary counterpart (27). These three metrics are given in closed form by introducing suitable coordinates.

The superposition of counter–moving pencils of light implies a rotational effect which compensates for the gravitational attraction to prevent mutual focusing of the rays. An important point is the fact that the propagation null vectors of the two null dust components are both *geodesic* with respect to the *total* metric arising from the nonlinear superposition.

The integration of the field equations requires, in addition to the space-time symmetries, an a priori ansatz for the metric functions. This ansatz implies a particular radial dependence of the energy density which in general cannot be prescribed appropriately. In the solution (18) the calculated energy density has a very nice behaviour.

The metric (18) describing counter-moving beams of light can be smoothly matched to the static cylindrically symmetric vacuum solution (24) at any distance from the regular axis.

References

[1] P. C. Vaidya,, *Proc. Indian Acad. Sci.* **A 33** (1951) 264.

[2] W. Kinnersley, *Phys. Rev.* **186** (1969) 1335.

[3] W. B.Bonnor, *Commun. Math. Phys.* **13** (1969) 163.

[4] P.R. Holvorcem, P.S. Letelier, and A. Wang, *J. Math. Phys.* **36** (1995) 3663.

[5] G. Date, *Gen. Rel. Grav* **29** (1997) 953.

[6] D. Kramer, *Class. Quantum Grav.* **15** (1998) L31.

[7] D. Kramer, *Class. Quantum Grav.* **15** (1998) L73.

[8] U. v.d.Gönna, and D. Kramer, *Gen. Rel. Grav.* **31** (1999) 349.

[9] U. v.d.Gönna and D. Kramer, *Annalen Physik* **9** (2000) Spec. Issue *Proceedings Journées Relativistes 1999, Weimar*) pp. 50.

SOLVING EQUILIBRIUM PROBLEM FOR THE DOUBLE–KERR SPACETIME

Vladimir S. Manko
Departamento de Física,
Centro de Investigación y de Estudios Avanzados del IPN,
A.P. 14-740, 07000 México D.F., Mexico

Eduardo Ruiz
Area de Física Teórica, Universidad de Salamanca,
37008 Salamanca, Spain

Abstract The double–Kerr equilibrium problem, its history and recent resolution are briefly reviewed.

Keywords: Exact solutions, Kerr spacetime, superposition of solutions.

1. INTRODUCTION

The discovery by D. Kramer and G. Neugebauer of their famous double–Kerr solution [1] dates to 1980, and the whole history of the equilibrium problem of two Kerr–NUT particles is therefore 20 year old since precisely this year 2000 the problem has received its final resolution. In our short review we are going to give a chronological analysis of the research related to the double–Kerr equilibrium problem and discuss the general analytic formulas recently obtained for the equilibrium states, individual Komar masses and angular momenta of the constituents which have permitted a rigorous proof of the non–existence of equilibrium configurations of two black hole constituents with positive masses.

2. A SHORT HISTORY OF THE PROBLEM

The original double–Kerr solution of Kramer–Neugebauer describes the non–linear superposition of two Kerr–NUT black holes, and it was claimed in [1] that for a special choice of the parameters the two constituents could be in a state of equilibrium due to the balance of the gravitational

Exact Solutions and Scalar Fields in Gravity: Recent Developments
Edited by Macias *et al.*, Kluwer Academic/Plenum Publishers, New York, 2001

attraction and spin–spin repulsion forces. Although it was shown later that this claim was only partially correct in a sense that the black hole equilibrium configurations necessarily implied the negative mass of one of the constituents, the Kramer–Neugebauer paper gave a tremendous impetus to the study of different two–body equilibrium problems within the framework of general relativity.

A detailed analysis of the physical properties of the double–Kerr solution had been probably started by the paper of Oohara and Sato [2] in which the existence of two separated stationary limit surfaces associated with each of the two black hole constituents was demonstrated, and it was shown that the Tomimatsu–Sato $\delta = 2$ solution [3] was a special case of the double–Kerr spacetime. The balance conditions were first derived and analyzed by Kihara and Tomimatsu [4, 5] and Tomimatsu produced formulas [6] for the calculation of the individual masses and angular momenta of the black holes; he also claimed the possibility of equilibrium between two extreme black holes, but later on it was demonstrated by Hoenselaers [7] that Tomimatsu's equilibrium case involved a ring singularity arising from the negative mass of one of the constituents. The main result of the paper [7] was, however, the derivation of the expressions for the Komar masses of the black hole constituents in equilibrium and the numerical check that these masses could not simultaneously assume positive values; this demonstrated (but a rigorous proof still was needed) that two normal Kerr black holes with positive individual masses could not be in gravitational equilibrium.

The non–existence of the black hole equilibrium configurations motivated a search for the equilibrium states of the superextreme Kerr constituents. Dietz and Hoenselaers used a complex continuation of the parameters in order to pass from the subextreme case to the superextreme one, and they found equilibrium positions of two equal superextreme spinning masses [8]. Dietz and Hoenselaers passed over the possibility to consider equilibrium configurations between a black hole and a superextreme object, and in the paper [9] the existence of such configurations was claimed for the first time, however without any analysis of the positivity of the Komar masses of the constituents.

3. A NEW APPROACH TO THE DOUBLE–KERR EQUILIBRIUM PROBLEM

A new approach to solving the equilibrium problem for the double–Kerr solution consists in using the extended version of this solution and searching for the *analytic* balance formulas equally applicable for treating equilibrium of any combination of the subextreme (black hole) and superextreme constituents. The complex Ernst potential $\mathcal{E}$ [10] and corresponding metric functions f, γ and ω entering the Papapetrou axisymmetric line element

$$ds^2 = f^{-1}[e^{2\gamma}(d\rho^2 + dz^2) + \rho^2 d\varphi^2] - f(dt - \omega d\varphi)^2 \tag{1}$$

(ρ, z, φ, t are the Weyl–Papapetrou cylindrical coordinates and time) for the extended double–Kerr solution have the form [11, 12]

$$\begin{aligned}
E &= E_+/E_-, \quad f = \frac{E_+\bar{E}_- + \bar{E}_+E_-}{2E_-\bar{E}_-}, \\
e^{2\gamma} &= \frac{E_+\bar{E}_- + \bar{E}_+E_-}{2\lambda\bar{\lambda}\prod_{n=1}^4 r_n}, \quad \omega = \frac{2i(G\bar{E}_- - \bar{G}E_-)}{E_+\bar{E}_- + \bar{E}_+E_-}, \\
E_\pm &= \Lambda \pm \Gamma, \quad \Lambda = \sum_{1\le i<j\le 4} \lambda_{ij} r_i r_j, \quad \Gamma = \sum_{i=1}^4 \nu_i r_i, \\
G &= -\sigma\Lambda + \bar{\sigma}\Gamma + z\Gamma + \sum_{1\le i<j\le 4} \lambda_{ij}(\alpha_i + \alpha_j) r_i r_j \\
&- \sum_{i=1}^4 \nu_i(\alpha_{i'} + \alpha_{j'} + \alpha_{k'}) r_i, \quad (i', j', k' \neq i;\ i' < j' < k') \\
\lambda_{ij} &:= (-1)^{i+j}(\alpha_i - \alpha_j)(\alpha_{i'} - \alpha_{j'})X_iX_j, \quad (i', j' \neq i, j;\ i' < j'), \\
\nu_i &:= (-1)^i(\alpha_{i'} - \alpha_{j'})(\alpha_{i'} - \alpha_{k'})(\alpha_{j'} - \alpha_{k'})X_i, \\
&\quad (i', j', k' \neq i;\ i' < j' < k') \\
X_i &:= \frac{(\alpha_i - \bar{\beta}_1)(\alpha_i - \bar{\beta}_2)}{(\alpha_i - \beta_1)(\alpha_i - \beta_2)}, \quad r_i := \sqrt{\rho^2 + (z - \alpha_i)^2}, \\
\sigma &:= [\nu + \sum_{1\le i<j\le 4} \lambda_{ij}(\alpha_1 - \alpha_j)]/\lambda, \ \nu := \sum_{i=1}^4 \nu_i, \ \lambda := \sum_{1\le i<j\le 4} \lambda_{ij}
\end{aligned} \tag{2}$$

where a bar over a symbol means complex conjugation, and the subindices i, j, i', j', k' vary from 1 to 4.

The arbitrary parameters entering the formulas (2) are β_1, β_2 which assume arbitrary complex values, and α_i, $i = 1, 2, 3, 4$, which can assume arbitrary real values or occur in complex conjugate pairs. From (2) follows that instead of the parameters β_1 and β_2 one can use a set of the constant objects X_1, X_2, X_3, X_4 that facilitates the solution of the balance equations. Without any lack of generality we can assume $\text{Re}\alpha_1 \ge \text{Re}\alpha_2 \ge \text{Re}\alpha_3 \ge \text{Re}\alpha_4$, then in the case of the real–valued α_n the parts of the symmetry axis $\alpha_2 \le z \le \alpha_1$ and $\alpha_4 \le z \le \alpha_3$ represent the Killing horizons of two black holes, while a pair of complex conjugate αs, say α_1 and $\alpha_2 = \bar{\alpha}_1$, represents a superextreme Kerr constituent.

The two constituents of the double–Kerr solution will be in equilibrium if the metric functions γ and ω satisfy the following conditions on the symmetry axis ($\rho = 0$)

$$\exp\{\gamma^{(+,0,-)}\} = 1, \quad \omega^{(+,0,-)} = 0, \tag{3}$$

where '+', '0' and '-' denote the $z > \mathrm{Re}\alpha_1$, $\mathrm{Re}\alpha_3 < z < \mathrm{Re}\alpha_2$ and $z < \mathrm{Re}\alpha_4$ parts of the symmetry axis, respectively.

The balance equations have been obtained in the explicit form and solved in the paper [12]. It turns out that for any choice of the parameters α_n, there always exists a set of the constants X_n determining an equilibrium configuration of the two constituents:

$$\begin{aligned}
&X_1 = -\frac{v_1 - \phi}{v_1 - \phi^{-1}}, \quad X_2 = -\frac{1 - v_1\phi}{1 - v_1\phi^{-1}}, \\
&X_3 = \frac{1 + iv_4\phi}{1 - iv_4\phi^{-1}}, \quad X_4 = \frac{-\phi + iv_4}{\phi^{-1} + iv_4}, \\
&v_1 := \epsilon_1\sqrt{\frac{(\alpha_1 - \alpha_3)(\alpha_1 - \alpha_4)}{(\alpha_2 - \alpha_3)(\alpha_2 - \alpha_4)}}, \ v_4 := \epsilon_4\sqrt{\frac{(\alpha_1 - \alpha_4)(\alpha_2 - \alpha_4)}{(\alpha_1 - \alpha_3)(\alpha_2 - \alpha_3)}}, \\
&\epsilon_1 := \pm 1, \quad \epsilon_4 := \pm 1,
\end{aligned} \tag{4}$$

where the complex parameter ϕ is subjected to the constraint

$$\phi\bar{\phi} = 1. \tag{5}$$

The above very concise balance formulas for X_n have been used in [12] for the analysis of of various particular equilibrium states involving either black holes or superextreme objects.

4. THE KOMAR MASSES AND ANGULAR MOMENTA

Although formulas (4) give a mathematical solution of the double–Kerr equilibrium problem, they still need to be complemented by the expressions of the individual Komar masses of the constituents in order to be able to judge whether a particular equilibrium configuration is physical or not, i.e. whether both masses of the constituents are positive. In [12] the individual masses were calculated anew for each particular equilibrium state. Recently, however, we have been able to obtain the general elegant expressions [13] for the Komar masses and angular momenta of each balancing constituent corresponding to the equilibrium formulas (4). The details of the derivation of these expressions can be found in Ref. [13], here we only give their explicit form:

$$M_u = \frac{ps(p - c_1)}{D}, \quad M_d = \frac{qs(q - c_4)}{D},$$

$$
\begin{aligned}
J_u &= \frac{\epsilon p^2 s^2 (p - c_1)(pc_1 - \epsilon q c_1 + \epsilon pq - 1)}{(c_1 + \epsilon c_4) D^2}, \\
J_d &= \frac{q^2 s^2 (q - c_4)(qc_4 - \epsilon p c_4 + \epsilon pq - 1)}{(c_1 + \epsilon c_4) D^2}, \\
D &= pc_1 + qc_4 + \epsilon pq - 1, \\
s &:= (\alpha_1 + \alpha_2 - \alpha_3 - \alpha_4)/2, \quad \epsilon := \epsilon_1 \epsilon_4, \qquad (6)
\end{aligned}
$$

where the subindices 'u' and 'd' denote the mass (M) and angular momentum (J) of the upper and lower constituents, respectively, and the constants p, q, c_1 and c_4 are introduced in the following way:

$$
\begin{aligned}
&\phi = p + iq, \quad p^2 + q^2 = 1, \\
&c_1 := (v_1 + v_1^{-1})/2, \quad c_4 := (v_4 + v_4^{-1})/2. \qquad (7)
\end{aligned}
$$

It can be readily proven *analytically* using formulas (6) and (7) that M_u and M_d cannot assume simultaneously positive values [13], so the equilibrium of two Kerr black holes possessing positive Komar masses is impossible. However, the equilibrium states with positive Komar masses of both constituents are possible when at least one of the constituents is a superextreme object.

Formulas (4)–(7) give a complete exact solution of the double–Kerr equilibrium problem.

5. TOWARDS THE ANALYSIS OF THE MULTI–BLACK HOLE EQUILIBRIUM STATES

It is very surprising that although the double–Kerr solution does not admit equilibrium states of two Kerr black holes, already in the case of three aligned Kerr black holes the gravitational equilibrium is possible for the constituents with positive Komar masses [14]. The results recently reported in [14] open some new horizons for the study of multi–black hole equilibrium configurations, and one may expect new interesting future findings in this area.

Acknowledgments

This work has been partially supported by CONACyT of Mexico (project 34222–E) and by DGICyT of Spain (project PB96-1306).

References

[1] D. Kramer and G. Neugebauer, *Phys. Lett.* **A75** (1980) 259.

[2] K. Oohara and H. Sato, *Prog. Theor. Phys.* **65** (1981) 1891.

[3] A. Tomimatsu and H. Sato, *Phys. Rev. Lett.* **29** (1972) 1344.

[4] M. Kihara and A. Tomimatsu, *Prog. Theor. Phys.* **67** (1982) 349.

[5] A. Tomimatsu and M. Kihara, *Prog. Theor. Phys.* **67** (1982) 1406.

[6] A. Tomimatsu, *Prog. Theor. Phys.* **70** (1983) 385.

[7] C. Hoenselaers, *Prog. Theor. Phys.* **72** 761 (1984).

[8] W. Dietz and C. Hoenselaers, *Ann. Phys. (N.Y.)* **165** (1985) 319.

[9] M.N. Zaripov, N. R. Sibgatullin and A. Chamorro, *Vestnik, Ser. Matem. Mech.* **6** (1994) 61.

[10] F.J. Ernst, *Phys. Rev.* **167** (1968) 1175.

[11] V.S. Manko and E. Ruiz, *Class. Quantum Grav.* **15** (1998) 2007.

[12] V.S. Manko, E. Ruiz and J.D. Sanabria–Gómez, *Class. Quantum Grav.* **17** (2000) 3881.

[13] V.S. Manko and E. Ruiz, *Class. Quantum Grav.* **18** (2001) L11.

[14] V.S. Manko, E. Ruiz and O.V. Manko, *Phys. Rev. Lett.* **85** (2000) 5504.

ROTATING EQUILIBRIUM CONFIGURATIONS IN EINSTEIN'S THEORY OF GRAVITATION

Reinhard Meinel*

Friedrich-Schiller-Universität Jena

Theoretisch-Physikalisches Institut

Max-Wien-Platz 1, D-07743 Jena, Germany

Abstract The theory of figures of equilibrium of rotating fluids arose from geophysical and astrophysical considerations. Within Newton's theory of gravitation, Maclaurin (1742) found the axially symmetric and stationary solution to the problem in the case of constant mass-density: a sequence of oblate spheroids. In the limit of maximum ellipticity a rotating disk is obtained. By applying analytical solution techniques from soliton theory this Maclaurin disk has been "continued" to Einstein's theory of gravitation [1, 2, 3].

After an introduction into these developments, the parametric collapse to a rotating black hole and possible generalizations are discussed.

Keywords: Rotation configurations, exact solutions.

1. FIGURES OF EQUILIBRIUM OF ROTATING FLUID MASSES

The theory of figures of equilibrium of rotating, self-gravitating fluids was developed in the context of questions concerning the shape of the Earth and of celestial bodies. Many famous physicists and mathematicians have contributed: Newton, Maclaurin, Jacobi, Liouville, Dirichlet, Dedekind, Riemann, Roche, Poincaré, H. Cartan, Lichtenstein, Chandrasekhar, and others. The shape of the rotating body can be obtained from the requirement that the force arising from pressure, the gravitational force, and the centrifugal force (in the corotating frame) be in equilibrium, cf. Lichtenstein [4] and Chandrasekhar [5]. Newton [6]

*E-mail: meinel@tpi.uni-jena.de

derived an approximate formula for the 'ellipticity' of the Earth:

$$\frac{\text{equatorial radius} - \text{polar radius}}{\text{mean radius}} \approx \frac{5\Omega^2 R^3}{4GM} \approx 0.0043, \tag{1}$$

a value not very much different from the actual value of about 0.0034 known today [Ω: angular velocity, R: (mean) radius, M: mass of the Earth, G: Newton's constant of gravitation]. However, the first empirical proof of the flattening of the Earth at the poles was provided by geodetic measurements in Lapland performed by Maupertius and Clairaut in 1736/1737. This was the end of a controversy between the schools of Newton and the Cassini family, the latter had claimed that the Earth is flattened at the equator.

It should also be mentioned that Kant [7] interpreted the Milky Way system as a rotating disk in balance between centrifugal and gravitational forces.

The first exact solution describing a rotating fluid body in Newton's theory of gravitation was found by Maclaurin [8]. The *Maclaurin spheroids* are characterized by the following relation between the (constant) angular velocity Ω, the (constant) mass–density ϱ_M, and the eccentricity

$$\epsilon = \sqrt{1 - \frac{a_3^2}{a_1^2}}, \quad a_1\text{: major semi-axis}, \quad a_3\text{: minor semi-axis}, \tag{2}$$

$$\frac{\Omega^2}{\pi G \varrho_M} = \frac{2}{\epsilon^3}\sqrt{1-\epsilon^2}(3-2\epsilon^2)\arcsin\epsilon - \frac{6}{\epsilon^2}(1-\epsilon^2). \tag{3}$$

It is interesting to note, that this reduces to Newton's formula (1) for small ϵ ($\epsilon \ll 1$). In the opposite limit, $\epsilon \to 1$, with $a_3 \to 0$, $a_1 = \varrho_0$ (fixed),

$$M = \frac{4\pi}{3} a_1^2 a_3 \varrho_M \quad \text{fixed, i.e. } \varrho_M \to \infty, \tag{4}$$

the *Maclaurin disk* with radius ϱ_0 is obtained. It has a surface mass–density σ depending on the distance ϱ from the center,

$$\sigma \propto \sqrt{1 - \frac{\varrho^2}{\varrho_0^2}}, \tag{5}$$

and satisfies the following relation between angular velocity, mass, and radius:

$$\Omega^2 = \frac{3\pi G M}{4\varrho_0^3}. \tag{6}$$

2. ROTATING DUST CONFIGURATIONS – THE MACLAURIN DISK

In Newton's theory of gravitation, it can easily be shown that a rotating, axially symmetric and stationary dust (Here "dust" means a perfect fluid with vanishing pressure.) configuration of finite mass must be an infinitesimally thin disk.

Therefore, the basic equations to be solved are the Poisson equation

$$\Delta U = 4\pi G \varrho_M, \quad \varrho_M = \sigma(\varrho)\delta(\zeta) \tag{7}$$

and the equation expressing the equilibrium between centrifugal and gravitational forces in the disk

$$\Omega^2 \varrho = \frac{\partial U}{\partial \varrho} \quad (\zeta = 0,\ \varrho \leq \varrho_0). \tag{8}$$

Here U is the gravitational potential, ϱ and ζ are cylindrical coordinates. It should be noted that for a prescribed rotation law $\Omega(\varrho)$, the surface mass–density $\sigma(\varrho)$ cannot be chosen arbitrarily but has to be calculated from the solution of the resulting boundary–value problem for the Laplace equation $\Delta U = 0$, according to

$$\sigma = \frac{1}{4\pi G}\left[\frac{\partial U}{\partial \zeta}\right]_{\zeta=0^-}^{\zeta=0^+}. \tag{9}$$

We assume

$$\Omega = const., \tag{10}$$

leading to the boundary condition

$$V \equiv U - \frac{1}{2}\Omega^2\varrho^2 = V_0 = const. \quad (\zeta = 0,\ \varrho \leq \varrho_0). \tag{11}$$

In addition to (11) we require $U \to 0$ at infinity and a bounded surface mass–density at the rim of the disk $\varrho = \varrho_0$. The latter requirement leads to a relation between the parameters Ω, ϱ_0, and V_0, see Eq. (15). This boundary–value problem has the unique solution

$$U = -\frac{\Omega^2\varrho_0^2}{\pi}\left\{\frac{4}{3}\operatorname{arccot}\xi + \left[\xi - (\xi^2 + \frac{1}{3})\operatorname{arccot}\xi\right](1 - 3\eta^2)\right\}, \tag{12}$$

with oblate elliptic coordinates ξ, η related to ϱ, ζ by

$$\varrho = \varrho_0\sqrt{(1+\xi^2)(1-\eta^2)}, \quad \zeta = \varrho_0\xi\eta \quad (0 \leq \xi < \infty,\ -1 \leq \eta \leq 1). \tag{13}$$

Note that $\xi = 0$ represents the disk. On the disk, Eq. (12) reduces to

$$\xi = 0: \quad U = -\Omega^2 \rho_0^2 + \frac{1}{2}\Omega^2 \varrho^2, \tag{14}$$

i.e. we may identify the constant V_0 from (11) as

$$V_0 = -\Omega^2 \varrho_0^2. \tag{15}$$

The solution depends on *two parameters*, e.g. Ω and ϱ_0. The surface mass–density calculated according to (9) is given by

$$\sigma = \frac{3M}{2\pi\varrho_0^2}\sqrt{1 - \frac{\varrho^2}{\varrho_0^2}}, \tag{16}$$

and the mass of the disk turns out to be

$$M = \frac{4}{3\pi G}\Omega^2 \varrho_0^3. \tag{17}$$

A comparison with Eqs. (5), (6) shows that this is exactly the Maclaurin disk.

It is interesting to insert numerical values characteristic for the Milky Way system into the last equation. Of course, Ω is not constant in the Galaxy. For a value of Ω corresponding to the revolution time of the sun about the galactic center (approximately 240 million years) and $\varrho_0 \approx 50,000$ light years, the quite reasonable mass value of $M \approx 2.3 \times 10^{11} M_\odot$ is obtained ($M_\odot$: mass of the sun).

3. GENERAL–RELATIVISTIC CONTINUATION

In Einstein's theory of gravitation, the Poisson equation $\Delta U = 4\pi G \varrho_M$ and the Euler equation $\nabla p = -\varrho_M \nabla V$ [p: pressure, V: "corotating" potential, cf. (11)] have to be replaced by Einstein's field equations $R_{ik} - \frac{1}{2}R g_{ik} = \kappa T_{ik}$ with the Ricci tensor R_{ik}, the Ricci scalar R, the metric tensor g_{ik}, the energy–momentum tensor T_{ik}, and $\kappa = 8\pi G/c^4$ (c: velocity of light). The advantage of the disk limit remains the same: The global (interior + exterior + matching) problem can be formulated as a boundary–value problem for the vacuum equations ($\Delta U = 0$ resp. $R_{ik} = 0$). For axially symmetric and stationary configurations, the Einstein vacuum equations $R_{ik} = 0$ are equivalent to the well–known Ernst equation

$$(\Re f)\Delta f = (\nabla f)^2 \tag{18}$$

for the (complex) Ernst potential $f(\rho, \zeta)$. The metric g_{ik} can be calculated from f. In the Newtonian limit, the Ernst potential becomes

$$f = 1 + \frac{2U}{c^2}, \quad |U| \ll c^2, \quad \Delta U = 0. \tag{19}$$

The Newtonian boundary condition (11) on the disk has to be replaced in Einstein's theory by (see [1])

$$\tilde{f} = e^{2V_0/c^2} \quad (\zeta = 0, \, \varrho \leq \varrho_0), \tag{20}$$

where $\tilde{f}$ denotes the Ernst potential in the corotating system and V_0 is a real (negative) constant. The asymptotic condition $U \to 0$ has to be replaced by $f \to 1$ at infinity. The Einsteinian solution depends on two parameters as well. It turns out that a useful parameter combination is given by

$$\mu = \frac{2\Omega^2 \rho_0^2}{c^2} e^{-2V_0/c^2}. \tag{21}$$

In terms of the two parameters μ and ρ_0, the solution is given by the following expression [3]:

$$f = \exp\left\{ \int_{K_1}^{K_a} \frac{K^2 dK}{W} + \int_{K_2}^{K_b} \frac{K^2 dK}{W} - w \right\}, \tag{22}$$

$$W = \sqrt{[(K-\zeta)^2 + \rho^2](K^2 - K_1^2)(K^2 - K_2^2)}\,, \tag{23}$$

$$K_1 = \varrho_0 \sqrt{\frac{\mathrm{i} - \mu}{\mu}} \quad (\Re K_1 < 0); \quad K_2 = -K_1^*. \tag{24}$$

The upper integration limits K_a and K_b have to be determined from the following Jacobi inversion problem:

$$\int_{K_1}^{K_a} \frac{dK}{W} + \int_{K_2}^{K_b} \frac{dK}{W} = u, \quad \int_{K_1}^{K_a} \frac{K dK}{W} + \int_{K_2}^{K_b} \frac{K dK}{W} = v, \tag{25}$$

where the functions u, v, and w are given by

$$u = \int_{-\mathrm{i}\varrho_0}^{\mathrm{i}\varrho_0} \frac{H dK}{W_1}, \quad v = \int_{-\mathrm{i}\varrho_0}^{\mathrm{i}\varrho_0} \frac{H K dK}{W_1}, \quad w = \int_{-\mathrm{i}\varrho_0}^{\mathrm{i}\varrho_0} \frac{H K^2 dK}{W_1}, \tag{26}$$

$$H = \frac{\mu \ln\left[\sqrt{1 + \mu^2(1 + K^2/\varrho_0^2)^2} + \mu(1 + K^2/\varrho_0^2)\right]}{\pi \mathrm{i} \varrho_0^2 \sqrt{1 + \mu^2(1 + K^2/\varrho_0^2)^2}}, \tag{27}$$

$$W_1 = \sqrt{(K-\zeta)^2+\varrho^2} \quad (\Re H = 0,\ \Re W_1 < 0). \tag{28}$$

The integration in (26) is along the imaginary K–axis, whereas the integration paths in (22) have to be the same as in (25). Alternatively, the Ernst potential (and the metric) can be expressed in terms of ultraelliptic theta functions [9], see also [10].

The formulae simplify considerably in the disk plane ($\zeta = 0$) and on the symmetry axis ($\varrho = 0$). In both cases the ultraelliptic integrals reduce to elliptic ones. In particular, this leads to a relatively simple expression for the surface mass–density [2].

It turns out that the parameter μ is restricted to the range

$$0 < \mu < \mu_0 = 4.62966\ldots, \tag{29}$$

where $\mu \ll 1$ represents the Newtonian limit (the Maclaurin disk!) and $\mu \to \mu_0$ leads to the black–hole limit to be discussed in the next section.

It should be noted that an approximate solution to this problem had already been found by Bardeen and Wagoner [11, 12] on the basis of a post–Newtonian expansion scheme including numerical methods. For a completely analytical version of this expansion see Petroff and Meinel [13].

4. BLACK–HOLE LIMIT

For $\mu \to \mu_0$, the following parameter limits are reached:

$$V_0 \to -\infty, \quad \Omega\varrho_0 \to 0, \quad \frac{G\Omega M}{c^3} \to \frac{1}{2}. \tag{30}$$

Consequently, for finite values of the total (gravitational) mass M,

$$\varrho_0 \to 0. \tag{31}$$

If the origin of the (Weyl–) coordinate system (where the disk shrinks to) is excluded, one obtains

$$f = \frac{2\Omega r/c - 1 - \mathrm{i}\cos\vartheta}{2\Omega r/c + 1 - \mathrm{i}\cos\vartheta}, \quad \varrho = r\sin\vartheta,\ \zeta = r\cos\vartheta \quad (r > 0,\ 0 \le \vartheta \le \pi). \tag{32}$$

This is exactly the Ernst potential of the extreme Kerr metric describing a maximally rotating black hole. Note, that in the coordinates used here, $r = 0$ represents the horizon. More details on this black–hole limit of the solution can be found in Refs. [14, 15].

5. DISCUSSION

The analytical solution for the rigidly rotating disk of dust in general relativity described above has been found by applying soliton–theoretical solution techniques to the Ernst equation. These methods only work for the vacuum equations however. Therefore, genuine perfect fluid configurations, e.g. models of rotating neutron stars, can only be treated numerically so far. However, there are some possible analytical generalizations of the disk solution considered that might be interesting:
(i) Differentially rotating disks of dust and (ii) black holes surrounded by dust rings. The first item has already been handled succesfully by a mixture of analytical and numerical methods [16, 17].

Acknowledgments

The support from the DFG is gratefully acknowledged.

References

[1] G. Neugebauer and R. Meinel, *Astrophys. J.* **414** (1993) L97.

[2] G. Neugebauer and R. Meinel, *Phys. Rev. Lett.* **73** (1994) 2166.

[3] G. Neugebauer and R. Meinel, *Phys. Rev. Lett.* **75** (1995) 3046.

[4] L. Lichtenstein, "*Gleichgewichtsfiguren rotierender Flüssigkeiten*", (Springer, Berlin, 1933).

[5] S. Chandrasekhar, "*Ellipsoidal figures of equilibrium*", (Dover, New York 1969).

[6] I. Newton, "*Philosophiae naturalis principia*", Book III, Propositions XVIII–XX. (1686).

[7] I. Kant, "*Allgemeine Naturgeschichte und Theorie des Himmels*" (1755).

[8] C. Maclaurin, "*A Treatise on Fluxions*", (1742).

[9] G. Neugebauer, A. Kleinwächter and R. Meinel, *Helv. Phys. Acta* **69** (1996) 472.

[10] A. Kleinwächter (2001), this volume.

[11] J.M. Bardeen and R.V. Wagoner, *Astrophys. J.* **158** (1969) L65.

[12] J.M. Bardeen and R.V. Wagoner, *Astrophys. J.* **167** (1971) 359.

[13] D. Petroff and R. Meinel *Phys. Rev. D.* (2000), submitted.

[14] R. Meinel, in: *Recent Developments in Gravitation and Mathematical Physics*, Eds. A. Garcia et al., (Science Network Publishing Konstanz, 1998). [gr-qc/9703077].

[15] R. Meinel, *Ann. Phys.* (Leipzig) **9** (2000) 335.

[16] M. Ansorg and R. Meinel, *Gen. Rel. Grav.* **32** (2000) 1365.

[17] M. Ansorg, *Gen. Rel. Grav.* (2001) in the press [gr-qc/0006045].

INTEGRABILITY OF SDYM EQUATIONS FOR THE MOYAL BRACKET LIE ALGEBRA

M. Przanowski*
Departamento de Fisica, CINVESTAV, 07000 Mexico D.F., Mexico
Institute of Physics, Technical University of Łódź,
Wólczańska 219. 93–005 Łódź, Poland.

J.F. Plebański†
Departamento de Fisica, CINVESTAV, 07000 Mexico D.F., Mexico.

S. Formański‡
Institute of Physics, Technical University of Łódź,
Wólczańska 219. 93–005 Łódź, Poland.

Abstract Integrability of the the *master equation* (ME) which is an especial form of the SDYM equations for the Moyal bracket Lie algebra is studied.
Infinite number of conservation laws and linear systems for ME are found. The twistor construction is also given.

Keywords: Integrable systems, SDYM equations, Moyal algebra, self–dual gravity.

1. INTRODUCTION

As has been pointed out by R.S. Ward [1]–[4] many of the integrable systems in mathematical physics arise as reductions of the self–dual Yang–Mills (SDYM) equations. In this way one can obtain the Bogomolny equation, the KdV and NLS equations, the Toda equations, the Ernst equation in general relativity, the Heisenberg ferromagnet equation, the top equations, the Nahm equation etc. (for review see

*E–mail:przan@fis.cinvestav.mx
†E–mail:pleban@fis.cinvestav.mx
‡E–mail:sforman@ck-sg.p.lodz.pl

[5]). Hence, in a sence the SDYM equations constitute a universal integrable system. Hovwever, it is known that some integrable equations, as for example the Kadomtsev–Petviashvili (KP) and Davey–Stewartson equations do not seem to be any reductions of the SDYM equations for a finite Lie group. For that reason L.J.Mason [6] has suggested that the heavenly equation might be a universal integrable equation. In 1992 I.A.B.Strachan [7] found the Moyal deformation of Plebański's first heavenly equation. Then K.Takasaki [8] and J.F.Plebański et al [9] considered the Moyal deformation of Plebański's second heavenly equation. Further generalization of heavenly equations has been given by J.F.Plebański and M.Przanowski [10, 11], where the SDYM system for the Moyal bracket Lie algebra has been written in the form of one nonlinear equation for a formal series in $\hbar$ of functions depending on six variables. This equation appears also to be a natural lift to six dimensions of the Moyal deformation of the Husain-Park heavenly equation. In the work by M.Przanowski et al [12] it has been argued that this equation (called the *master equation* (ME)) is a universal integrable system. In particular, it has been shown how ME can be reduced to the KP equation. However, we were not able to provide some evidence for the integrability of ME. We are going to accomplish the proof in the present paper.

In the next section (Sec.2) it is shown that ME admits an infinite number of nonlocal conservation laws. In Sec.3 we get the linear systems for ME i.e., the linear systems (Lax pairs) such that ME is the integrability condition for these systems. The section 4 is devoted to the twistor construction for ME in the case when the solution Θ of ME is a power series in $\hbar$, $\Theta = \sum_{n=0}^{\infty} \hbar^n \Theta_n$, $\frac{\partial \Theta_n}{\partial \hbar} = 0$. The case of $\Theta = \sum_{n=-N<0}^{\infty} \hbar^n \Theta_n$, $\frac{\partial \Theta_n}{\partial \hbar} = 0$ is analysed in Sec.5, where the integrability of ME is shown by using the recent result of D.B.Fairlie and A.N.Leznov [13].

Let us remind how one can arrive at ME (see [10, 11, 12] for details). We deal with C^4 endowed with the flat metric $ds^2 = 2(dx \otimes_s d\tilde{x} + dy \otimes_s d\tilde{y})$. Consider the SDYM equations for a Lie group G on some $U \subset C^4$

$$F_{xy} = 0\,, \quad F_{\tilde{x}\tilde{y}} = 0 \quad \text{and} \quad F_{x\tilde{x}} + F_{y\tilde{y}} = 0 \tag{1}$$

where $F \in \Gamma(g \otimes \Lambda^2 U)$ is the curvature 2-form defining the Yang-Mills field and g denotes the Lie algebra of G. In terms of the connection 1-form $A \in \Gamma(g \otimes T^* U)$ (the Yang-Mills potential) after suitable choice of gauge i.e. such that $A_x = 0 = A_y$ Eqs. (1) read

$$\partial_{\tilde{x}} A_{\tilde{y}} - \partial_{\tilde{y}} A_{\tilde{x}} + [A_{\tilde{x}}, A_{\tilde{y}}] = 0, \quad \partial_x A_{\tilde{x}} + \partial_y A_{\tilde{y}} = 0. \tag{2}$$

The second equation of (2) we interpret as an integrability condition. It means that there exist $\Theta \in \Gamma(g \otimes C^{\infty}(U))$ such that $A_{\tilde{x}} = -\partial_y \Theta$ and $A_{\tilde{y}} = \partial_x \Theta$ In terms of this function the self-duality conditions may be written in the form

$$\partial_x \partial_{\tilde{x}} \Theta + \partial_y \partial_{\tilde{y}} \Theta + [\partial_x \Theta, \partial_y \Theta] = 0 \tag{3}$$

Assume now that g is the *Moyal bracket Lie algebra.* Therefore, (3) gives an equation for a function Θ in six dimensions depending on a parameter $\hbar \in R$, $\Theta = \Theta(\hbar; x, y, \tilde{x}, \tilde{y}, p, q)$ which we call the *master equation* (ME)

$$\partial_x \partial_{\tilde{x}} \Theta + \partial_y \partial_{\tilde{y}} \Theta + \{\partial_x \Theta, \partial_y \Theta\}_M = 0 \tag{4}$$

where $\{\cdot, \cdot\}_M$ stands for the Moyal bracket

$$\begin{aligned}
\{f, g\}_M &:= \tfrac{1}{i\hbar}(f * g - g * f) = f \tfrac{2}{\hbar} sin(\tfrac{\hbar}{2} \overleftrightarrow{P}) g; \quad \hbar \epsilon R \\
f * g &:= \textstyle\sum_{n=0}^{\infty} \frac{1}{n!} (\frac{i\hbar}{2})^n \omega^{i_1 j_1} \ldots \omega^{i_n j_n} \frac{\partial^n f}{\partial X^{i_1} \ldots \partial X^{i_n}} \frac{\partial^n g}{\partial X^{j_1} \ldots \partial X^{j_n}} \\
&= f \exp(\tfrac{i\hbar}{2} \overleftrightarrow{P}) g, \quad i_1, \ldots j_1, \ldots = 1, 2, \\
(X^1, X^2) &= (q, p), (\omega^{ij}) = \begin{pmatrix} 0 & 1 \\ -1 & 0 \end{pmatrix}, \quad \overleftrightarrow{P} := \frac{\overleftarrow{\partial}}{\partial q} \frac{\overrightarrow{\partial}}{\partial p} - \frac{\overleftarrow{\partial}}{\partial p} \frac{\overrightarrow{\partial}}{\partial q}
\end{aligned} \tag{5}$$

This ME can be also obtained from the variational principle [10] for the action

$$S = \int_{U \times R^2} dv[-\frac{1}{3} \Theta * \{\partial_x \Theta, \partial_y \Theta\}_M + \frac{1}{2}(\partial_x \Theta * \ \partial_{\tilde{x}} \Theta + \partial_y \Theta * \partial_{\tilde{y}} \Theta)]. \tag{6}$$

[The "Planck constant" $\hbar$ plays the role of a *deformation parameter.* Nowadays it is well known that the Moyal deformation of differential equations is of great importance in the theory of integrable systems (see also [14]–[17])]

2. NONLOCAL CONSERVATION LAWS

In this section we show that ME admits an infinite number of nonlocal conservation laws. Our considerations follow E.Brezin et al [18], M.K.Prasad et al [19], L.L.Chau et al [20] and L.L.Chau [21], where nonlocal conservation laws for some nonlinear fields in two dimensions and for SDYM fields have been found.

Define the 0-order "currents" $j_{\tilde{x}}^{(0)} := -\partial_y \Theta^{(0)}$ and $j_{\tilde{y}}^{(0)} := \partial_x \Theta^{(0)}$ where $\Theta^{(0)} = \Theta$. Then of course due to commutativity of partial derivatives
$\partial_x j_{\tilde{x}}^{(0)} + \partial_y j_{\tilde{y}}^{(0)} = 0$. In the next step we put

$$j_{\tilde{x}}^{(1)} := \mathcal{D}_{\tilde{x}} * \Theta^{(0)} = \mathcal{D}_{\tilde{x}} * \Theta, \quad j_{\tilde{y}}^{(1)} := \mathcal{D}_{\tilde{y}} * \Theta^{(0)} = \mathcal{D}_{\tilde{y}} * \Theta \tag{7}$$

where $\mathcal{D}_{\tilde{x}} := \partial_{\tilde{x}} - \frac{1}{i\hbar}\partial_y\Theta$, $\mathcal{D}_{\tilde{y}} := \partial_{\tilde{y}} + \frac{1}{i\hbar}\partial_x\Theta$. Simple calculation gives the conservation law

$$\partial_x j_{\tilde{x}}^{(1)} + \partial_y j_{\tilde{y}}^{(1)} = \partial_x\partial_{\tilde{x}}\Theta + \partial_y\partial_{\tilde{y}}\Theta + \{\partial_x\Theta, \partial_y\Theta\}_M \overset{by ME}{=} 0 \tag{8}$$

Hence, there exists function $\Theta^{(1)}$ such that $j_{\tilde{x}}^{(1)} := -\partial_y\Theta^{(1)}$ $j_{\tilde{y}}^{(1)} := \partial_x\Theta^{(1)}$ Thus we are led to an iteration procedure. Given the nth *currents* $j_{\tilde{x}}^{(n)}$ and $j_{\tilde{y}}^{(n)}$ satisfying the equation $\partial_x j_{\tilde{x}}^{(n)} + \partial_y j_{\tilde{y}}^{(n)} = 0$ for $n \geq 1$ one defines the function $\Theta^{(n)}$ such that $j_{\tilde{x}}^{(n)} := -\partial_y\Theta^{(n)}$ and $j_{\tilde{y}}^{(n)} := \partial_x\Theta^{(n)}$. Then the (n+1)th currents are defined by

$$j_{\tilde{x}}^{(n+1)} := \mathcal{D}_{\tilde{x}} * \Theta^{(n)}, \quad j_{\tilde{y}}^{(n+1)} := \mathcal{D}_{\tilde{y}} * \Theta^{(n)} \tag{9}$$

and we get the following conservation law

$$\begin{aligned} \partial_x j_{\tilde{x}}^{(n+1)} + \partial_y j_{\tilde{y}}^{(n+1)} &= (\mathcal{D}_{\tilde{x}} * \mathcal{D}_{\tilde{y}} - \mathcal{D}_{\tilde{y}} * \mathcal{D}_{\tilde{x}}) * \Theta^{(n-1)} \\ &= \frac{1}{i\hbar}(\partial_x\partial_{\tilde{x}}\Theta + \partial_y\partial_{\tilde{y}}\Theta + \{\partial_x\Theta, \partial_y\Theta\}_M) * \Theta^{(n-1)} \overset{by ME}{=} 0 \end{aligned} \tag{10}$$

Thus one arrives at an infinite number of nonlocal conservation laws for n=1,2,.... The term *nonlocal* stands for the fact that the currents $j_{\tilde{x}}^{(n)}$ and $j_{\tilde{y}}^{(n)}$ for $n \geq 2$ are defined by integration and differentiation of lower currents. For the nth *charge* $\Theta^{(n)}(\hbar, x, y, \tilde{x}, \tilde{y}, p, q)$, $n \geq 1$ we get

$$\begin{aligned} \Theta^{(n)} &= \int_{x_0}^{x} dx'\mathcal{D}_{\tilde{y}} * \Theta^{(n-1)} = \int_{x_0}^{x} dx'\mathcal{D}_{\tilde{y}} * \int_{x_0}^{x'} dx''\mathcal{D}_{\tilde{y}} * \Theta^{(n-2)} \\ &= \int_{x_0}^{x} dx'\mathcal{D}_{\tilde{y}} * \int_{x_0}^{x'} dx''\mathcal{D}_{\tilde{y}} * \int_{x_0}^{x''} dx'''\mathcal{D}_{\tilde{y}} * \ldots * \Theta. \end{aligned} \tag{11}$$

So the (n+1)th currents read

$$\begin{aligned} j_{\tilde{x}}^{(n+1)} &= \mathcal{D}_{\tilde{x}} * \int_{x_0}^{x} dx'\mathcal{D}_{\tilde{y}} * \int_{x_0}^{x'} dx''\mathcal{D}_{\tilde{y}} * \int_{x_0}^{x''} dx'''\mathcal{D}_{\tilde{y}} * \ldots * \Theta \\ j_{\tilde{y}}^{(n+1)} &= \mathcal{D}_{\tilde{y}} * \int_{x_0}^{x} dx'\mathcal{D}_{\tilde{y}} * \int_{x_0}^{x'} dx''\mathcal{D}_{\tilde{y}} * \int_{x_0}^{x''} dx'''\mathcal{D}_{\tilde{y}} * \ldots * \Theta. \end{aligned} \tag{12}$$

[The natural question arises what the relations are between the above conservations laws and the ones obtained by C.P.Boyer and J.F.Plebański [22, 23] for heavenly equation in 4 dimensions.]

3. LINEAR SYSTEMS FOR ME

Here we show how one can adopt some results obtained for SDYM equations (L.L.Chau et al [20], L.L.Chau [21]) to get the linear systems connected with ME. Define

$$\tilde{\Psi} := 1 + \frac{1}{i\hbar}\sum_{n=1}^{\infty} \lambda^n \Theta^{(n-1)} \tag{13}$$

where λ is a complex parameter and $\Theta^{(n-1)}$, $n = 1, 2, \ldots$, are the charges defined in the previous section. Then we easily get

$$\partial_x \tilde{\Psi} = \lambda \mathcal{D}_{\tilde{y}} * \tilde{\Psi}\,, \quad -\partial_y \tilde{\Psi} = \lambda \mathcal{D}_{\tilde{x}} * \tilde{\Psi} \tag{14}$$

Differentiating the first equation of (14) with respect to y and the second one with respect to x then employing Eq. (14) once more one obtains

$$\lambda(\mathcal{D}_{\tilde{x}} * \mathcal{D}_{\tilde{y}} - \mathcal{D}_{\tilde{y}} * \mathcal{D}_{\tilde{x}}) * \tilde{\Psi} = 0 \tag{15}$$

As (15) is satisfied for every $\lambda \epsilon U$ $(0 \epsilon U \subset C)$ it follows that

$$\begin{aligned} \mathcal{D}_{\tilde{x}} * \mathcal{D}_{\tilde{y}} - \mathcal{D}_{\tilde{y}} * \mathcal{D}_{\tilde{x}} \equiv \frac{1}{i\hbar}(\partial_x\partial_{\tilde{x}}\Theta + \partial_y\partial_{\tilde{y}}\Theta + \{\partial_x\Theta, \partial_y\Theta\}_M) = 0 \\ \Rightarrow \partial_x\partial_{\tilde{x}}\Theta + \partial_y\partial_{\tilde{y}}\Theta + \{\partial_x\Theta, \partial_y\Theta\}_M = 0 \end{aligned} \tag{16}$$

i.e. Θ fulfills ME. Thus substituting the definition of $\mathcal{D}_{\tilde{x}}$ and $\mathcal{D}_{\tilde{y}}$ into (14) and changing $\tilde{\Psi}$ by the general solution $\Psi(\lambda)$ we arrive at the linear system for ME

$$\begin{aligned} (\partial_x - \lambda\partial_{\tilde{y}})\Psi(\lambda) &= \lambda \tfrac{1}{i\hbar}\partial_x\Theta * \Psi(\lambda) \\ (\partial_y + \lambda\partial_{\tilde{x}})\Psi(\lambda) &= \lambda \tfrac{1}{i\hbar}\partial_y\Theta * \Psi(\lambda)\,, \quad \lambda \epsilon \overline{C} - \{\infty\}. \end{aligned} \tag{17}$$

Integrability condition of the system (17) is given by ME. The solution $\Psi(\lambda)$ is analytic in $\lambda \epsilon \overline{C} - -\{\infty\}$ and is assumed to be of the form

$$\Psi := 1 + \lambda \tfrac{1}{i\hbar}\Theta + O(\lambda^2) = \exp_*\{\tfrac{1}{i\hbar}\textstyle\sum_{n=1}^{\infty} \lambda^n \Xi_n\} \tag{18}$$

$$\Xi_n = \Xi_n(\hbar; x, y, \tilde{x}, \tilde{y}, p, q) = \textstyle\sum_{k=0}^{\infty} \hbar^k \Xi_{nk}(x, y, \tilde{x}, \tilde{y}, p, q)\,; \;\; \Xi_1 = \Theta$$

where $\exp_*\{\cdot\}$ stands for the exponential in associative Moyal $*$-algebra, i.e.

$$\exp_*\{\alpha\} := 1 + \sum_{n=1}^{\infty} \frac{1}{n!} \underbrace{\alpha * \ldots * \alpha}_{n-\text{times}}$$

Analogously, we can consider another linear system for ME

$$\begin{aligned} (\partial_{\tilde{y}} - \tfrac{1}{\lambda}\partial_x)\Phi(\lambda) &= -\tfrac{1}{i\hbar}\partial_x\Theta * \Phi(\lambda) \\ (\partial_{\tilde{x}} + \tfrac{1}{\lambda}\partial_y)\Phi(\lambda) &= \tfrac{1}{i\hbar}\partial_y\Theta * \Phi(\lambda)\,, \quad \lambda \epsilon \overline{C} - \{0\}. \end{aligned} \tag{19}$$

As before, ME is the integrability condition of (18). The solution $\Phi(\lambda)$ has the Laurent series expansion

$$\Phi = J_*^{-1} + O(\frac{1}{\lambda}) = \exp_*\{\frac{1}{i\hbar}\sum_{n=0}^{\infty} \lambda^{-n}\Upsilon_n\} \tag{20}$$

$$\begin{aligned} \Upsilon_n &= \Upsilon_n(\hbar; x, y, \tilde{x}, \tilde{y}, p, q) \\ &= \sum_{k=0}^{\infty} \hbar^k \Upsilon_{nk}(x, y, \tilde{x}, \tilde{y}, p, q)\,; \exp_*\{\frac{1}{i\hbar}\Upsilon_0\} = J_*^{-1} \end{aligned} \tag{21}$$

Inserting $\Phi(\lambda)$ given by(20) into (18) one quickly finds the following relations

$$\frac{1}{i\hbar}\partial_x\Theta = J_*^{-1} * \partial_{\tilde{y}}J\,, \quad -\frac{1}{i\hbar}\partial_y\Theta = J_*^{-1} * \partial_{\tilde{x}}J \tag{22}$$

where $J * J_*^{-1} = J_*^{-1} * J = 1$.

[The linear system for SDYM equations was first obtained by R.S.Ward [24] and then generalized to ME by J.F.Plebański et al. [9, 11] and M.Przanowski et al [12]. For SDYM equations on noncommutative spacetime this linear system was considered by K.Takasaki [25]. Note that our solutions $\Psi(\lambda)$ (18) and $\Phi(\lambda)$ (20) correspond to f and $\tilde{f}$, respectively, given in the monograph by L.J.Mason and N.M.J.Woodhouse [5] p.175 for the SDYM case. The exponential form of $\Psi(\lambda)$ and $\Phi(\lambda)$ is proposed following K.Takasaki [26].]

4. TWISTOR CONSTRUCTION

Now we are in a position to generalize the twistor construction for SDYM equations [24, 5] to the case of ME.

To this end consider the following function

$$H(\lambda) := \Phi_*^{-1}(\lambda) * \Psi(\lambda) \tag{23}$$

where $\lambda \epsilon (\overline{C} - \{0\}) \cap (\overline{C} - \{\infty\})$ and $\Phi_*^{-1}(\lambda) * \Phi(\lambda) = \Phi * \Phi_*^{-1}(\lambda)(\lambda) = 1$ The function $H(\lambda)$ is analytic in λ for $\lambda \epsilon (\overline{C} - \{0\}) \cap (\overline{C} - \{\infty\})$. Straightforward calculations show that by (17) and (18) one gets

$$(\lambda\partial_{\tilde{y}} - \partial_x)H(\lambda) = 0 \text{ and } (\lambda\partial_{\tilde{x}} + \partial_y)H(\lambda) = 0 \tag{24}$$

It means that $H = H(\lambda)$ is of the following form

$$H = H(\lambda) = H(\hbar; \tilde{y} + \lambda x, \tilde{x} - \lambda y, \lambda, p, q). \tag{25}$$

But the equations

$$\tilde{y} + \lambda x =: w^1 = \text{const.}, \;\; \tilde{x} - \lambda y =: w^2 = \text{const.}, \;\; \lambda =: w^3 = \text{const.} \tag{26}$$

define a totally null anti–self–dual 2–surface in C^4 (the *twistor surface*) described by the anti–self–dual 2–form

$$\omega(\lambda) = (d\tilde{y}+\lambda dx)\wedge(d\tilde{x}-\lambda dy) = -d\tilde{x}\wedge d\tilde{y}+\lambda(dx\wedge d\tilde{x}+dy\wedge d\tilde{y})-\lambda^2(dx\wedge dy) \tag{27}$$

Finally, inserting (18) and (20) into (23), employing also (25) and the Campbell–Baker–Hausdorff formula we obtain

$$H(\lambda) \;\; = \;\; H(\hbar; \tilde{y} + \lambda x, \tilde{x} - \lambda y, \lambda, p, q)$$

$$
\begin{aligned}
&= \exp_*\{-\frac{1}{i\hbar}\sum_{n=0}^{\infty}\lambda^{-n}\Upsilon_n\} * \exp_*\{\frac{1}{i\hbar}\sum_{m=1}^{\infty}\lambda^m\Xi_m\} \\
&= \exp_*\{\frac{1}{i\hbar}[\sum_{n=1}^{\infty}\lambda^n\Xi_n \\
&- \sum_{n=0}^{\infty}\lambda^{-n}\Upsilon_n - \frac{1}{2}\{\sum_{n=0}^{\infty}\lambda^{-n}\Upsilon_n, \sum_{m=1}^{\infty}\lambda^m\Xi_m\}_M + \ldots]\} \\
&= \exp_*\{\frac{1}{i\hbar}\sum_{m=-\infty}^{\infty}\lambda^m\Lambda_m\}; \ \lambda\,\epsilon\,(\overline{C}-\{0\})\cap(\overline{C}-\{\infty\}) \\
\Lambda_m &= \Lambda_m(\hbar; x, y, \tilde{x}, \tilde{y}, p, q) = \sum_{k=0}^{\infty}\hbar^k\Lambda_{mk}(x, y, \tilde{x}, \tilde{y}, p, q). \qquad (28)
\end{aligned}
$$

Therefore, the function $H(\lambda)$ defined by (23) is for any fixed (p, q) a formal series with respect to $\hbar$ and $\hbar^{-1}$ of analytic functions in the projective twistor space CP_3. This formal series has an especial form given by the last part of (28) i.e.,

$$
\begin{gathered}
H(\lambda) = H(\hbar; \tilde{y}+\lambda x, \tilde{x}-\lambda y, \lambda, p, q) = \exp_*\{\tfrac{1}{i\hbar}\textstyle\sum_{m=-\infty}^{\infty}\lambda^m\Lambda_m\} \\
\lambda\,\epsilon\,(\overline{C}-\{0\})\cap(\overline{C}-\{\infty\}) \\
\Lambda_m = \Lambda_m(\hbar; x, y, \tilde{x}, \tilde{y}, p, q) = \textstyle\sum_{k=0}^{\infty}\hbar^k\Lambda_{mk}(x, y, \tilde{x}, \tilde{y}, p, q). \qquad (29)
\end{gathered}
$$

The above considerations suggest the twistor construction for ME which is analogous to the one for SDYM equations [24, 5]. The main idea of our construction was given by K.Takasaki [26].

Let $H(\lambda)$ be any analytic function of the form (29) and let $\Psi(\lambda)$ be a function of the form (18) analytic for $\lambda\,\epsilon\,(\overline{C}--\{\infty\})$ and $\Phi(\lambda)$ be a function of the form (20) analytic for $\lambda\,\epsilon\,(\overline{C}--\{0\})$. Assume that the factorization (23) holds (the *Riemann–Hilbert problem*). As $(\lambda\partial_{\tilde{y}}-\partial_x)H(\lambda)=0$ and $(\lambda\partial_{\tilde{x}}+\partial_y)H(\lambda)=0$ one quickly finds that

$$
\begin{gathered}
[(\lambda\partial_{\tilde{y}}-\partial_x)\Psi(\lambda)] * \Psi_*^{-1}(\lambda) = \lambda[(\partial_{\tilde{y}}-\tfrac{1}{\lambda}\partial_x)\Phi(\lambda)] * \Phi_*^{-1}(\lambda) \\
[(\lambda\partial_{\tilde{x}}+\partial_y)\Psi(\lambda)] * \Psi_*^{-1}(\lambda) = \lambda[(\partial_{\tilde{x}}+\tfrac{1}{\lambda}\partial_y)\Phi(\lambda)] * \Phi_*^{-1}(\lambda) \\
\lambda\,\epsilon\,(\overline{C}-\{0\})\cap(\overline{C}-\{\infty\}) \qquad (30)
\end{gathered}
$$

The left–hand side of Eq. (30) is analytic in λ for $\lambda\,\epsilon\,\overline{C}--\{\infty\}$ and the right–hand side devided by λ is analytic for $\lambda\,\epsilon\,\overline{C}-\{0\}$. It means that the left–hand side of (30) can be analytically extended on the whole Riemann sphere $\overline{C}$ to give

$$
\begin{gathered}
[(\lambda\partial_{\tilde{y}}-\partial_x)\Psi(\lambda)] * \Psi_*^{-1}(\lambda) = \lambda B_1 + C_1 \\
[(\lambda\partial_{\tilde{x}}+\partial_y)\Psi(\lambda)] * \Psi_*^{-1}(\lambda) = \lambda B_2 + C_2, \ \ \lambda\,\epsilon\,\overline{C} \qquad (31)
\end{gathered}
$$

where B_1, B_2, C_1 and C_2 are independent of λ. In the gauge (18) one gets

$$C_1 = 0 = C_2\,;\; B_1 = -\frac{1}{i\hbar}\partial_x\Theta\,;\; B_2 = \frac{1}{i\hbar}\partial_y\Theta \tag{32}$$

Finally, substituting (32) into (31) we obtain the linear system (17) and inserting (31) under (32) into (30) one recovers the linear system (18). But, as we know ME is the integrability condition of these systems. Thus one arrives at the twistor construction for ME. [*Remark* : The factorization (23) is gauge invariant i.e., the transformation $\Phi(\lambda) \mapsto S * \Phi(\lambda)$, $\Psi(\lambda) \mapsto S * \Psi(\lambda)$ does not change the function $H(\lambda)$. Consequently, one can choose the gauge such that $\Psi(\lambda)$ is of the form (18)]

5. THE CASE OF $\Theta = \sum_{K=-N<0}^{\infty} \hbar^K \Theta_K$

In previous sections we dealt with the analytic in $\hbar$ solution of ME. Now consider the case that it might be singular in $\hbar = 0$ but its pole is of rank not grater than $N < \infty$ so assume solution of the form

$$\Theta = \sum_{k=-N}^{\infty} \hbar^k \Theta_k\,, \qquad 1 \leq N \leq \infty \tag{33}$$

Thus defining new function Ω

$$\Omega := \hbar^N \Theta = \sum_{k=0}^{\infty} \hbar^k \Omega_k\,, \quad \text{where} \quad \Omega_k = \Theta_{k-N} \tag{34}$$

In terms of which the ME might be written as:

$$\hbar^N(\partial_x\partial_{\tilde{x}}\Omega + \partial_y\partial_{\tilde{y}}\Omega) + \{\partial_x\Omega\,,\,\partial_y\Omega\}_M = 0 \tag{35}$$

In an explicit form (compare with Eq. (4).)

$$\sum_{k=0}^{\infty} \hbar^{k+N}(\partial_x\partial_{\tilde{x}}\Omega_k + \partial_y\partial_{\tilde{y}}\Omega_k) + \sum_{n,m,l=0}^{\infty} \hbar^{2n+m+l}\frac{1}{(2n+1)!}\left(\frac{i}{2}\right)^{2n}$$
$$\times \quad \omega^{i_1 j_1}\ldots\omega^{i_{2n+1}j_{2n+1}}\frac{\partial^{2n+1}(\partial_x\Omega_m)}{\partial X^{i_1}\ldots\partial X^{i_{2n+1}}}\frac{\partial^{2n+1}(\partial_y\Omega_l)}{\partial X^{j_1}\ldots\partial X^{j_{2n+1}}} = 0$$

Hence, equalizing to 0 each term in power of $\hbar$ one obtains:

$$\partial_x\partial_{\tilde{x}}\Omega_k + \partial_y\partial_{\tilde{y}}\Omega_k + \sum_{2n+m+l=k+N} \frac{(-1)^n}{4^n(2n+1)!}\omega^{i_1 j_1}\ldots\omega^{i_{2n+1}j_{2n+1}}$$
$$\times \quad \frac{\partial^{2n+1}(\partial_x\Omega_m)}{\partial X^{i_1}\ldots\partial X^{i_{2n+1}}}\frac{\partial^{2n+1}(\partial_y\Omega_l)}{\partial X^{j_1}\ldots\partial X^{j_{2n+1}}} = 0 \tag{36}$$

for $2n+m+l \geq N$ and

$$\sum_{2n+m+l=s} \frac{(-1)^n}{4^n(2n+1)!} \omega^{i_1j_1} \dots \omega^{i_{2n+1}j_{2n+1}} \times \frac{\partial^{2n+1}(\partial_x \Omega_m)}{\partial X^{i_1} \dots \partial X^{i_{2n+1}}} \frac{\partial^{2n+1}(\partial_y \Omega_l)}{\partial X^{j_1} \dots \partial X^{j_{2n+1}}} = 0 \tag{37}$$

for $s = 0, 1, \dots, N-1$.

Starting with $s = 0$ which implies $n = m = l = 0$, Eq. (36) reads:

$$\omega^{i_1j_1} \frac{\partial(\partial_x \Omega_0)}{\partial X^{i_1}} \frac{\partial(\partial_y \Omega_0)}{\partial X^{j_1}} = \partial_x \Omega_0 \overleftrightarrow{P} \partial_y \Omega_0 = 0 \tag{38}$$

where $\overleftrightarrow{P} := \frac{\overleftarrow{\partial}}{\partial q}\frac{\overrightarrow{\partial}}{\partial p} - \frac{\overleftarrow{\partial}}{\partial p}\frac{\overrightarrow{\partial}}{\partial q}$.

Suppose there exist a solution of this nonlinear equation. Then for $1 \leq s < N$ we arrive at the iterative system of linear equations for functions Ω_s:

$$\partial_x \Omega_s \overleftrightarrow{P} \partial_y \Omega_0 + \partial_x \Omega_0 \overleftrightarrow{P} \partial_y \Omega_s = F(\Omega_m, \Omega_l) \quad \text{where } m < s, l < s \tag{39}$$

and inhomogeneity F is of the form:

$$F(\Omega_m, \Omega_l) := -\sum_{\substack{m+l=s \\ m,l<s}} \partial_x \Omega_m \overleftrightarrow{P} \partial_y \Omega_l - \sum_{\substack{2n+m+l=s \\ n\neq 0}} \frac{(-1)^n}{4^n(2n+1)!} \times \omega^{i_1j_1} \dots \omega^{i_{2n+1}j_{2n+1}} \frac{\partial^{2n+1}(\partial_x \Omega_m)}{\partial X^{i_1} \dots \partial X^{i_{2n+1}}} \frac{\partial^{2n+1}(\partial_y \Omega_l)}{\partial X^{j_1} \dots \partial X^{j_{2n+1}}} \tag{40}$$

from which we can find functions Ω_s, $s = 1, \dots, N-1$ under the condition that there exist solution of Eq. (38). Remaining functions Ω_s for $s = N, N+1, \dots$ have to be computed from equations (35) written as equations for Ω_s in the following form:

$$\begin{aligned} \partial_x \Omega_s \overleftrightarrow{P} \partial_y \Omega_0 + \partial_x \Omega_0 \overleftrightarrow{P} \partial_y \Omega_s &= -\partial_x \partial_{\tilde{x}} \Omega_{s-N} - \partial_y \partial_{\tilde{y}} \Omega_{s-N} \\ &+ F(\Omega_m, \Omega_l) \end{aligned} \tag{41}$$

The general solution of (38) exist due to the following theorem, which is an especial 4–dimensional case of more general result:

Theorem (Fairlie & Leznov [13]) The homogenous complex Monge–Ampere equation:

$$\det \begin{bmatrix} \frac{\partial^2 \eta}{\partial x \partial q} & \frac{\partial^2 \eta}{\partial x \partial p} \\ \frac{\partial^2 \eta}{\partial y \partial q} & \frac{\partial^2 \eta}{\partial y \partial p} \end{bmatrix} = 0$$

posses a general solution $\eta(x, y, p, q)$ i.e. solution given in terms of two arbitrary function each of which depends on three variables $\zeta = \zeta(x, y, \beta)$ and $\xi = \xi(q, p, \beta)$ and is given by

$$\partial_x \eta = \partial_x \zeta(x, y, \beta(x, y, q, p)), \quad \partial_y \eta = \partial_x \zeta(x, y, \beta(x, y, q, p)),$$
$$\partial_q \eta = \partial_q \xi(x, y, \beta(x, y, q, p)), \quad \partial_p \eta = \partial_p \xi(x, y, \beta(x, y, q, p))$$

where β is given in implicit form $\frac{\partial \zeta}{\partial \beta} = -\frac{\partial \xi}{\partial \beta}$.

Summarizing, the above construction leads to a singular in $\hbar$ solution of ME.

Acknowledgments

This work was partially supported by the CONACyT and CINVESTAV (México) and by KBN (Poland). One of us (M.P.) is indebted to the staff of Departamento de Fisica, CINVESTAV, Mexico D.F., for warm hospitality.

References

[1] R.S. Ward, *Nucl. Phys. B* **236** (1984) 381.

[2] R.S. Ward, *Philos. Trans. R.Soc.* **A315** (1985) 451.

[3] R.S. Ward, in: *Field theory, quantum gravity and strings*, eds. H.J. de Vega and N. Sanchez, Lecture Notes in Physics **280** (Springer, Berlin 1986).

[4] R.S. Ward, in: *Twistors in mathematics and physics*, eds. T.N. Bailey and R.J. Baston (Cambridge University Press, Cambridge 1990).

[5] L.J. Mason and N.M.J.Woodhouse, *Integrability, self–duality, and twistor theory* (Clarendon Press, Oxford 1996).

[6] L.J. Mason, *Twistor Newsletter* **30** (1990) 14.

[7] I.A.B. Strachan, *Phys. Lett.* **B282** (1992) 63.

[8] K. Takasaki, *J. Geom. Phys.* **14** (1994) 332.

[9] J.F. Plebański, M. Przanowski, B. Rajca and J. Tosiek , *Acta Phys. Pol.* **B26** (1995) 889.

[10] J.F. Plebański and M. Przanowski, *Phys. Lett.* **A212** (1996) 22.

[11] J.F. Plebański and M. Przanowski, in: *Gravitation, electromagnetism and geometrical structures*, ed. G.Farrarese (Pitagora Editrice, Bologna 1996).

[12] M. Przanowski and S. Formański, *Acta Phys. Pol.* **B30** (1999) 863.

[13] D.B. Fairlie and A.N. Leznov, *J. Phys. A: Math. Gen.* **33** (2000) 4657.

[14] B.A. Kupershmidt, *Lett. Math. Phys.* **20** (1990) 19.

[15] I.A.B. Strachan, *J. Geom. Phys.* **21** (1997) 255.

[16] M. Przanowski, S. Formański and F.J. Turrubiates, *Mod. Phys. Lett.* **A13** (1998) 3193.

[17] C. Castro and J.F. Plebański, *J. Math. Phys.* **40** (1999) 3738.

[18] E. Brezin, C. Itzykson, J. Zinn-Justin and J.B. Zuber, *Phys. Lett.* **B82** (1979) 442.

[19] M.K. Prasad, A. Sinha and L.L. Chau Wang, *Phys. Lett.* **B87** (1979) 237.

[20] L.L. Chau, M.K. Prasad and A.Sinha, *Phys. Rev.* **D24** (1981) 1574.

[21] L.L. Chau, in: *Nonlinear phenomena*, ed. K.B. Wolf, Lecture Notes in Physics **189** (Springer, New York 1983).

[22] C.P. Boyer and J.F. Plebański, *J. Math. Phys.* **18** (1977) 1022.

[23] C.P. Boyer and J.F. Plebański, *J. Math. Phys.* **26** (1985) 229; see also: C.P. Boyer, in: *Nonlinear phenomena*, ed. K.B. Wolf, Lecture Notes in Physics **189** (Springer, New York 1983).

[24] R.S. Ward, *Phys. Lett.* **A61** (1977) 81.

[25] K. Takasaki, *Anti-self-dual Yang-Mills equations on noncommutative spacetime*, hep-th/0005194 v2.

[26] K. Takasaki, *J. Geom. Phys.* **14** (1994) 111.

II

ALTERNATIVE THEORIES AND SCALAR FIELDS

THE FRW UNIVERSES WITH BAROTROPIC FLUIDS IN JORDAN–BRANS–DICKE THEORY

P. Chauvet–Alducin *

Departamento de Física , Universidad Autónoma Metropolitana–Iztapalapa
P. O. Box. 55–534, México D. F., C. P. 09340 México

Abstract

The description of homogeneous and isotropic models for the expansion of the Universe represented by a perfect, barotropic, fluid pose mathematical difficulties that have been overcome when the space is non–flat in scalar–tensor theories only if the energy–stress tensor is vacuous or describes stiff matter, incoherent radiation or ultrarelativistic matter. This paper reveals the dynamics for these and the other ideal fluid models, in non–flat spaces mainly, for the Jordan–Brans–Dicke cosmological theory by showing that its field equations, when written in terms of reduced variables, allow their straightforward partial integration.

Keywords: Jordan–Brans–Dicke theory, FRW universes, exact solutions.

1. INTRODUCTION

The current evolution of the Universe in terms of Friedmann–Robertson–Walker (FRW) metrics seems to give an acceptable description of it over large scales even if the apparent homogeneity and isotropy on those scales still demands justification, like possibly including indispensable initial conditions before the onset of an inflationary epoch. Nowadays, for theoretical and also observational [1] reasons scalar–tensor theories that can generalize in different ways Einstein's general relativity (GR) theory have captured new attention, and it is worthwhile to extend the number of allowed exact cosmological solutions where to choose from to describe its mean evolution.

*E–mail: pcha@xanum.uam.mx

Jordan–Brans–Dicke (JBD) [2] cosmology is the most simple scalar–tensor theory that under FRW metrics with perfect fluids its nonlinear field equations comprise two unknown variables which makes their integration more difficult, and its solutions hard to come by in contrast with GR which only has a single function –the scale factor– to determine. Consequently, a fruitful avenue exploited to obtain cosmological solutions in JBD, first done in [3] and later elsewhere [4], has been to find the way to combine the two aforementioned variables into a single one. This or any other method employed to obtain barotropic, fluid solutions in this, and other similar but more general scalar–tensor theories [5] has been fully accomplished in non flat spaces only for the vacuum, incoherent radiation, and stiff matter [6] so this paper shows how to advance the "single function" method further in order to integrate the field equations for other perfect fluids.

The Lagrangian for scalar–tensor gravity theories with no self interaction can be expressed as

$$L = \phi R - \frac{\omega(\phi)}{\phi}\phi_{;\alpha}\phi^{;\alpha} + 16\pi L_M, \qquad \alpha = 0, ..., 3 \tag{1}$$

where R is the spacetime Ricci curvature scalar, ϕ is the scalar field, $\omega(\phi)$ a dimensionless coupling function while G, Newton's gravitational constant together with c, the speed of light are put equal to 1. Finally, L_M is the Lagrangian for the matter fields. JBD assumes $\omega = const.$, and the variation of its action with respect to $g_{\mu\nu}$, and ϕ generates the following field equations

$$\begin{aligned} R_{\mu\nu} - \frac{1}{2}g_{\mu\nu}R &= 8\pi\frac{T_{\mu\nu}}{\phi} + \frac{1}{\phi}[\phi_{;\mu\nu} - g_{\mu\nu}\phi_{;\alpha}\phi^{;\alpha}] \\ &+ \frac{\omega}{\phi}[\phi_{;\mu}\phi_{;\nu} - \frac{1}{2}g_{\mu\nu}\phi_{;\alpha}\phi^{;\alpha}] \ , \end{aligned} \tag{2}$$

and

$$\phi_{;\mu}{}^{;\mu} = \left(\frac{8\pi}{3+2\omega}\right)T \ . \tag{3}$$

$$T^{\mu\nu} = (\rho + p)u^\mu u^\nu \ + \ pg^{\mu\nu}$$

is the energy–momentum tensor for a perfect fluid where p is the pressure, and ρ its energy density. u^β is its four velocity, $T \equiv T_\mu{}^\mu$ its trace, and its covariant derivative is

$$T^{\nu}_{\mu;\nu} = 0. \tag{4}$$

The line element for the FRW metric in (t, r, θ, φ) coordinates is

$$ds^2 = -dt^2 + a(t)^2[(1 - kr^2)^{-1}dr^2 + r^2(d\theta^2 + sin^2\theta d\varphi)] \, , \tag{5}$$

where a is the scale factor, and $k = +1, 0\, or\, -1$ stands for positive, vanishing or negative space curvature, respectively.

For a perfect fluid in a FRW cosmology its ordinary, but nonlinear field equations comprise the two unknown variables $a = a(t)$ and $\phi = \phi(t)$:

$$\frac{\ddot{\phi}}{\phi} + 3\frac{\dot{a}\dot{\phi}}{a\phi} = \frac{8\pi}{3+2\omega}\frac{(\rho - 3p)}{\phi a^3} \quad , \tag{6}$$

is the ϕ field equation, together with

$$\frac{\ddot{a}}{a} + 2\frac{\dot{a}^2}{a^2} + \frac{\dot{a}\dot{\phi}}{a\phi} + 2\frac{k}{a^2} = \frac{8\pi}{3+2\omega}\frac{[(1+\omega)\rho - \omega p]}{\phi} \quad , \tag{7}$$

and

$$3\frac{\ddot{a}}{a} + \omega\frac{\dot{\phi}^2}{\phi^2} + \frac{\ddot{\phi}}{\phi} = \frac{-8\pi}{3+2\omega}\frac{[(2+\omega)\rho + 3(1+\omega)p]}{\phi} \, , \tag{8}$$

where overdots stand for time derivatives.

2. BAROTROPIC FLUIDS

The equation $p = n\rho$ with $-1 < n < 1$, distinguish barotropic fluids. The conservation equation

$$\dot{\rho} = -3(\rho + p)(\ln a)^{\cdot}$$

yields

$$\rho = M_n a^{-3(1+n)} . \qquad M_n = \text{const.} \tag{9}$$

However instead of "t", I use in this section the "time parameter η", defined by $dt = a^{3n}d\eta$, with $(\,)' = \partial\eta$. With it, the ϕ field equation is once formally integrated $(n \neq \frac{1}{3}, \pm 1)$

$$\frac{\phi'}{\phi} = \frac{(1-3n)m_n\eta + \eta_0}{q} \quad , \tag{10}$$

where $q \equiv \phi a^{3(1-n)}$, η_0 is an integration constant, and

$$3(1-n)\frac{a'}{a} = \frac{q'}{q} - \frac{(1-3n)m_n\eta + \eta_0}{q} \quad . \quad m_n = \frac{8\pi M_n}{3+2\omega} \tag{11}$$

q obeys the following "dynamic" equation

$$3(1-n)q\,\frac{d^3q}{d\eta^3} = \left\{q'' - [2(2-3n)+3(1-n)^2\omega]m_n\right\} \times$$
$$[(3n+1)q' + 2(1-3n)(1-3n)m_n\eta + \eta_0] \quad , \tag{12}$$

together with its constraint

$$\frac{q''}{q} - \frac{2}{3(1-n)}\left(\frac{q'}{q}\right)^2 - \frac{2(1-3n)}{3(1-n)}\frac{q'}{q}\left[\frac{(1-3n)m_n\eta+\eta_0}{q}\right]$$
$$+ \frac{1}{3}\left[2(2-3n)+3(1-n)^2\omega\right]\left[\frac{(1-3n)m_n\eta+\eta_0}{q}\right]^2$$
$$+ \frac{[2+(1-n)(1+3n)\omega]m_n}{q} = 0. \tag{13}$$

Its solutions determine the way these models expand via the scale factor function "a" which, from Eq.(11), is given by

$$a^{3(1-n)} = q\exp\left(-\int\frac{(1-3n)m_n\eta+\eta_0}{q}d\eta\right) . \tag{14}$$

Eq.(12) whose integral is

$$\frac{q''}{q} - \frac{[2(2-3n)+3(1-n)^2\omega]m_n}{q} = \frac{-6(1-n)k}{a^{2(1-3n)}} \quad , k=0, \ \pm 1 \tag{15}$$

together with Eq.(13) turn out the following solutions. For $k=0$,

$$q = A_0\eta^2 + B_0\eta + C_0 \tag{16}$$

with

$$2A_0 = \left[2(2-3n)+3(1-n)^2\omega\right] m_n \quad , \tag{17}$$

while B_0, and C_0 remain undetermined. Therefore, for the FRW flat space three different cosmic solutions exist, distinguished by the sign of the discriminant $\Delta_0 \equiv 4A_0C_0 - {B_0}^2$ which depends the coupling parameter ω, and through n on the equation of state. For the discriminants $\Delta_0 > 0$, and $\Delta_0 < 0$, the scalar field ϕ has two branches: one with ϕ an increasing function of time, and the other with ϕ a decreasing function of time [7].

For $k=\pm 1$ spaces one obtains a similar solution:

$$q = A_{\pm}\eta^2 + B_{\pm}\eta + C_{\pm} \tag{18}$$

where, on this occasion

$$\begin{aligned} A_{\pm} &= -\frac{(3n-1)^2}{3n+1} m_n \\ B_{\pm} &= \frac{2(3n-1)}{3n+1} \eta_0 \\ C_{\pm} &= -\frac{\eta_0{}^2}{(3n+1)m_n} \quad , \end{aligned} \tag{19}$$

and since $\Delta\pm \equiv 4A_{\pm}C_{\pm} - B_{\pm}^2 = 0$, without loosing any in generality one can put $\eta_0 = 0$, and give the scale factor "a" in terms of "cosmic time t" :

$$a = \left[-\frac{2(3n-1)^2\, k}{2+(1-n)(3n+1)\omega} \right]^{\frac{1}{2(1-n)}} t\,. \tag{20}$$

3. RADIATION

Radiation or ultrarelativistic fluids where $p = \frac{1}{3}\rho$, imply

$$\rho = M_r a^{-4}. \qquad M_r = const.$$

Such fluids cannot generate a scalar field that if present must then be sourceless

$$\dot{\phi} a^3 = \gamma = const. \neq 0 \ , \tag{21}$$

Here, $r \equiv \phi a^2$ plays the role of dependent variable on "η–time" defined by $dt \equiv a d\eta$. With them one obtains the following equations

$$\frac{\phi'}{\phi} = \frac{\gamma}{r} \ , \tag{22}$$

and

$$2\frac{a'}{a} = \frac{r'}{r} - \frac{\gamma}{r} \tag{23}$$

so that

$$a^2 = r \exp\left(-\int \frac{\gamma}{r} dt \right) . \tag{24}$$

The dynamic equation is

$$\frac{r''}{r} = \frac{2}{3}\frac{m_r}{r} - 4k \ , \quad k = 0, \ \pm 1 \tag{25}$$

and its constriction is give by

$$\left(\frac{r'}{r}\right)^2 - \frac{3+2\omega}{3}\left(\frac{\gamma}{r}\right)^2 = \frac{4m_r}{3r} - 4k \ . \quad m_r \equiv 8\pi M_r \tag{26}$$

Even if $k \neq 0$ these last two equations do not contain the scale factor "a" that with other fluids is always present. This fact makes it unnecessary recourse to a third order differential equation in order to eliminate the aforementioned term. For $k = 0$ Eq.(25), and Eq.(26) obtain

$$r = R\eta^2 + Q\eta + P \ , \tag{27}$$

with $R = \frac{1}{3}m_r$, while $3Q^2 = 4m_r P + (3 + 2\omega)\gamma^2$, while P remains free. The discriminant is

$$\Delta_0 = -\frac{(3+2\omega)}{3}\gamma.$$

In all, three different cosmic solutions for this FRW flat space ($k = 0$) model exist distinguished by the discriminant whose sign depends on the sign of ω. The solutions have been analyzed in [7]. For $k \neq 0$ Eq.(25) and Eq.(26) have the following solutions

$$2(\eta + \eta_0) = \begin{cases} \ln\left[2\left(16k^2r^2 + \frac{16}{3}m_r r + \frac{(3+2\omega)}{3}\gamma^2\right)^{\frac{1}{2}} - 8kr + \frac{4}{3}m_r\right] , \\ arcsh\frac{-8kr+\frac{4}{3}m_r}{\Delta^{\frac{1}{2}}} , \\ -\arcsin\frac{-8kr+\frac{4}{3}m_r}{(-\Delta)^{\frac{1}{2}}} , \\ \ln(-8kr + \frac{4}{3}m_r) \end{cases} \tag{28}$$

the first is for $k = -1$, and with $\Delta = -\frac{16}{9}(m_r + 3(3+2\omega)\gamma^2$ the second is for $\Delta > 0$ and $k = -1$, $\Delta < 0$ and $k = 1$ for the third, the last is for $\Delta = 0$ and $k = -1$. Moreover, when $\Delta \neq 0$ depending on the sign of ω for each type of determinant one can get two solutions.

Other particular solutions for $k \neq 0$ are found when

$$r = r_0 = const. \tag{29}$$

with $r_0 = \frac{1}{6k}m_r$, so that

$$a = \left(\frac{1}{6k}m_r\right)^{\frac{1}{2}} \exp\left(-\frac{k\gamma}{3m_r}\eta\right) . \tag{30}$$

This last expression can be given in terms of the "cosmic time t . The result is a linear expansion

$$a = \left(\frac{6k}{m_r}\right)^{\frac{1}{2}} t = \left(-\frac{3k}{3+2\omega}\right)^{\frac{1}{2}} t \ , \tag{31}$$

because, from the constriction one gets

$$m_r{}^2 = -3(3+2\omega)k\gamma^2 \ , \tag{32}$$

Another set, obtained when $\gamma = 0$, correspond to Einstein's models solutions.

4. VACUUM

With the notation $v \equiv \phi a^3$ the vacuum field equations, where the scalar field is also sourceless, are

$$\frac{\dot{\phi}}{\phi} = \frac{\nu}{v}, \qquad \nu = const. \tag{33}$$

from which the relation between a and v

$$3\frac{\dot{a}}{a} = \frac{\dot{v}}{v} - \frac{\nu}{v}, \tag{34}$$

obtains

$$a^3 = v \exp\left(\frac{-\nu}{3}\int\frac{dt}{v}\right). \tag{35}$$

For $\rho = p = 0$, the "dynamic" equation is

$$\frac{1}{v}\frac{d^3v}{dt^3} - \frac{1}{3}\frac{\ddot{v}}{v}\left(\frac{\dot{v}}{v} + 2\frac{\nu}{v}\right) = 0 , \qquad k = 0, \ \pm 1 \tag{36}$$

and its constriction

$$3\frac{\ddot{v}}{v} - 2\left(\frac{\dot{v}}{v}\right)^2 - 2\left(\frac{\nu}{v}\right)\frac{\dot{v}}{v} + (4+3\omega)\left(\frac{\nu}{v}\right)^2 = 0 . \tag{37}$$

For $k = 0$

$$v = ct + t_0 , \qquad c, t_0 = const. \tag{38}$$

where

$$2c = \left[-1 \pm (3(3+2\omega))^{\frac{1}{2}}\right]\nu .$$

With τ, the use of $dt = v d\tau$ hands out for $k = 1$

$$v = \left[\cosh\left(\frac{(-\Delta)^{\frac{1}{2}}}{6}\tau\right)\right] e^{-\frac{\nu}{2}(\tau+\tau_0)} , \ \omega > -3/2 \tag{39}$$

and for $k = -1$

$$v = \left[\cos\left(\frac{(\Delta)^{\frac{1}{2}}}{6}\tau\right)\right] e^{-\frac{\nu}{2}(\tau+\tau_0)} , \ \omega < -3/2 \tag{40}$$

where $\Delta \equiv -12(3+2\omega)\nu^2$.

The corresponding scale factors are

$$a = \left[e^{-\frac{\nu}{2}(\tau+\tau_0)} \cosh \left(\frac{(-\Delta)^{\frac{1}{2}}}{-6} (\tau + \tau_0) \right) \right]^{\frac{1}{2}} \tag{41}$$

for $k = 1$, and $\omega > -3/2$.

$$a = \left[e^{\frac{\nu}{2}(\tau+\tau_0)} \cos \left(\frac{(\Delta)^{\frac{1}{2}}}{6} (\tau + \tau_0) \right) \right]^{\frac{1}{2}} \tag{42}$$

for $k = -1$, and $\omega < -3/2$.

5. CONCLUSIONS

For $k = \pm 1$ spaces, perfect fluid solutions imply that given k, and n the numerical value of ω is determined by the amount of "matter" present while also its sign depends on n. In other words, when space is non flat one can not freely choose the sign, or the degree of coupling between the tensor and the scalar modes. Note the fact that in such spaces these fluids expand in the same way, that is, expansion ignores the particular nature of "matter". Even more disturbing is the possibility that this evolution turns out to be the only kind of expansion which can be given in terms of polynomial, transcendental, or other fairly well known functions which may furthermore be also time invertible. For the open space, $k = -1$, a linear expansion is known to give a Milne Universe where space–time is, actually, flat Minkowski space. However, a linear expansion in $k = 1$ is not a Milne solution.

Acknowledgments

This work was supported by CONACyT Grant No. 5–3672–E9312

References

[1] P.M. Garnavich *et al.*, *Astrophys. J.* **509** (1998) 74; A G. Reiss *et al.*, *Astron. J.* **116** (1998) 1009; S. Perlmutter *et al.*, *Astrophys. J.* **517** (1999) 565.

[2] C. Brans and R. H. Dicke, *Phys. Rev.* **124** (1961) 925;

[3] P. Chauvet, *Astrophys. Space Sci.* **90** (1983) 51.

[4] P. Chauvet, and J.L. Cervantes-Cota, *Phys. Rev.* **D52** (1995) 3416; P. Chauvet and O. Pimentel, *Gen. Rel. Grav.* **24** (1992) 243; P. Chauvet, J.L. Cervantes–Cota, and H.N. Núñez–Yépez, *Class and Quantum Grav.* **9** (1992) 1923.

[5] P.G. Bergmann, *Int. J. Theor. Phys.* **1** (1968) 25; R.V. Wagoner, *Phys. Rev. D* **1** (1970) 3209; K. Nordvedt, *Astrophys. J.* **161** (1970) 1059.

[6] J.P. Mimoso and D. Wands, *Phys. Rev.* **D51** (1995) 477; J.D. Barrow, *Phys. Rev.* **D47**, 5329 (1993); P. Chauvet and O. Obregón, *Astrophys. Space Sci.* **66**

(1979) 515; H. Dehnen and O. Obregón, *Astrophys. Space Sci.* **14** (1971) 454; R.E. Morganstern, *Phys. Rev.* **D4** (1971) 278.

[7] L.E. Gurevich, A.M. Finkelstein, and V.A. Ruban, *Astrophys. Space Sci.* **22** (1973) 231; V.A. Ruban, and A.M. Finkelstein, *Gen. Rel. Grav.* **6** (1975) 601;

HIGGS–FIELD AND GRAVITY

Heinz Dehnen *
Physics Department. University of Konstanz
Box M 677. D-78457 Konstanz. Germany

Abstract The only scalar field of concrete physical importance is the Higgs-field of the elementary particle physics, especially after the possible detection of the Higgs-particle with a mass of 114 GeV at CERN some weeks ago in September 2000. Therefore our first question is: Which kind of interaction is mediated by the exchange of the Higgs–particle? This question can be answered classically by considering the field equations following from the Lagrangian of the electro-weak interaction.

Keywords: Higgs field, gravity, gauge theories.

1. HIGGS–FIELD GRAVITY

Without any details the general structure of the electro-weak Lagrangian is given by ($c = 1$, $\hbar = 1$, $\eta_{\mu\nu} = diag(1, -1, -1, -1)$):

$$L = \frac{1}{2} i\bar{\psi}\gamma^{\mu} D_{\mu}\psi + h.c. - \frac{1}{16\pi} F^{a}{}_{\lambda\mu} F_{a}{}^{\lambda\mu} +$$
$$+ \quad \frac{1}{2}(D_{\mu}\phi)^{\dagger} D^{\mu}\phi - \frac{\mu^2}{2}\phi^{\dagger}\phi - \frac{\lambda}{4!}(\phi^{\dagger}\phi)^2 - k\bar{\psi}\phi^{\dagger}\hat{x}\psi + h.c. \qquad (1)$$

($\mu^2 < 0, \lambda > 0$ are real parameters of the Higgs-potential). Herein D_{μ} represents the covariant derivative with respect to the localized group $SU(2) \times U(1)$

$$D_{\mu} = \partial_{\mu} + igA_{\mu} \qquad (2)$$

(g gauge coupling constants, $A_{\mu} = A^{a}_{\mu}\tau_{a}$ gauge potentials, τ_{a} generators of the group $SU(2) \times U(1)$) and the gauge field strength $F_{\mu\nu}$ is determined by its commutator according to $F_{\mu\nu} = (1/ig)[D_{\mu}, D_{\nu}] = F^{a}{}_{\mu\nu}\tau_{a}$; furthermore $\hat{x}$ is the Yukawa coupling-matrix of the Yukawa coupling term for generation of the fermionic masses ($k > 0$ is a real valued constant, the only non-dimensionless constant is μ^2 with the dimension of

*E–mail: Heinz.Dehnen@uni-konstanz.de

mass square). ψ summarizes the left- and right-handed lepton- and quark-wave functions.

From (1) we get immediately the field equations for the spinorial matter fields (ψ-fields):

$$i\gamma^\mu D_\mu \psi - k(\phi^\dagger \hat{x} + \hat{x}^\dagger \phi)\psi = 0, \tag{3}$$

for the scalar isospinorial Higgs-field ϕ

$$D^\mu D_\mu \phi + \mu^2 \phi + \frac{\lambda}{3!}(\phi^\dagger \phi)\phi = -2k\overline{\psi}\hat{x}\psi \tag{4}$$

and for the gauge-fields $F^{a\mu\lambda}$

$$\partial_\mu F^{a\mu\lambda} + igf^a{}_{bc}A^{b\mu}F^{c\lambda}_\mu = 4\pi j^{a\lambda} \tag{5}$$

with the gauge-current densities

$$j^{a\lambda} = g(\overline{\psi}\gamma^\lambda \tau^a \psi + \frac{i}{2}[\phi^\dagger \tau^a D^\lambda \phi - (D^\lambda \phi)^\dagger \tau^a \phi]). \tag{6}$$

Herein $f^a{}_{bc}$ are the totally skew symmetric structure constants of the group. The gauge invariant canonical energy momentum tensor reads with the use of (3)

$$\begin{aligned} T^\mu_\lambda &= \frac{i}{2}[\overline{\psi}\gamma^\mu D_\lambda \psi - (\overline{D_\lambda \psi})\gamma^\mu \psi] - \frac{1}{4\pi}\left[F^a{}_{\lambda\nu}F_a{}^{\lambda\mu} - \frac{1}{4}\delta_\lambda{}^\mu F^a{}_{\alpha\beta}F_a{}^{\alpha\beta}\right] \\ &+ \frac{1}{2}\Big[(D_\lambda \phi)^\dagger D^\mu \phi + (D^\mu \phi)^\dagger D_\lambda \phi \\ &- \delta_\lambda{}^\mu \left\{(D_\alpha \phi)^\dagger D^\alpha \phi - \mu^2 \phi^\dagger \phi - \frac{2\lambda}{4!}(\phi^\dagger \phi)^2\right\}\Big] \end{aligned} \tag{7}$$

and fulfills the conservation law

$$\partial_\mu T_\lambda{}^\mu = 0. \tag{8}$$

Obviously, the current-density (6) has a gauge-covariant matter-field and Higgs-field part, i.e. $j^{a\lambda}(\psi)$ and $j^{a\lambda}(\phi)$ respectively, whereas the energy-momentum tensor (7) consists of a sum of three gauge-invariant parts:

$$T_\lambda{}^\mu = T_\lambda{}^\mu(\psi) + T_\lambda{}^\mu(F) + T_\lambda{}^\mu(\phi), \tag{9}$$

represented by the brackets on the right hand side of (7).

For demonstrating the Higgs-field interaction explicitly we perform now the spontaneous symmetry breaking. According to (4) and (7) the state of lowest energy (ground state), which fulfils the field equations, is defined by $\psi \equiv 0$, $A_{\mu a} \equiv 0$, $F_{\mu\nu a} \equiv 0$ and the non–trivial Higgs–field condition

$$\phi_0^\dagger \phi_0 = v^2 = \frac{-6\mu^2}{\lambda}, \tag{10}$$

which we resolve as

$$\phi_0 = vN \tag{11}$$

with the isospinorial unit vector N:

$$N^\dagger N = 1, \quad \partial_\lambda N \equiv 0. \tag{12}$$

The general Higgs–field ϕ is different form (11) by a local unitary transformation:

$$\phi = \rho U N, \quad U^\dagger U = 1 \tag{13}$$

with

$$\phi^\dagger \phi = \rho^2, \quad \rho = v(1+\varphi), \tag{14}$$

where φ represents the real valued excited Higgs-field normalized with respect to the ground-state value v. Now we use the possibility of a unitary gauge transformation which is inverse to (13), so that Goldstone–bosons are avoided and (13) goes over in

$$\phi = \rho N. \tag{15}$$

In the following all calculations are performed in the gauge (15) (unitary gauge).

Using (11) up to (15) the field equations (3) through (5) take the form

$$i\gamma^\mu D_\mu \psi - \hat{m}(1+\varphi)\psi = 0, \tag{16}$$

$$\partial_\mu F_a^{\mu\lambda} + igf_{abc}A^{b\mu}F^c{}_\mu{}^\lambda + M_{ab}^2(1+\varphi)^2 A^{b\lambda} = 4\pi j_a^\lambda(\psi), \tag{17}$$

$$\partial^\mu\partial_\mu\varphi + M^2\varphi + \frac{1}{2}M^2(3\varphi^2+\varphi^3) = -\frac{1}{v^2}\left[\bar{\psi}\hat{m}\psi - \frac{1}{4\pi}M_{ab}^2 A^a{}_\lambda A^{b\lambda}(1+\varphi)\right], \tag{18}$$

wherein

$$\hat{m} = kv(N^\dagger\hat{x} + \hat{x}^\dagger N) \tag{19}$$

is the mass–matrix of the matter-fields (ψ-fields),

$$M^2{}_{ab} = 4\pi g^2 v^2 N^\dagger \tau_{(a}\tau_{b)} N \tag{20}$$

the symmetric matrix of the mass-square of the gauge-fields (A^a_μ-fields) and

$$M^2 = -2\mu^2, \quad (\mu^2 < 0) \tag{21}$$

the square of the mass of the excited Higgs–field (φ–field) corresponding to the Higgs–particle. The diagonalization procedure of (20) is not given here explicitly but results in the usual Weinberg–mixing with the masses $M_W = \pi^{1/2} g_2 v$, $M_Z = v\sqrt{\pi(g_1^2 + g_2^2)}$ and $M_A = 0$ of the electro–weak gauge bosons (g_1, g_2 gauge coupling constants belonging to the group $U(1) \times SU(2)$).

Obviously in the field equations (16) up to (18) the Higgs–field φ plays the role of an attractive scalar gravitational potential between the massive particles [1]: According to (18) the source of φ is the mass of the fermions and of the gauge–bosons. (The second term in the bracket on the right hand side of Eq.(18) is positive with respect to the signature of the metric.), whereby this equation linearized with respect to φ results in a potential equation of Yukawa–type. Accordingly the potential φ has finite range

$$l = M^{-1} \tag{22}$$

given by the mass of the Higgs-particle and v^{-2} has the meaning of a gravitational constant, so that

$$v^{-2} = 4\pi G\alpha \tag{23}$$

is valid, where G is the Newtonian gravitational constant and α a dimensionless factor, which compares the strength of the Newtonian gravity with that of the Higgs–field. On the other hand, the gravitational potential φ acts back on the mass of the fermions and of the gauge–bosons according to the field equations (16) and (17). Simultaneously the equivalence between inertial and passive as well as active gravitational mass is guaranteed. This feature results from the fact that by the symmetry breaking only one type of mass is introduced. We mention that this Higgs–gravity fulfils an old idea of Einstein known as "Mach's principle of relativity of inertia" [2], according to which the masses should be produced by the interaction with a gravitational field.

Finally the factor α in (23) can be estimated as follows: Taking into consideration the unified theory of electroweak interaction the value of v is correlated according to (20) with the mass M_W of the $W^\pm$-bosons according to $v^{-2} = \pi g_2^2/M_W^2$. Combination with (23) results in

$$\alpha = \frac{g_2^2}{2}\left(\frac{M_P}{M_W}\right)^2 = 2 \times 10^{32} \tag{24}$$

(M_P Planck-mass). Consequently the Higgs–gravity represents a relatively strong attractive scalar gravitational interaction between massive elementary particles, however with extremely short range and with the essential property of quantizability. If any Higgs–field exists in nature, this gravity is present. It seems that the Higgs–field has to do with quantum gravity. The expression (24) shows namely, that in the case of a symmetry breaking where the bosonic mass is of order of the Planck–mass, the Higgs–gravity approaches the Newtonian gravity, if the mass of the Higgs–particle is sufficiently small in view of the long range. In this connection the question arises if it is possible to construct a tensorial quantum theory of gravity with the use of the Higgs–mechanism, leading at last to Einstein's gravitational theory in the classical macroscopic limit.

2. HIGGS–FIELD SCALAR TENSOR THEORY OF GRAVITY

We will not solve this question in the following. But a first step in this direction consists in a stronger connection of the Higgs–field gravity with the actual gravity. This can be realized by construction of a scalar–tensor theory of gravity with the usual Higgs–field discussed above as scalar field. Then not only the inertial and gravitational masses but also the gravitational constant are generated simultaneously by spontaneous symmetry breaking.

The associated Lagrangian which unifies gravity and the other known interactions with a minimum of effort, takes even in more detail as before the very simple form [3]

$$\mathcal{L} = \left\{ \frac{1}{4}\alpha\phi^\dagger\phi R + \frac{1}{2}\phi^\dagger_{||\mu}\phi^{||\mu} - V(\phi) + L_M \right\} \sqrt{-g} \tag{25}$$

with the Higgs-potential

$$V(\phi) = \frac{\mu^2}{2}\phi^\dagger\phi + \frac{\lambda}{4!}\left(\phi^\dagger\phi\right)^2 + \frac{3}{2}\frac{\mu^4}{\lambda}, \tag{26}$$

the ground-state value of which is normalized to zero in contrast to (1); otherwise a cosmological constant would appear. But the Higgs–potential will now produce a cosmological function. R is the Ricci–scalar and α a dimensionless factor which must be determined empirically as above. The symbol $||\mu$ means the covariant derivative with respect to all external and internal gauge groups. The usual partial derivative with respect to the coordinate x^μ is denoted by $|\mu$. ϕ is the $SU(2) \times U(1)$–Higgs–field generating all the fermionic and the W– and Z^0– masses and the matter Lagrangian reads explicitly (L, R means summation of the left and right handed terms) as in (1):

$$L_M = \frac{i}{2}\overline{\psi}\gamma^{\mu}_{L,R}\psi_{||\mu} + h.c. - \frac{1}{16\pi}F^a_{\mu\nu}F^{\mu\nu}_a - k\overline{\psi}_R\phi^{\dagger}\hat{x}\psi_L + h.c.. \qquad (27)$$

For the case of applying the Lagrange density (25) to the GUT SU(5)–model, the Yukawa coupling term does not appear in (27) and the wave function ψ, the generators τ_a and the Higgs–field ϕ must be specified correspondigly.

The field equations following from the action principle associated with (25) are generalized Einstein equations

$$\begin{aligned} R_{\mu\nu} &- \frac{1}{2}Rg_{\mu\nu} + \frac{2}{\alpha\phi^{\dagger}\phi}V(\phi)g_{\mu\nu} = -\frac{2}{\alpha\phi^{\dagger}\phi}T_{\mu\nu} - \frac{2}{\alpha\phi^{\dagger}\phi}\left[\phi^{\dagger}_{(||\mu}\phi_{||\nu)} - \right. \\ &- \left. \frac{1}{2}\phi^{\dagger}_{||\lambda}\phi^{||\lambda}g_{\mu\nu}\right] - \frac{1}{\phi^{\dagger}\phi}\left[(\phi^{\dagger}\phi)_{|\mu||\nu} - (\phi^{\dagger}\phi)^{|\lambda}{}_{||\lambda}g_{\mu\nu}\right], \qquad (28) \end{aligned}$$

the Higgs–field equation

$$\phi^{||\mu}{}_{||\mu} - \frac{\alpha}{2}R\phi + (\mu^2 + \frac{\lambda}{6}\phi^{\dagger}\phi)\phi = -2k\overline{\psi}_R\hat{x}\psi_L, \qquad (29)$$

the Dirac equations

$$i\gamma^{\mu}_{\binom{L}{R}}\psi_{||\mu} - k\begin{pmatrix}\hat{x}^{\dagger}\phi\psi_R \\ \phi^{\dagger}\hat{x}\psi_L\end{pmatrix} = 0 \qquad (30)$$

and the inhomogeneous Yang–Mills equations for the gauge field strengths

$$F^{\nu\mu}_a{}_{||\nu} = 4\pi j^{\mu}_a \qquad (31)$$

with the gauge currents

$$j^{\mu}_a = j^{\mu}_a(\psi) + j^{\mu}_a(\phi) = g\overline{\psi}\gamma^{\mu}_{L,R}\tau_a\psi + \frac{ig}{2}\phi^{\dagger}\tau_a\phi^{||\mu} + h.c. \qquad (32)$$

belonging to the fermions and the Higgs–field respectively.

The energy–momentum tensor $T_{\mu\nu}$ in (28) is the symmetric metrical one belonging to (27) and reads with the use of the Dirac equations (30):

$$T^{\mu\nu} = \frac{i}{2}\overline{\psi}\gamma^{(\mu}_{L,R}\psi^{||\nu)} + h.c. - \frac{1}{4\pi}\left(F^{\mu a}{}_{\lambda}F^{\nu\lambda}_a - \frac{1}{4}F^a_{\alpha\beta}F^{\alpha\beta}_a g^{\mu\nu}\right). \qquad (33)$$

In consequence of the coupling with the Higgs–field the energy–momentum law is modified in comparison with the pure Einstein theory; it holds in contrast to (8):

$$T_{\mu}{}^{\nu}{}_{||\nu} = k\overline{\psi}_R\phi^{\dagger}_{||\mu}\hat{x}\psi_L + h.c. + F^a_{\nu\mu}j^{\nu}_a(\phi). \qquad (34)$$

From the field equation (29) it follows for the non-trivial Higgs–field ground–state, c.f. (10) and (11) ($\mu^2 < 0$):

$$\phi_0 = vN,\ N^\dagger N = 1, \quad v^2 = \phi_0^\dagger \phi_0 = -\frac{6\mu^2}{\lambda}, \tag{35}$$

for which the Higgs–potential (26) is minimized simultaneously; its corresponding ground–state value is zero and herewith the field equations (28) for the metric are fulfilled identically with the Minkowski–metric as the metrical ground–state. Insertion of (35) into the Dirac equation (30) and into the Higgs–field gauge current (32) of the Yang–Mills equation (31 yields the fermionic mass–matrix (19 and the matrix of the mass square of the gauge–bosons (20). In the unitary gauge the Higgs–field ϕ takes the form (15).

Insertion of (14) and (15) into (28) gives the field equation of gravitation:

$$\begin{aligned} & R_{\mu\nu} - \frac{1}{2} R g_{\mu\nu} + \frac{3}{\alpha v^2} \frac{\mu^4}{\lambda} \frac{\left[(1+\varphi)^2 - 1\right]^2}{(1+\varphi)^2} g_{\mu\nu} \\ = & -\frac{2}{\alpha v^2 (1+\varphi)^2} \left[T_{\mu\nu} + \frac{1}{4\pi}(1+\varphi)^2 M^2_{ab} \left(A^a_\mu A^b_\nu - \frac{1}{2} A^a_\lambda A^{b\lambda} g_{\mu\nu} \right) + \right. \\ + & \left. v^2 \left(\varphi_{|\mu} \varphi_{|\nu} - \frac{1}{2} \varphi_{|\lambda} \varphi^{|\lambda} g_{\mu\nu} \right) \right] - \frac{1}{(1+\varphi)^2} \left[(1+\varphi)^2{}_{|\mu||\nu} \right. \\ & \left. -(1+\varphi)^{2|\lambda}{}_{||\lambda} g_{\mu\nu} \right], \end{aligned} \tag{36}$$

with the trace

$$\begin{aligned} R = & \frac{2}{\alpha v^2 (1+\varphi)^2} \left[T - \frac{1}{4\pi}(1+\varphi)^2 M^2_{ab} A^a_\lambda A^{b\lambda} - \right. \\ - & \left. v^2 \varphi_{|\lambda} \varphi^{|\lambda} + 6\frac{\mu^4}{\lambda} \left((1+\varphi)^2 - 1 \right)^2 \right] - \frac{3}{(1+\varphi)^2} (1+\varphi)^{2\,|\lambda}{}_{||\lambda} \end{aligned} \tag{37}$$

and the equation of motion

$$\begin{aligned} & \left[T_\mu{}^\nu + \tfrac{(1+\varphi)^2}{4\pi} M^2_{ab} (A^a_\mu A^{b\nu} - \tfrac{1}{2} \delta_\mu{}^\nu A^a_\lambda A^{b\lambda}) \right]_{||\nu} = \\ & = \varphi_{|\mu} \left[\overline{\psi} \hat{m} \psi - \tfrac{1+\varphi}{4\pi} M^2_{ab} A^a_\lambda A^{b\lambda} \right]. \end{aligned} \tag{38}$$

According to (38) the excited Higgs–field φ generates a gravitation like potential force acting on the massive particles as in section 1.

Obviously the Newtonian gravitational constant is defined as in (23) only after symmetry breaking by the ground–state value v. Simultaneously there exists a variability of the gravitational constant described by

$(1+\varphi)^{-2}$. The value of the parameter α is the same as in (24). The cosmological function originated by the Higgs-potential is necessarily positive and possesses the value:

$$\Lambda = 12\pi G \frac{\mu^4}{\lambda} \frac{\left[(1+\varphi)^2 - 1\right]^2}{(1+\varphi)^2}. \tag{39}$$

It vanishes for the ground–state ($\varphi = 0$). The first bracket on the right hand side of (36) represents the effective energy–momentum tensor of fermions and gauge bosons taking additionally into account the masses of the gauge bosons and energy and momentum of the excited Higgs–field.

In the same way it follows from (29) after insertion of (37) and (23)

$$\begin{aligned} &\left[(1+\varphi)^2 - 1\right]^{|\mu}{}_{||\mu} + M^2\left[(1+\varphi)^2 - 1\right] = \\ &= \frac{1}{1+\frac{1}{3\alpha}} \frac{8\pi G}{3} \left[T - (1+\varphi)\overline{\psi}\,\hat{m}\psi\right], \end{aligned} \tag{40}$$

with the square of the Higgs–mass:

$$M^2 = \frac{16\pi G \frac{\mu^4}{\lambda}}{(1+\frac{1}{3\alpha})}. \tag{41}$$

Accordingly the value of (41) is smaller than the usual one (see (21)) by the very small factor α^{-1} and of the same order of magnitude as Λ, see (39). In consequence of the non-minimal coupling of the gravitational and the Higgs field in (25), the trace T appears as an additional source in (40, which results later in a total cancellation of the right hand side.

Finally we give the Dirac equation (30) and the Yang–Mills equation (31) after symmetry breaking. One obtains with (19)

$$i\gamma^{\mu}_{\binom{L}{R}}\psi_{||\mu} - (1+\varphi)\,\hat{m}\psi_{\binom{R}{L}} = 0 \tag{42}$$

and with (20)

$$F_a{}^{\nu\mu}{}_{||\nu} + (1+\varphi)^2 M^2_{ab} A^{\mu b} = 4\pi j^{\mu}_a(\psi). \tag{43}$$

Now we are able to calculate the source of the excited Higgs-field φ according to (40). From (33) and (42) we get for the trace of $T^{\mu\nu}$:

$$T = \frac{i}{2}\overline{\psi}\gamma^{\mu}_{L,R}\psi_{||\mu} + h.c. = (1+\varphi)\overline{\psi}\,\hat{m}\psi. \tag{44}$$

Evidently by insertion of (44) into (40) the source of the excited Higgs–field φ vanishes identically; it obeys exactly the homogeneous non-linear Klein–Gordon wave–equation [4]:

$$\varphi^{|\mu}{}_{||\mu} + M^2 \varphi \frac{(1+\frac{\varphi}{2})}{1+\varphi} + \frac{\varphi_{|\mu}\varphi^{|\mu}}{1+\varphi} \equiv 0 \,. \tag{45}$$

We note here explicitly that not only the fermionic masses but also those of the gauge–bosons do no more appear in (45) as source for the excited Higgs–field; it is coupled only to the very weak gravitational field contained in the only space–time covariant derivative. In case of several Higgs–fields the cancellation discussed above takes place only in the field equation for that particular Higgs–field generating the gravitational constant and the masses of the fermions. Thus in case of the SU(5) Higgs–field the cancellation of the right–hand side of Eq. (40) does not happen.

In any case the φ–field can exist as cosmological background field and perhaps solve the dark matter problem. There exists the possibility, that the φ–field interacts only gravitationally, where matter fields and gauge fields are absent ($\psi_{L,R} \equiv 0, A_\mu{}^b \equiv 0$). On the other hand also the solution $\varphi \equiv 0$ is always possible. Herewith we obtain from (36) and (38) the usual Einstein gravity (without cosmological constant) with fermionic and bosonic energy-momentum tensor as source, and from (42) and (43) the Dirac–equations and the Yang–Mills equations for the $SU(2) \times U(1)$ standard model follow without any influence of the excited Higgs–field φ. Instead of this there exists the only very weak gravitational interaction, where the gravitational constant is generated by the Higgs–field ground–state.

In this context there remains finally an interesting question: Is it possible to introduce a flavour–dependent Higgs–field resulting in a flavour–dependent gravitational constant after symmetry breaking perhaps connected with the Cabbibo–mixing? Then neutrino–oscillations are possible with massless neutrinos interacting differently with the gravitational field [5]. In this case a modification of the standard model of the elementary particles by massive inert right–handed neutrinos would not be necessary.

Acknowledgments

This research was supported by the joint German–Mexican project DLR–Conacyt MXI 010/98 OTH — E130–1148.

References

[1] H. Dehnen, H. Frommert and F. Ghaboussi, *Int. Jour. Theor. Phys.* **29** (1990) 537.

[2] A Einstein, *Sitzungsber. Preu. Akad. d. Wiss.*, Berlin, **142** 2 (1917).

[3] H. Dehnen, H. Frommert and F. Ghaboussi, *Int. Jour. Theor. Phys.* **31** (1992) 109.

[4] H. Dehnen and H. Frommert, *Int. Jour. Theor. Phys.* **32** (1993) 1135.

[5] M. Gasperini, *Phys. Rev.* **D8**, (1988), 2635.
J. Pantaleone, *Phys. Rev.* **D10** (1993) 4199.

THE ROAD TO GRAVITATIONAL S–DUALITY

H. Garcia–Compeán*
Departamento de Física, Centro de Investigación y Estudios Avanzados del IPN
P.O. Box 14-740, 07000, México D.F., México.

O. Obregón†
Instituto de Física de la Universidad de Guanajuato
P.O. Box E–143, 37150, León Gto., México.

C. Ramirez‡
Facultad de Ciencias Físico Matemáticas,
Universidad Autónoma de Puebla
P.O. Box 1364, 72000, Puebla, México.

M. Sabido§
Instituto de Física de la Universidad de Guanajuato
P.O. Box E–143, 37150, León Gto., México.

Abstract We overview the road to defining S–duality analogues for non–supersymmetric theories of gravity. The case of pure topological gravity in four dimensions, and, MacDowell–Mansouri gauge theory of gravity are discussed. Three–dimensional dimensional reductions from the topological gravitational sector in four dimensions, enable to recuperate the $2+1$ Chern–Simons gravity and the corresponding S–dual theory, from the notion of self–duality in the four–dimensional theory.

Keywords: S–duality, MacDowell–Mansouri gauge theories, Chern–Simons gravity.

*E–mail: compean@fis.cinvestav.mx
†E–mail: octavio@ifug3.ugto.mx
‡E–mail: cramirez@fcfm.buap.mx
§E–mail: msabido@ifug3.ugto.mx

Exact Solutions and Scalar Fields in Gravity: Recent Developments
Edited by Macias *et al.*, Kluwer Academic/Plenum Publishers, New York, 2001

1. INTRODUCTION

The study of symmetries in physics, has become a subject of most importance, for example in particle physics, gauge symmetry is the central idea to understanding the fundamental forces, Lorentz invariance is crucial in quantum field theory, general coordinate covariance in General Relativity, etc.

In recent years a new wave of dualities has emerge, as a most important technique in the understanding of non–perturbative aspects of quantum field theory, and string theory.

In superstring theory, duality shows the equivalence among different types of perturbative string theories and gives insight of a new underlying theory known as M–theory. Specially interesting are T–duality, and S–duality; the first one gives a new view of space time physics, revealing the existence of a minimal length.

Duality is an old known type of symmetry which by interchanging the electric, and magnetic fields leaves invariant the vacuum Maxwell equations. It was extended by Dirac to include sources, with the well known price of the predictions of monopoles, which appear as the dual particles of the electrically charged ones, and whose existence has not been confirmed. Dirac obtained that the charges (couplings) of the magnetical and electrical charged particles are inverse to each other, so if the electrical force is *"weak"*, then the magnetic force between monopoles will be *"strong"*, then one description would be treated perturbatively, and other one not, but of course we could always use the duality to go from one theory to the other.

The Strong/weak coupling duality (S–duality) in superstring and supersymmetric gauge theories in various dimensions has been, in the last five years, the major tool to study the strong coupling dynamics of these theories. Much of these results require supersymmetry through the notion of BPS state. These states describe the physical spectrum and they are protected of quantum corrections leaving the strong/coupling duality under control to extract physical information. In particular, the exact Wilson effective action of $\mathcal{N} = 2$ supersymmetric gauge theories has been computed, showing the duality symmetries of these effective theories. It turns out that the dual description is quite adequate to address standard non-perturbative problems of Yang–Mills theory. In the non–supersymmetric case there are no BPS states and the situation is much more involved. This latter case is an open question and it is still under current investigation.

In the specific case of non–supersymmetric gauge theories in four dimensions, the subject has been explored recently in the Abelian as well as in the non–Abelian cases [1, 2] (for a review see [3]). In the Abelian case, one considers CP non–conserving Maxwell theory on a curved compact four–manifold X with Euclidean signature or, in other words, U(1) gauge theory with a θ vacuum coupled to four–dimensional gravity. The manifold X is basically described by its associated classical topological invariants: the Euler characteristic $\chi(X) = \frac{1}{16\pi^2}\int_X \mathrm{tr}R\wedge\tilde{R}$ and the signature $\sigma(X) = -\frac{1}{24\pi^2}\int_X \mathrm{tr}R\wedge R$. In the Maxwell theory, the partition function $Z(\tau)$ transforms as a modular form under a finite index subgroup $\Gamma_0(2)$ of $\mathrm{SL}(2,\mathbf{Z})$ [1], $Z(-1/\tau) = \tau^u\bar{\tau}^v Z(\tau)$, with the modular weight $(u,v) = (\frac{1}{4}(\chi+\sigma), \frac{1}{4}(\chi-\sigma))$. In the above formula $\tau = \frac{\theta}{2\pi} + \frac{4\pi i}{g^2}$, where g is the U(1) electromagnetic coupling constant and θ is the usual theta angle.

In order to cancel the modular anomaly in Abelian theories, it is known that one has to choose certain holomorphic couplings $B(\tau)$ and $C(\tau)$ in the topological gravitational (non-dynamical) sector, through the action

$$I^{TOP} = \int_X \Big(B(\tau)\mathrm{tr}R\wedge\tilde{R} + C(\tau)\mathrm{tr}R\wedge R\Big), \tag{1}$$

i.e., which is proportional to the appropriate sum of the Euler characteristic $\chi(X)$ and the signature $\sigma(X)$.

2. S–DUALITY IN TOPOLOGICAL GRAVITY

We will first show our procedure to define a gravitational "S–dual" Lagrangian, by beginning with the non–dynamical topological gravitational action of the general form

$$I^{TOP} = \frac{\Theta_G^E}{2\pi}\int_X \mathrm{tr}R\wedge\tilde{R} + \frac{\Theta_G^P}{2\pi}\int_X \mathrm{tr}R\wedge R, \tag{2}$$

X is a four dimensional closed lorentzian manifold, *i.e* compact, without boundary $\partial X = 0$ and with lorentzian signature. In this action, the coefficients are the gravitational analogues of the θ–vacuum in QCD [4].

This action can be written in terms of the self–dual and anti–self–dual parts of the Riemann tensor as follows

$$I^{TOP} = \int_X \Big(\tau^+\mathrm{tr}R^+\wedge R^+ - \tau^-\mathrm{tr}R^-\wedge R^-\Big), \tag{3}$$

with $\tau^\pm = \frac{1}{2\pi}(\Theta_G^E \mp \Theta_G^P)$. In local coordinates on X, this action is written as

$$I^{TOP} = \int_X d^4x\epsilon^{\mu\nu\rho\sigma}\Big(\tau^+ R^{+ab}_{\mu\nu}R^+_{\rho\sigma ab} - \tau^- R^{-ab}_{\mu\nu}R^-_{\rho\sigma ab}\Big), \tag{4}$$

where ${R^{\pm}}_{\mu\nu}{}^{ab} = \frac{1}{2}({R_{\mu\nu}}^{ab} \mp \frac{i}{2}\epsilon^{ab}{}_{cd}{R_{\mu\nu}}^{cd})$ and satisfies

$$\epsilon^{ab}{}_{cd}{R^{\pm}}_{\mu\nu}{}^{cd} = \pm 2i{R^{\pm}}_{\mu\nu}{}^{ab}. \qquad (5)$$

Self–dual (anti–self–dual) Riemann tensors are defined as well in terms of the self–dual (anti–self–dual) component of the spin connection $\omega_{\mu}^{\pm ab} := \frac{1}{2}(\omega_{\mu}^{ab} \mp \frac{i}{2}\epsilon^{ab}{}_{cd}\omega_{\mu}^{cd})$ as

$${R^{\pm}}_{\mu\nu}{}^{ab} = \partial_{\mu}{\omega^{\pm}}_{\nu}{}^{ab} - \partial_{\nu}{\omega^{\pm}}_{\mu}{}^{ab} + \frac{1}{2}f^{[ab]}_{[cd][ef]}\omega^{\pm cd}{}_{\mu}\,\omega^{\pm ef}{}_{\nu}, \qquad (6)$$

with

$$f^{[ab]}_{[cd][ef]} = \frac{1}{2}\Big(\eta_{ce}\delta^{a}_{d}\delta^{b}_{f} - \eta_{cf}\delta^{a}_{d}\delta^{b}_{e} + \eta_{df}\delta^{a}_{c}\delta^{b}_{e} - \eta_{de}\delta^{a}_{c}\delta^{b}_{f} - (a \leftrightarrow b)\Big). \qquad (7)$$

The Lagrangian (4) can be written obviously as $\mathcal{L} = \mathcal{L}_{+} + \mathcal{L}_{-}$, where $\mathcal{L}_{\pm} = \epsilon^{\mu\nu\rho\sigma}(\tau^{\pm}{R^{\pm}}_{\mu\nu}{}^{ab}{R^{\pm}}_{\rho\sigma ab})$.

The Euclidean partition function is defined as [5]

$$Z(\tau) = Z(\tau^{+},\tau^{-}) = \int \mathcal{D}\omega^{+}\mathcal{D}\omega^{-} exp\Big(-\int_{X}(\mathcal{L}_{+} + \mathcal{L}_{-})\Big), \qquad (8)$$

which satisfies a factorization $Z(\tau^{+},\tau^{-}) = Z_{+}(\tau^{+})Z_{-}(\tau^{-})$ where

$$Z_{\pm}(\tau^{\pm}) = \int \mathcal{D}\omega^{\pm} exp\Big(-\int_{X}\mathcal{L}_{\pm}\Big). \qquad (9)$$

It is an easy matter to see from action (4) that the partition function (8) is invariant under combined shifts of Θ^{E}_{G} and Θ^{P}_{G} for $\tau^{\pm} \to \tau^{\pm} \mp 2$, in the case of spin manifolds and for non-spin manifolds $\tau^{\pm} \to \tau^{\pm} \mp 5$.

In order to find a S–dual theory to (4) we follow references [2]. The procedures are similar to those presented in the Introduction for Yang–Mills theories. We begin by proposing a parent Lagrangian

$$\begin{aligned} L &= L_{+} + L_{-} = \epsilon^{\mu\nu\rho\sigma}\Big[\tau^{+}{R^{+}}_{\mu\nu}{}^{ab}{R^{+}}_{\rho\sigma ab} \\ &+ iG^{+ab}_{\mu\nu}\Big(R^{+}_{\rho\sigma ab} - (\partial_{\rho}\omega^{+}_{\sigma ab} - \partial_{\sigma}\omega^{+}_{\rho ab} + [\omega^{+}_{\rho},\omega^{+}_{\sigma}]_{ab})\Big)\Big] \\ &+ \epsilon^{\mu\nu\rho\sigma}\Big[\tau^{-}{R^{-}}_{\mu\nu}{}^{ab}{R^{-}}_{\rho\sigma ab} \\ &+ iG^{-ab}_{\mu\nu}\Big(R^{-}_{\rho\sigma ab} - (\partial_{\rho}\omega^{-}_{\sigma ab} - \partial_{\sigma}\omega^{-}_{\rho ab} + [\omega^{-}_{\rho},\omega^{-}_{\sigma}]_{ab})\Big)\Big], \end{aligned} \qquad (10)$$

where R^+ is of course treated as a field independent on ω^+.

For simplicity, we focus on the partition function of the self–dual part of this Lagrangian, for the anti–self–dual part L_- one can follow the same procedure, the corresponding partition function is

$$Z_+^*(\tau^+) = \int \mathcal{D}R^+\mathcal{D}\omega^+\mathcal{D}G^+ exp\biggl(-\int_X L_+\biggr). \tag{11}$$

First of all we would like to regain the original Lagrangian (4). Thus, we should integrate over the Lagrange multiplier G^+

$$exp\biggl(-\int_X L_+^{**}\biggr) = \int \mathcal{D}G^+ exp\biggl(-\int_X L_+\biggr). \tag{12}$$

The relevant part of the integral is

$$\int \mathcal{D}G^+ exp\biggl\{-i\int_X \epsilon^{\mu\nu\rho\sigma}G_{\mu\nu}^{+ab}\Bigl(R_{\rho\sigma ab}^+ - (\partial_\rho\omega_{\sigma ab}^+ - \partial_\sigma\omega_{\rho ab}^+ + [\omega_\rho^+,\omega_\sigma^+]_{ab})\Bigr)\biggr\}. \tag{13}$$

Using the formula $\delta(x) = \int \frac{de}{2\pi} exp(iex)$ we find

$$Z_+^*(\tau^+) = -2\pi \int \mathcal{D}R^+\mathcal{D}\omega^+\delta[\epsilon(R^\pm - \ldots)]exp(-\int_X \epsilon^{\mu\nu\rho\sigma}\tau^+ R_{\mu\nu}^{+ab}R_{\rho\sigma ab}^+). \tag{14}$$

Thus we get the following original Lagrangian $L^{**} = L_+^{**} + L_-^{**} = \epsilon^{\mu\nu\rho\sigma}(\tau^+ R_{\mu\nu}^{+ab}R_{\rho\sigma ab}^+ + \tau^- R_{\mu\nu}^{-ab}R_{\rho\sigma ab}^-)$, where now $R_{\mu\nu}^{\pm ab}$ depend on $\omega_\mu^{\pm ab}$ as usual.

Obviously, as before, factorization holds, $Z^*(\tau^+,\tau^-) = Z_+^*(\tau^+)Z_-^*(\tau^-)$.

Now we would like to find the dual Lagrangian $\tilde{L}^{**}$. To do this we have first to integrate out the variables R^+ and ω^+, thus

$$exp\biggl(-\int_X \tilde{L}_+^{**}\biggr) = \int \mathcal{D}R^+\mathcal{D}\omega^+ exp\biggl(-\int_X L_+\biggr). \tag{15}$$

We consider first the integration over R^+

$$exp\biggl(-\int_X \tilde{L}_{\omega+}^{**}\biggr) = \int \mathcal{D}R^+ exp\biggl(-\int_X \epsilon^{\mu\nu\rho\sigma}(\tau^+ R_{\mu\nu}^{+ab}R_{ab\rho\sigma}^+ + iR_{\mu\nu}^{+\ ab}G_{\rho\sigma ab}^+)\biggr). \tag{16}$$

The functional integral over R^+ is of the Gaussian type and defines the following Lagrangian

$$L_+ = \epsilon^{\mu\nu\rho\sigma}\biggl[\frac{1}{4\tau^+}G_{\mu\nu}^{+\ ab}G_{\rho\sigma ab}^+ - iG_{\mu\nu}^{+\ ab}R_{\rho\sigma ab}^+(\omega^+)\biggr], \tag{17}$$

where $R^{+}_{\rho\sigma ab}(\omega^{+})$ is given by (6). Therefore, another intermediate Lagrangian emerges, which otherways would result from adding the degrees of freedom $G^{\pm ab}_{\mu\nu}$.

Now we integrate in the variable ω^{+}, and after some manipulations, the relevant part of the above Lagrangian necessary for the integration in ω^{+} is

$$\tilde{L}^{**}_{\omega+} = \ldots - 2i\epsilon^{\mu\nu\rho\sigma}\omega^{+ab}_{\mu}\partial_{\nu}G^{+}_{\rho\sigma ab} + \frac{i}{2}\epsilon^{\mu\nu\rho\sigma}f^{[ab]}_{[cd][ef]}\omega^{+cd}_{\mu}\omega^{+ef}_{\nu}G^{+}_{\rho\sigma ab} + \ldots \quad . \tag{18}$$

Inserting (18) into (15) and integrating out with respect to ω^{+}, we finally find

$$\tilde{L}^{**}_{+} = +2\epsilon^{\mu\nu\rho\sigma}\partial_{\nu}G^{+}_{\rho\sigma ab}(M^{+})^{-1\ abcd}_{\mu\lambda}\epsilon^{\lambda\theta\alpha\beta}\partial_{\theta}G^{+}_{\alpha\beta cd} + \ldots \quad . \tag{19}$$

Therefore, the complete dual Lagrangian is

$$\begin{aligned}
\tilde{L}^{**} &= \tilde{L}^{**}_{+} + \tilde{L}^{**}_{-} \\
&= \epsilon^{\mu\nu\rho\sigma}\Big[- \frac{1}{4\tau^{+}}G^{+ab}_{\mu\nu}G^{+}_{\rho\sigma ab} + \frac{1}{4\tau^{-}}G^{-ab}_{\mu\nu}G^{-}_{\rho\sigma ab} \\
&+ 2\partial_{\nu}G^{+}_{\rho\sigma ab}(M^{+})^{-1\ abcd}_{\mu\lambda}\epsilon^{\lambda\theta\alpha\beta}\partial_{\theta}G^{+}_{\alpha\beta cd} \\
&- 2\partial_{\nu}G^{-}_{\rho\sigma ab}(M^{-})^{-1\ abcd}_{\mu\lambda}\epsilon^{\lambda\theta\alpha\beta}\partial_{\theta}G^{-}_{\alpha\beta cd}\Big].
\end{aligned} \tag{20}$$

Where $M^{+\rho\sigma}_{\ \ cdef} \equiv i\epsilon^{\mu\nu\rho\sigma}f^{[ab]}_{[cd][ef]}G^{+}_{\mu\nu ab}$. Of course, the condition of factorization for the above Lagrangian still holds.

3. S–DUALITY IN MACDOWELL–MANSOURI GAUGE THEORY OF GRAVITY

Let us briefly review the MacDowell–Mansouri (MM) proposal [6]. The starting point for the construction of this theory is to consider an SO(3,2) gauge theory with a Lie algebra–valued gauge potential A^{AB}_{μ}, where the indices $\mu = 0,1,2,3$ are space–time indices and the indices $A, B = 0,1,2,3,4$. From the gauge potential $A_{\mu}^{\ AB}$ we may introduce the corresponding field strength $F_{\mu\nu}^{\ \ AB} = \partial_{\mu}A_{\nu}^{\ AB} - \partial_{\nu}A^{AB}_{\mu} + \frac{1}{2}f^{AB}_{CDEF}A^{CD}_{\mu}A^{EF}_{\nu}$, where f^{AB}_{CDEF} are the structure constants of SO(3,2). MM choose $F^{a4}_{\mu\nu} \equiv 0$ and as an action

$$S_{MM} = \int d^4x\epsilon^{\mu\nu\alpha\beta}\epsilon_{abcd}F_{\mu\nu}^{\ \ ab}F_{\alpha\beta}^{\ \ cd}, \tag{21}$$

where $a, b, \ldots$etc. $= 0, 1, 2, 3$.

On the other hand, by considering the self–dual (or anti–self–dual) part of the connection, a generalization has been proposed [7]. The extension to the supergravity case is considered in [8].

One can then search whether the construction of a linear combination of the corresponding self–dual and anti–self–dual parts of the MacDowell–Mansouri action can be reduced to the standard MM action plus a kind of Θ–term and, moreover, if by this means one can find the "dual–theory" associated with the MM theory. This was showed in [9] and the corresponding extension to supergravity is given at [10]. In what follows we follow Ref. [9]. Let us consider the action

$$S = \int d^4x \epsilon^{\mu\nu\alpha\beta} \epsilon_{abcd} \Big({}^+\tau\, {}^+F_{\mu\nu}{}^{ab}\, {}^+F_{\alpha\beta}{}^{cd} - {}^-\tau\, {}^-F_{\mu\nu}{}^{ab}\, {}^-F_{\alpha\beta}{}^{cd} \Big), \qquad (22)$$

where ${}^{\pm}F_{\mu\nu}{}^{ab} = \frac{1}{2}\Big(F_{\mu\nu}{}^{ab} \pm \tilde{F}_{\mu\nu}{}^{ad} \Big)$ and $\tilde{F}_{\mu\nu}{}^{ab} = -\frac{1}{2} i \epsilon^{ab}{}_{cd} F_{\mu\nu}{}^{cd}$. It can be easily shown [9], that this action can be rewritten as

$$S = \frac{1}{2} \int d^4x \epsilon^{\mu\nu\alpha\beta} \epsilon_{abcd} \Big[({}^+\tau - {}^-\tau) F_{\mu\nu}{}^{ab} F_{\alpha\beta}{}^{cd} + ({}^+\tau + {}^-\tau) F_{\mu\nu}{}^{ab} \tilde{F}_{\alpha\beta}{}^{cd} \Big]. \qquad (23)$$

In their original paper, MM have shown [6] that the first term in this action reduces to the Euler topological term plus the Einstein–Hilbert action with a cosmological term. This was achieved after identifying the components of the gauge field $A_\mu{}^{AB}$ with the Ricci rotation coefficients and the vierbein. Similarly, the second term can be shown to be equal to $i\theta P$, where P is the Pontrjagin topological term [7] . Thus, it is a genuine θ term, with θ given by the sum ${}^+\tau + {}^-\tau$.

Our second task is to find the "dual theory", following the same scheme as for Yang–Mills theories [2]. For that purpose we consider the parent action

$$\begin{aligned} I &= \int d^4x \epsilon^{\mu\nu\alpha\beta} \epsilon_{abcd} \Big(c_1\, {}^+G_{\mu\nu}{}^{ab}\, {}^+G_{\alpha\beta}{}^{cd} + c_2\, {}^-G_{\mu\nu}{}^{ab}\, {}^-G_{\alpha\beta}{}^{cd} \\ &+ c_3\, {}^+F_{\mu\nu}{}^{ab}\, {}^+G_{\alpha\beta}{}^{cd} + c_4\, {}^-F_{\mu\nu}{}^{ab}\, {}^-G_{\alpha\beta}{}^{cd} \Big). \end{aligned} \qquad (24)$$

From which the action (3) can be recovered after integration on ${}^+G$ and ${}^-G$.

In order to get the "dual theory" one should start with the partition function

$$Z = \int \mathcal{D}^+G\, \mathcal{D}^-G\, \mathcal{D}A\, e^{-I}. \qquad (25)$$

To proceed with the integration over the gauge fields we observe that $F_{\mu\nu}{}^{ab} = \partial_\mu A_\nu{}^{ab} - \partial_\nu A_\mu{}^{ab} + \frac{1}{2} f^{ab}_{CDEF} A_\mu{}^{CD} A_\nu{}^{EF}$. Taking into account the explicit expression for the structure constants, the second term of $F_{\mu\nu}{}^{ab}$ will naturally split in four terms given by $A_\mu{}^{ad} A_{\nu d}{}^{b} - A_\nu{}^{ad} A_{\mu d}{}^{b}$ $-\lambda^2 \left(A_\mu{}^{a4} A_\nu{}^{b4} - A_\nu{}^{a4} A_\mu{}^{b4} \right)$. The integration over the components $A_\mu{}^{a4}$ is given by a Gaussian integral, which turns out to be $det\, \mathbf{G}^{-1/2}$, where $\mathbf{G}$ is a matrix given by $\mathbf{G}_{ab}{}^{\mu\nu} = 8i\lambda^2 \epsilon^{\mu\nu\alpha\beta} \left(c_3{}^{+}G_{\alpha\beta ab} - c_4{}^{-}G_{\alpha\beta ab} \right)$.

Thus, the partition function (6) can be written as

$$Z = \int \mathcal{D}^{+}G\, \mathcal{D}^{-}G\, \mathcal{D}A_\mu^{ab}\; det\, \mathbf{G}^{-1/2}\; e^{-\mathbb{I}}, \tag{26}$$

where

$$\begin{aligned} \mathbb{I} &= 2i \int d^4x \epsilon^{\mu\nu\alpha\beta} \Big[c_1{}^{+}G_{\mu\nu}{}^{ab+}G_{\alpha\beta ab} - c_2{}^{-}G_{\mu\nu}{}^{ab-}G_{\alpha\beta ab} \\ &+ 2H_{\mu\nu}{}^{ab}(c_3{}^{+}G_{\alpha\beta ab} - c_4{}^{-}G_{\alpha\beta ab}) \Big], \end{aligned} \tag{27}$$

and $H_{\mu\nu}{}^{ab} = \partial_\mu A_\nu{}^{ab} - \partial_\nu A_\mu{}^{ab} + \frac{1}{2} f^{ab}_{cdef} A_\mu{}^{cd} A_\nu{}^{ef}$ is the SO(3,1) field strength.

Our last step to get the dual action is to integrate over $A_\mu{}^{ab}$. This kind of integration is well known and has been performed in previous works [2, 9, 11]. The result is

$$Z = \int \mathcal{D}^{+}G\, \mathcal{D}^{-}G\; det\, \mathbf{G}^{-1/2}\; det({}^{+}M)^{-1/2} det({}^{-}M)^{-1/2}\; e^{-\int d^4x \tilde{L}}, \tag{28}$$

with

$$\begin{aligned} \tilde{L} &= \epsilon^{\mu\nu\rho\sigma} \Big[-\frac{1}{4^{+}\tau}{}^{+}G_{\mu\nu}{}^{ab+}G_{\rho\sigma ab} + \frac{1}{4^{-}\tau}{}^{-}G_{\mu\nu}{}^{ab-}G_{\rho\sigma ab} \\ &+ 2\partial_\nu{}^{+}G_{\rho\sigma ab}({}^{+}M)^{-1\;abcd}_{\mu\lambda} \epsilon^{\lambda\theta\alpha\beta} \partial_\theta{}^{+}G_{\alpha\beta cd} \\ &- 2\partial_\nu{}^{-}G_{\rho\sigma ab}({}^{-}M)^{-1\;abcd}_{\mu\lambda} \epsilon^{\lambda\theta\alpha\beta} \partial_\theta G^{-}_{\alpha\beta cd} \Big], \end{aligned} \tag{29}$$

where ${}^{\pm}M^{\mu\nu}{}_{ab}{}^{cd} = \frac{1}{2}\epsilon^{\mu\nu\alpha\beta} \left(-\delta^{c\pm}_a G_{\alpha\beta b}{}^{d} + \delta^{c\pm}_b G_{\alpha\beta a}{}^{d} + \delta^{d\pm}_a G_{\alpha\beta b}{}^{c} - \delta^{d\pm}_b G_{\alpha\beta a}{}^{c} \right)$ and ${}^{+}\tau = -\frac{1}{4c_1}$, ${}^{-}\tau = -\frac{1}{4c_2}$, $c_3 = c_4 = 1$.

The non–dynamical model considered in the previous section, [11] results in a kind of non–linear sigma model [2] of the type considered by Freedman and Townsend [12], as in the usual Yang–Mills dual models. The dual to the dynamical gravitational model (9) considered here, results in a Lagrangian of the same structure. However, it differs from the non–dynamical case by the features discussed above.

4. (ANTI)SELF–DUALITY OF THE THREE–DIMENSIONAL CHERN–SIMONS GRAVITY

It is well known that the $2+1$ Einstein–Hilbert action with nonvanishing cosmological constant λ is given by the "standard" and "exotic" Einstein actions [13]. It is well known that for $\lambda > 0$ (and $\lambda < 0$), these actions are equivalent to a Chern–Simons actions in $2+1$ dimensions with gauge group $\mathcal{G}$ to be SO(3,1) (and SO(2,2)).

In this section we will work out the Chern–Simons Lagrangian for (anti)self–dual gauge connection with respect to duality transformations of the internal indices of the gauge group $\mathcal{G}$, in the same philosophy of MM [6], and that of [7]

$$L^{\pm}{}_{CS} = \int_{\mathcal{M}} \varepsilon^{ijk}\left({}^{\pm}A_i^{AB}\partial_j\, {}^{\pm}A_{kAB} + \frac{2}{3}{}^{\pm}A_{iA}^{B}\, {}^{\pm}A_{jB}^{C}\, {}^{\pm}A_{kC}^{A}\right), \tag{30}$$

where $A,B,C,D = 0,1,2,3$, $\eta_{AB} = diag(-1,+1,+1,+1)$ and the complex (anti) self-dual connections are ${}^{\pm}A_i^{AB} = \frac{1}{2}(A_i^{AB} \mp \frac{i}{2}\varepsilon^{AB}{}_{CD}A_i^{CD})$, which satisfy the relation $\varepsilon^{AB}{}_{CD}\,{}^{\pm}A_i^{CD} = \pm i\,{}^{\pm}A^{AB}$.

Thus using the above equations, the action (9) can be rewritten as

$$\begin{aligned} L^{\pm}{}_{CS} &= \int_{\mathcal{M}} \frac{1}{2}\varepsilon^{ijk}\left(A_i^{AB}\partial_j A_{kAB} + \frac{2}{3}A_{iA}{}^{B}A_{jB}{}^{C}A_{kC}{}^{A}\right) \\ &\mp \frac{i}{4}\varepsilon^{ijk}\varepsilon^{ABCD}\left(A_{iAB}\partial_j A_{kCD} + \frac{2}{3}A_{iA}{}^{E}A_{jEB}A_{kCD}\right). \end{aligned} \tag{31}$$

In this expression the first term is the Chern–Simons action for the gauge group $\mathcal{G}$, while the second term appears as its corresponding "θ–term". The same result was obtained in $3+1$ dimensions when we considered the (anti)self–dual MM action [9], or the (anti)self-dual $3+1$ pure topological gravitational action [11].

One should remark that the two terms in the action (10) are the Chern–Simons and the corresponding "θ–term" for the gauge group $\mathcal{G}$ under consideration. After imposing the particular identification $A_i^{AB} = (A_i^{ab}, A_i^{3a}) = (\omega_i^{ab}, \sqrt{\lambda}e_i^a)$ and $\omega_i^{ab} = \varepsilon^{abc}\omega_{ic}$, the "exotic" and "standard" actions for the gauge group SO(3,1) are given respectively by

$$\begin{aligned} L_{CS}{}^{\pm} = \int_X \frac{1}{2}\varepsilon^{ijk}\Big(&\omega_i^a(\partial_j\omega_{ka} - \partial_k\omega_{ja}) + \frac{2}{3}\varepsilon_{abc}\omega_i^a\omega_j^b\omega_k^c \\ +\ &\lambda e_i^a(\partial_j e_{ka} - \partial_k e_{ia}) - 2\lambda\varepsilon_{abc}e_i^a e_j^b\omega_k^c\Big) \\ \pm\ i\sqrt{\lambda}\varepsilon^{ijk}\Big(&e_i^a(\partial_j\omega_{ka} - \partial_k\omega_{ja}) - \varepsilon_{abc}e_i^a\omega_j^b\omega_k^c + \frac{1}{3}\lambda\varepsilon_{abc}e_i^a e_j^b e_k^c\Big) \end{aligned} \tag{32}$$

plus surface terms. It is interesting to note that the above action (32) can be obtained from action (1) (for a suitable choice of $B(\tau)$ and $C(\tau)$) by dimensional reduction from X to its boundary $\mathcal{M} = \partial X$. Thus the "standard" action come from the Euler characteristic $\chi(X)$, while the "exotic" action come from the signature $\sigma(X)$.

5. CHERNSIMONS GRAVITY DUAL ACTION IN THREE DIMENSIONS

This section is devoted to show that a "dual" action to the Chern–Simons gravity action can be constructed following [14]. Essentially we will repeat the procedure to find the the "dual" action to MM gauge theory given in Sec. II.

We begin from the original non–Abelian Chern–Simons action given by

$$L = \int_{\mathcal{M}} d^3x \frac{g}{4\pi} \varepsilon^{ijk} A_i^{AB} \left(\partial_j A_{kAB} + \frac{1}{3} f_{ABCDEF} A_j^{CD} A_k^{EF} \right). \tag{33}$$

Now, as usual we propose a parent action in order to derive the dual action

$$L_D = \int_{\mathcal{M}} d^3x \varepsilon^{ijk} \left(a B_i^{AB} H_{jkAB} + b A_i^{AB} G_{jkAB} + c B_i^{AB} G_{jkAB} \right), \tag{34}$$

where $H_{jkAB} = \partial_j A_{kAB} + \frac{1}{3} f_{ABCDEF} A_j^{CD} A_k^{EF}$ and B_i^{AB} and G_{ij}^{AB} are vector and tensor fields on $\mathcal{M}$. It is a very easy matter to show that the action (13) can be derived from this parent action after integration of G fields

The "dual" action L_D^* can be computed as usually in the Euclidean partition function, by integrating first out with respect to the physical degrees of freedom A_i^{AB}. The resulting action is of the Gaussian type in the variable A and thus, after some computations, it is easy to find the "dual" action

$$\begin{aligned} L_D^* &= \int_{\mathcal{M}} d^3x \varepsilon^{ijk} \Big\{ -\frac{3}{4a} (a\partial_i B_{jAB} + bG_{ijAB}) [\mathbf{R}^{-1}]_{kn}^{ABCD} \\ &\times \ \varepsilon^{lmn} (a\partial_l B_{mCD} + bG_{lmCD}) + c\alpha_i^{AB} G_{jkAB} \Big\}, \end{aligned} \tag{35}$$

where $[\mathbf{R}]$ is given by $[\mathbf{R}]^{ij}_{ABCD} = \varepsilon^{ijk} f^{EF}{}_{ABCD} B_{kEF}$ whose inverse is defined by $[\mathbf{R}]^{ij}_{ABCD} [\mathbf{R}^{-1}]^{CDEF}_{jk} = \delta^i_k \delta^{EF}_{AB}$.

The partition function is finally given by

$$Z = \int \mathcal{D}G\mathcal{D}B\sqrt{det(\mathbf{M}^{-1})}exp(-L_D^*). \tag{36}$$

In this "dual action" the G field is not dynamical and can be integrated out. The integration of this auxiliary field gives

$$L_D^{**} = \int_{\mathcal{M}} d^3x \frac{4\pi}{g} \varepsilon^{lmn} \Big(B_l^{AB} \partial_m B_{nAB} - \frac{4\pi}{g} f_{ABCDEF} B_l^{AB} B_m^{CD} B_n^{EF} \Big). \tag{37}$$

The fields B cannot be rescaled if we impose "periodicity" conditions on them. Thus, this dual action has inverted coupling with respect to the original one (compare with [15] for the Abelian case).

6. CONCLUDING REMARKS

In this work we have shown the road to the formulation of an analog of S -duality in gravity, following similar procedures to those well known in non–Abelian nonsupersymmetric Yang-Mills theories. We constructed for dual theories for pure topological gravity, for MacDowell–Mansouri gauge theory of gravity, and for 2 + 1 Chern–Simons gravity. Work for the supersymmetric versions of these theories has been done.

References

[1] E. Witten, *Selecta Mathematica, New Series*, **1** (1995) 383.

[2] O. Ganor and J. Sonnenschein, *Int. J. Mod. Phys.* **A11** (1996) 5701; N. Mohammedi, hep-th/9507040; Y. Lozano, *Phys. Lett.* **B364** (1995) 19.

[3] F. Quevedo, *Nucl. Phys.* Proc. Suppl. **A61** (1998) 23.

[4] S. Deser, M.J. Duff and C.J. Isham, *Phys. Lett.* **B93**, 419 (1980).; A. Ashtekar, A.P. Balachandran and S. Jo, *Int. J. Mod. Phys.* **A4** (1989) 1493.

[5] M. Abe, A. Nakamichi and T. Ueno, *Mod. Phys. Lett.* **A9**, 895 (1994); *Phys. Rev.* **D50** (1994) 7323.

[6] S.W. Mac Dowell and F. Mansouri, *Phys. Rev. Lett.* **38** (1977) 739; F. Mansouri, *Phys. Rev.* **D16** (1977) 2456.

[7] J.A. Nieto, O. Obregón and J. Socorro, *Phys. Rev.* **D50** (1994) R3583.

[8] J.A. Nieto, J. Socorro and O. Obregón, *Phys. Rev. Lett.* **76** (1996) 3482.

[9] H. García–Compeán, O. Obregón and C. Ramírez, *Phys. Rev.* **D58** (1998) 104012-1; *ibid. Chaos, Solitons and Fractals* **10** (1999) 373.

[10] H. García–Compeán, J.A. Nieto, O. Obregón and C. Ramírez, *Phys. Rev.* **D59** (1999) 124003.

[11] H. García–Compeán, O. Obregón, J.F. Plebański and C. Ramírez, *Phys. Rev.* **D57** (1998) 7501.

[12] D.Z. Freedman and P.K. Townsend, *Nucl. Phys.* **B177** (1981) 282; P.C. West, *Phys. Lett.* **B76** (1978) 569; A.H. Chamseddine, *Ann. Phys.* (N.Y.) **113** (1978) 219; S. Gotzes and A.C. Hirshfeld, *Ann. Phys.* (N.Y.) **203** (1990) 410.

[13] E. Witten, *Nucl. Phys.* **B311** (1988) 46; *Nucl. Phys.* **B323** (1989) 113.

[14] H. García–Compeán, O. Obregón, C. Ramírez and M. Sabido, "*Remarks on* 2+1 *Self–dual Chern–Simons Gravity*", hep-th/9906154, to appear in Phys. Rev. D.

[15] A.P. Balachandran, L. Chandar and B. Sathiapalan, Int. J. *Mod. Phys.* **A11** (1996).

EXACT SOLUTIONS IN MULTIDIMENSIONAL GRAVITY WITH P–BRANES AND PPN PARAMETERS

V. D. Ivashchuk
Center for Gravitation and Fundamental Metrology, VNIIMS,
3–1 M. Ulyanovoy Str., Moscow, 117313, Russia
Institute of Gravitation and Cosmology, Peoples' Friendship University,
6 Miklukho–Maklaya St., Moscow 117198, Russia.

V. N. Melnikov*
Center for Gravitation and Fundamental Metrology, VNIIMS,
3–1 M. Ulyanovoy Str., Moscow, 117313, Russia
Institute of Gravitation and Cosmology, Peoples' Friendship University,
6 Miklukho–Maklaya St., Moscow 117198, Russia.

Abstract

Multidimensional gravitational model containing several dilatonic scalar fields and antisymmetric forms is considered. The manifold is chosen in the form $M = M_0 \times M_1 \times \ldots \times M_n$, where M_i are Einstein spaces ($i \geq 1$). The sigma-model representation for p–branes ansatz and exact solutions in the model are given (e.g. solutions depending on harmonic function, spherically symmetric ones and black holes). PPN parameters for 4–dimensional section of the metric are also presented.

Keywords: Multidimensional gravity, p–branes, exact solutions.

1. INTRODUCTION

More than 30 years passed after the pioneering paper [1] of Dietrich Kramer on spherically symmetric solution in 5–dimensional model (suggested by E. Schmutzer). Later this solution was rediscovered by several

*Visiting Professor at Depto.de Fisica, CINVESTAV, A.P.14–740, Mexico 07000, D.F.

authors and became a starting point for different multidimensional generalizations.

Here we consider a more general class of exact solutions in a D-dimensional model with scalar fields and fields of forms, inspired by modern unified models. Such generalization is intended to study general gravitational and cosmological properties of these models. Clear, that most of their symmetries are not preserved in arbitrary dimensions, but are recovered in 10 or 11. Among our solutions, there are solutions with harmonic functions obtained by the null–geodesic method suggested by G. Neugebauer and D. Kramer in [2]. We also consider exact spherically symmetric, cosmological and black hole configurations with p–branes. All of them have a general structure since minor restrictions on parameters of the model are imposed. We also present PPN parameters for 4–dimensional section of the metric with the aim of studying new observational windows to unified models and extra dimensions [3, 4, 5].

2. THE MODEL

We consider the model governed by the action

$$\begin{aligned} S &= \int_M d^D z \sqrt{|g|}\{R[g] - 2\Lambda - h_{\alpha\beta} g^{MN} \partial_M \varphi^\alpha \partial_N \varphi^\beta \\ &- \sum_{a\in\Delta} \frac{\theta_a}{n_a!} \exp[2\lambda_a(\varphi)](F^a)^2_g\}, \end{aligned} \tag{1}$$

where $g = g_{MN} dz^M \otimes dz^N$ is the metric on the manifold M, $\dim M = D$, $\varphi = (\varphi^\alpha) \in \mathbf{R}^l$ is a vector from dilatonic scalar fields, $(h_{\alpha\beta})$ is a non–degenerate $l \times l$ matrix ($l \in \mathbf{N}$), $\theta_a \neq 0$, $F^a = dA^a = \frac{1}{n_a!} F^a_{M_1 \dots M_{n_a}} dz^{M_1} \wedge \dots \wedge dz^{M_{n_a}}$ is a n_a-form ($n_a \geq 2$) on a D–dimensional manifold M, Λ is a cosmological constant and λ_a is a 1–form on $\mathbf{R}^l$: $\lambda_a(\varphi) = \lambda_{a\alpha}\varphi^\alpha$, $a \in \Delta$, $\alpha = 1, \dots, l$. In (0) we denote $|g| = |\det(g_{MN})|$, $(F^a)^2_g = F^a_{M_1 \dots M_{n_a}} F^a_{N_1 \dots N_{n_a}} g^{M_1 N_1} \dots g^{M_{n_a} N_{n_a}}$, $a \in \Delta$, where Δ is some finite set. In the models with one time all $\theta_a = 1/2$ when the signature of the metric is $(-1, +1, \dots, +1)$.

2.1. ANSATZ FOR COMPOSITE P-BRANES

Let us consider the manifold and metric as

$$M = M_0 \times M_1 \times \dots \times M_n, \tag{2}$$

$$g = e^{2\gamma(x)} \hat{g}^0 + \sum_{i=1}^{n} e^{2\phi^i(x)} \hat{g}^i, \tag{3}$$

where $g^0 = g^0_{\mu\nu}(x)dx^\mu \otimes dx^\nu$ is an arbitrary metric with any signature on the manifold M_0 and $g^i = g^i_{m_i n_i}(y_i)dy_i^{m_i} \otimes dy_i^{n_i}$ is a metric on M_i satisfying the equation $R_{m_i n_i}[g^i] = \xi_i g^i_{m_i n_i}$, $m_i, n_i = 1, \ldots, d_i$; $\xi_i =$ const, $i = 1, \ldots, n$. The functions $\gamma, \phi^i : M_0 \to \mathbf{R}$ are smooth. We denote $d_\nu = \dim M_\nu$; $\nu = 0, \ldots, n$; $D = \sum_{\nu=0}^{n} d_\nu$. We put any manifold M_ν, $\nu = 0, \ldots, n$, to be oriented and connected. Then the volume d_i–form $\tau_i \equiv \sqrt{|g^i(y_i)|}\, dy_i^1 \wedge \ldots \wedge dy_i^{d_i}$, and signature parameter $\varepsilon(i) \equiv \mathrm{sign}(\det(g^i_{m_i n_i})) = \pm 1$ are correctly defined.

Let $\Omega = \Omega(n)$ be a set of all non–empty subsets of $\{1, \ldots, n\}$. The number of elements in Ω is $|\Omega| = 2^n - 1$. For any $I = \{i_1, \ldots, i_k\} \in \Omega$, $i_1 < \ldots < i_k$, we denote $\tau(I) \equiv \tau_{i_1} \wedge \ldots \wedge \tau_{i_k}$, $\varepsilon(I) \equiv \varepsilon(i_1) \ldots \varepsilon(i_k)$, $d(I) \equiv \sum_{i \in I} d_i$.

For fields of forms we consider the composite electromagnetic ansatz

$$F^a = \sum_{I \in \Omega_{a,e}} \mathcal{F}^{(a,e,I)} + \sum_{J \in \Omega_{a,m}} \mathcal{F}^{(a,m,J)}, \tag{4}$$

$$\mathcal{F}^{(a,e,I)} = d\Phi^{(a,e,I)} \wedge \tau(I), \tag{5}$$

$$\mathcal{F}^{(a,m,J)} = e^{-2\lambda_a(\varphi)} * (d\Phi^{(a,m,J)} \wedge \tau(J)) \tag{6}$$

with elementary forms of electric and magnetic types, $a \in \Delta$, $I \in \Omega_{a,e}$, $J \in \Omega_{a,m}$ and $\Omega_{a,v} \subset \Omega$, $v = e, m$. For scalar functions we put

$$\varphi^\alpha = \varphi^\alpha(x), \quad \Phi^s = \Phi^s(x), \tag{7}$$

$s \in S$. Thus φ^α and Φ^s are functions on M_0. Here and below $S = S_e \sqcup S_m$, $S_v = \sqcup_{a \in \Delta} \{a\} \times \{v\} \times \Omega_{a,v}$, $v = e, m$. The set S consists of elements $s = (a_s, v_s, I_s)$, where $a_s \in \Delta$, $v_s = e, m$ and $I_s \in \Omega_{a_s, v_s}$. Due to (5) and (6): $d(I) = n_a - 1$ and $d(J) = D - n_a - 1$, for $I \in \Omega_{a,e}$ and $J \in \Omega_{a,m}$.

2.2. THE SIGMA MODEL

Let $d_0 \neq 2$ and $\gamma = \gamma_0(\phi) \equiv \frac{1}{2-d_0} \sum_{j=1}^{n} d_j \phi^j$ (harmonic gauge).

It was proved in [6] that eqs. of motion for (0) and the Bianchi identities: $d\mathcal{F}^s = 0$, $s \in S_m$, for fields from (3), (4)–(7), when some restrictions are imposed, are equivalent to eqs. of motion for σ–model

$$\begin{aligned} S_{\sigma 0} = & \int d^{d_0}x \sqrt{|g^0|} \Big\{ R[g^0] - \hat{G}_{AB} g^{0\mu\nu} \partial_\mu \sigma^A \partial_\nu \sigma^B \\ & - \sum_{s \in S} \varepsilon_s \exp(-2U^s_A \sigma^A) g^{0\mu\nu} \partial_\mu \Phi^s \partial_\nu \Phi^s - 2V \Big\}, \end{aligned} \tag{8}$$

where $(\sigma^A) = (\phi^i, \varphi^\alpha)$,

$$V = V(\phi) = \Lambda e^{2\gamma_0(\phi)} - \frac{1}{2}\sum_{i=1}^{n} \xi_i d_i e^{-2\phi^i + 2\gamma_0(\phi)}, \tag{9}$$

$(\hat{G}_{AB}) = \mathrm{diag}(G_{ij}, h_{\alpha\beta})$ is the target space metric, $G_{ij} = d_i\delta_{ij} + d_i d_j (d_0 - 2)^{-1}$,

$$U^s_A = U^s_A \sigma^A = \sum_{i \in I_s} d_i \phi^i - \chi_s \lambda_{a_s}(\varphi), \quad (U^s_A) = (d_i \delta_{iI_s}, -\chi_s \lambda_{a_s \alpha}), \tag{10}$$

$s = (a_s, v_s, I_s)$. Here $\chi_e = +1$ and $\chi_m = -1$; $\delta_{iI} = \sum_{j \in I} \delta_{ij}$ and $\varepsilon_s = (-\varepsilon[g])^{(1-\chi_s)/2} \varepsilon(I_s)\theta_{a_s}$, $s \in S$, $\varepsilon[g] \equiv \mathrm{sign}\det(g_{MN})$.

3. TODA–TYPE SOLUTIONS FROM NULL–GEODESIC METHOD

It is known that geodesics of the target space equipped with some harmonic function on a three–dimensional space generate a solution to the σ–model equations [2]. Here we apply this null–geodesic method to our sigma-model and obtain a new class of solutions in multidimensional gravity. Action (8) may be also written in the form

$$S_{\sigma 0} = \int d^{d_0}x \sqrt{|g^0|} \{ R[g^0] - \mathcal{G}_{\hat{A}\hat{B}}(X) g^{0\mu\nu} \partial_\mu X^{\hat{A}} \partial_\nu X^{\hat{B}} - 2V \} \tag{11}$$

where $X = (X^{\hat{A}}) = (\phi^i, \varphi^\alpha, \Phi^s) \in \mathbf{R}^N$, and $\mathcal{G} = \mathcal{G}_{\hat{A}\hat{B}}(X) dX^{\hat{A}} \otimes dX^{\hat{B}}$ on $\mathcal{M} = \mathbf{R}^N$, $N = n + l + |S|$ is defined as follows $(\mathcal{G}_{\hat{A}\hat{B}}(X)) = \mathrm{diag}(G_{ij}, h_{\alpha\beta}, \varepsilon_s \exp(-2U^s(\sigma))\delta_{ss'})$.

Consider exact solutions to field equations for (11). We put

$$X^{\hat{A}}(x) = F^{\hat{A}}(H(x)), \tag{12}$$

where $F : (u_-, u_+) \to \mathbf{R}^N$ is a smooth function, $H : M_0 \to \mathbf{R}$ is a harmonic function on M_0 (i.e. $\Delta[g^0]H = 0$), satisfying $u_- < H(x) < u_+$ for all $x \in M_0$.

Let all factor spaces are Ricci-flat and cosmological constant is zero. Then, the potential is zero and the field equations corresponding to (11) are satisfied identically if $F = F(u)$ obey the Lagrange equations for

$$L = \frac{1}{2}\mathcal{G}_{\hat{A}\hat{B}}(F)\dot{F}^{\hat{A}}\dot{F}^{\hat{B}}, \tag{13}$$

$$E = \frac{1}{2}\mathcal{G}_{\hat{A}\hat{B}}(F)\dot{F}^{\hat{A}}\dot{F}^{\hat{B}} = 0. \tag{14}$$

First, we integrate the Lagrange equations for Φ^s

$$\frac{d}{du}\left(\exp(-2U^s(\sigma))\dot{\Phi}^s\right) = 0 \Longleftrightarrow \dot{\Phi}^s = Q_s \exp(2U^s(\sigma)), \tag{15}$$

where Q_s are constants, $s \in S$. Here $(F^{\hat{A}}) = (\sigma^A, \Phi^s)$. We put $Q_s \neq 0$.

For fixed $Q = (Q_s, s \in S)$ equations for (13) corresponding to $(\sigma^A) = (\phi^i, \varphi^\alpha)$, when (15) are substituted, are equivalent to equations for

$$L_Q = \frac{1}{2}\hat{G}_{AB}\dot{\sigma}^A\dot{\sigma}^B - V_Q, \tag{16}$$

where $V_Q = \frac{1}{2}\sum_{s\in S}\varepsilon_s Q_s^2 \exp[2U^s(\sigma)]$. (14) reads $E_Q = \frac{1}{2}\hat{G}_{AB}\dot{\sigma}^A\dot{\sigma}^B + V_Q = 0$. Here we are interested in exact solutions for a case when the vectors U^s have non-zero length, i.e. $K_s = (U^s, U^s) \neq 0$ and the quasi-Cartan matrix

$$(A_{ss'}) = 2(U^s, U^{s'})/(U^{s'}, U^{s'}) \tag{17}$$

is a non–degenerate one. Here some ordering in S is assumed.

Exact solutions to Lagrange equations for (16) are

$$\sigma^A = \sum_{s\in S}\frac{U^{sA}}{(U^s, U^s)}q^s + c^A u + \bar{c}^A, \tag{18}$$

where q^s are solutions to Toda-type equations

$$\ddot{q}^s = -B_s \exp(\sum_{s'\in S} A_{ss'}q^{s'}), \tag{19}$$

with $B_s = 2K_s A_s$, $A_s = \frac{1}{2}\varepsilon_s Q_s^2$, $s \in S$. These equations correspond to

$$L_{TL} = \frac{1}{4}\sum_{s,s'\in S} h_s A_{ss'}\dot{q}^s\dot{q}^{s'} - \sum_{s\in S} A_s \exp(\sum_{s'\in S} A_{ss'}q^{s'}), \tag{20}$$

where we denote $h_s = K_s^{-1}$. $c = (c^A)$ and $\bar{c} = (\bar{c}^A)$ satisfy constraints

$$U^s(c) = U_A^s c^A = 0, \qquad U^s(\bar{c}) = U_A^s \bar{c}^A = 0. \tag{21}$$

Using (18) we obtain the solutions:

$$g = \left(\prod_{s\in S} f_s^{2d(I_s)h_s/(D-2)}\right)\Big\{\exp(2c^0 H + 2\bar{c}^0)\hat{g}^0 \tag{22}$$

$$+\sum_{i=1}^{n}\Big(\prod_{s\in S} f_s^{-2h_s\delta_{iI_s}}\Big)\exp(2c^i H + 2\bar{c}^i)\hat{g}^i\Big\},$$

$$\exp(\varphi^\alpha) = \left(\prod_{s\in S} f_s^{h_s\chi_s\lambda^\alpha_{a_s}}\right)\exp(c^\alpha H + \bar{c}^\alpha), \tag{23}$$

$$f_s = f_s(H) = \exp(-q^s(H)), \tag{24}$$

where $H = H(x)$ ($x \in M_0$) is a harmonic function on (M_0, g^0),

$$c^0 = \frac{1}{2-d_0}\sum_{j=1}^{n} d_j c^j, \quad \bar{c}^0 = \frac{1}{2-d_0}\sum_{j=1}^{n} d_j \bar{c}^j, \tag{25}$$

$$2E = 2E_{TL} + \hat{G}_{AB}c^A c^B = 0, \tag{26}$$

$$E_{TL} = \frac{1}{4}\sum_{s,s' \in S} h_s A_{ss'} \dot{q}^s \dot{q}^{s'} + \sum_{s \in S} A_s \exp(\sum_{s' \in S} A_{ss'} q^{s'}). \tag{27}$$

The following relations are valid for electric and magnetic forms

$$\mathcal{F}^s = Q_s \left(\prod_{s' \in S} f_{s'}^{-A_{ss'}}\right) dH \wedge \tau(I_s), \tag{28}$$

$$\mathcal{F}^s = Q_s(*_0 dH) \wedge \tau(\bar{I}_s). \tag{29}$$

4. COSMOLOGICAL SOLUTIONS

Here we study the case $d_0 = 1$, $M_0 = (u_-, u_+)$, i.e. with one–variable dependence. We consider $M = (u_-, u_+) \times M_1 \times \ldots \times M_n$ with the metric

$$g = w \exp(2\gamma(u)) du \otimes du + \sum_{i=1}^{n} \exp(2\phi^i(u)) \hat{g}^i. \tag{30}$$

4.1. SOLUTIONS WITH RICCI-FLAT SPACES

Let all spaces be Ricci–flat, $\xi_1 = \ldots = \xi_n = 0$. Since $H(u) = u$ is a harmonic function on (M_0, g^0) with $g^0 = w du \otimes du$, then the solutions from 3 contain cosmological ones [9] (for $\mathbf{A}_m$ Toda chain see [8]).

4.2. SOLUTIONS WITH ONE CURVED SPACE

This solution may be modified to the case: $\xi_1 \neq 0$, $\xi_2 = \ldots = \xi_n = 0$, when all "brane" submanifolds do not contain M_1. It is:

$$g = \left(\prod_{s \in S} [f_s(u)]^{2d(I_s)h_s/(D-2)}\right) \Big\{ [f_1(u)]^{2d_1/(1-d_1)} \exp(2c^1 u + 2\bar{c}^1) \tag{31}$$

$$\times[wdu \otimes du + f_1^2(u)\hat{g}^1] + \sum_{i=2}^{n}\Big(\prod_{s\in S}[f_s(u)]^{-2h_s\delta_{iI_s}}\Big)\exp(2c^iu + 2\bar{c}^i)\hat{g}^i\Big\}.$$

$$\exp(\varphi^\alpha) = \left(\prod_{s\in S} f_s^{h_s\chi_s\lambda^\alpha_{a_s}}\right)\exp(c^\alpha u + \bar{c}^\alpha), \quad (32)$$

$$F^a = \sum_{s\in S}\delta^a_{a_s}\mathcal{F}^s. \quad (33)$$

$$\mathcal{F}^s = Q_s\left(\prod_{s'\in S} f_{s'}^{-A_{ss'}}\right) du \wedge \tau(I_s), \quad (34)$$

$$\mathcal{F}^s = Q_s\tau(\bar{I}_s), \quad (35)$$

Forms above correspond to electric and magnetic p–branes.

Here $f_1(u) = R\sinh(\sqrt{C_1}(u - u_1))$ for $C_1 > 0$, $\xi_1 w > 0$; $f_1(u) = R\sin(\sqrt{|C_1|}(u - u_1))$ for $C_1 < 0$, $\xi_1 w > 0$; $f_1(u) = R\cosh(\sqrt{C_1}(u - u_1))$ for $C_1 > 0$, $\xi_1 w < 0$; and $f_1(u) = |\xi_1(d_1 - 1)|^{1/2}$ for $C_1 = 0$, $\xi_1 w > 0$, where u_1, C_1 are constants and $R = |\xi_1(d_1 - 1)/C_1|^{1/2}$ [9].

$$U^r(c) = U^r(\bar{c}) = 0, \qquad r = s, 1. \quad (36)$$

Here $U^1(c) = -c^1 + \sum_{i=1}^{n} d_i c^i$. The energy constraint reads

$$E = E_1 + E_{TL} + \frac{1}{2}\hat{G}_{AB}c^A c^B = 0, \quad (37)$$

5. BLACK HOLE SOLUTIONS

Now we come to spherically symmetric case of (31), i.e. we put $w = 1$, $M_1 = S^{d_1}$, $g^1 = d\Omega^2_{d_1}$, where $d\Omega^2_{d_1}$ is the metric on a unit sphere S^{d_1}, $d_1 \geq 2$. Then,$\xi^1 = d_1 - 1$. We put $M_2 = \mathbf{R}$, $g^2 = -dt \otimes dt$, i.e. M_2 is a time manifold. We also assume that $(U^s, U^s) \neq 0$ and $\det((U^s, U^{s'})) \neq 0$. These solutions contain the black hole ones [10, 11, 12]:

$$g = \Big(\prod_{s\in S} H_s^{2h_s d(I_s)/(D-2)}\Big)\Big\{F^{-1}dR \otimes dR + R^2 d\Omega^2_{d_1} \quad (38)$$

$$-\Big(\prod_{s\in S} H_s^{-2h_s}\Big)F dt \otimes dt + \sum_{i=3}^{n}\Big(\prod_{s\in S} H_s^{-2h_s\delta_{iI_s}}\Big)\hat{g}^i\Big\},$$

$$\exp(\varphi^\alpha) = \prod_{s\in S} H_s^{h_s\chi_s\lambda^\alpha_{a_s}}, \quad (39)$$

$$F^a = \sum_{s\in S}\delta^a_{a_s}\mathcal{F}^s, \quad (40)$$

$$\mathcal{F}^s = -\frac{Q_s}{R^{d_1}} \left(\prod_{s' \in S} H_{s'}^{-A_{ss'}} \right) dR \wedge \tau(I_s), \tag{41}$$

$$\mathcal{F}^s = Q_s \tau(\bar{I}_s), \tag{42}$$

Here $Q_s \neq 0$, $h_s = K_s^{-1}$. $H_s > 0$ obey the equations

$$R^{d_1} \frac{d}{dR} \left(R^{d_1} \frac{F}{H_s} \frac{dH_s}{dR} \right) = B_s \prod_{s' \in S} H_{s'}^{-A_{ss'}}, \tag{43}$$

$s \in S$, where $B_s = \varepsilon_s K_s Q_s^2 \neq 0$. Here

$$H_s(R = +\infty) = 1, \tag{44}$$

$$H_s \to H_{s0} \neq 0, \tag{45}$$

for $R^{\bar{d}} \to 2\mu$, $s \in S$. The metric (38) has a regular horizon at $R^{\bar{d}} = 2\mu$.

There exist solutions to eqs. (43)–(45) of polynomial type v.r.t. $z = R^{-\bar{d}}$ variable. The simplest example occurs in orthogonal case [5]: $(U^s, U^{s'}) = 0$, for $s \neq s'$, $s, s' \in S$. Then, $(A_{ss'}) = \mathrm{diag}(2, \ldots, 2)$ is a Cartan matrix for Lie algebra $\mathbf{A_1} \oplus \ldots \oplus \mathbf{A_1}$ and $H_s(z) = 1 + P_s z$ with $P_s \neq 0$, $s \in S$. In [7] this solution was generalized to block–orthogonal case .

6. POST–NEWTONIAN APPROXIMATION

Let $d_1 = 2$. Here we consider the 4–dimensional section of (38), namely,

$$g^{(4)} = U \left\{ \frac{dR \otimes dR}{1 - 2\mu/R} + R^2 d\Omega_2^2 - U_1 \left(1 - \frac{2\mu}{R} \right) dt \otimes dt \right\}, \tag{46}$$

$$U = \prod_{s \in S} H_s^{2d(I_s)h_s/(D-2)}, \qquad U_1 = \prod_{s \in S} H_s^{-2h_s}. \tag{47}$$

For PN approximation we take the first two powers of $1/R$, i. e.

$$H_s = 1 + \frac{P_s}{R} + \frac{P_s^{(2)}}{R^2} + o(\frac{1}{R^3}). \tag{48}$$

Introducing ρ by $R = \rho(1 + (\mu/2\rho))^2$, $(\rho > \mu/2)$, we get from (46)

$$g^{(4)} = U \left\{ -U_1 \frac{(1 - (\mu/2\rho))^2}{(1 + (\mu/2\rho))^2} dt \otimes dt + \left(1 + \frac{\mu}{2\rho} \right)^4 \delta_{ij} dx^i \otimes dx^j \right\}. \tag{49}$$

For possible physical applications, one should calculate β and γ using

$$g_{00}^{(4)} = -(1 - 2V + 2\beta V^2) + O(V^3), \tag{50}$$

$$g_{ij}^{(4)} = \delta_{ij}(1 + 2\gamma V) + O(V^2), \tag{51}$$

$i, j = 1, 2, 3$, where $V = GM/\rho$. From (47)–(51) we deduce [11, 13]:

$$GM = \mu + \sum_{s \in S} h_s P_s \left(1 - \frac{d(I_s)}{D-2}\right), \tag{52}$$

$$\beta - 1 = \frac{1}{2(GM)^2} \sum_{s \in S} (-\varepsilon_s) Q_s^2 \left(1 - \frac{d(I_s)}{D-2}\right), \tag{53}$$

$$\gamma - 1 = -\frac{1}{GM} \sum_{s \in S} h_s P_s \left(1 - 2\frac{d(I_s)}{D-2}\right). \tag{54}$$

Parameter β depends upon ratios of physical parameters: Q_s/GM and does not depend upon the quasi–Cartan matrix and, hence, upon intersections of p–branes. The parameter γ depends upon ratios P_s/GM, where P_s are functions of GM, Q_s and $A = (A_{ss'})$. The calculation of γ needs an exact solution for radial functions H_s. Relations (53) and (54) coincide with those obtained in [7] for block–orthogonal case, when charges Q_s (and P_s) are coinciding inside blocks. For most interesting p–brane solutions with $\varepsilon_s = -1$ and $d(I_s) < D - 2$ (for all $s \in S$) (53) implies $\beta > 1$.

Acknowledgments

This work was supported in part by DFG grant 436 RUS 113/236/O(R), by RFBR, grant 98–02–16414, Project SEE and CONACYT, Mexico. VNM is grateful to Depto. de Fisica, CINVESTAV for their hospitality.

References

[1] D. Kramer, *Acta Physica Polonica* **2** (1969) 807.

[2] G. Neugebauer and D. Kramer, *Ann. der Physik* (Leipzig) **24** (1969) 62.

[3] V.N.Melnikov, in: *Cosmology and Gravitation*, ed. M.Novello, (Editions Frontieres, Singapore, 1994) pp.147.

[4] V.N.Melnikov, in: *Cosmology and Gravitation II.* (Editions Frontieres, Singapore, 1996) pp.465.

[5] V.D.Ivashchuk and V.N.Melnikov, in: *Lecture Notes in Physics*, v.537. Eds. S.Cotsakis and G.Gibbons, (Springer, Berlin, 2000) pp.214.

[6] V.D. Ivashchuk and V.N. Melnikov, *Class. Quantum Grav.* **14** (1997) 3001.
[7] V.D.Ivashchuk and V.N.Melnikov, *J. Math. Phys.*, **40** (1999) 6558.
[8] V.R. Gavrilov and V.N. Melnikov, *TMP*, **123** (2000) 374.
[9] V.D.Ivashchuk and S.W. Kim, *J. Math. Phys.* **41** (2000) 444.
[10] V.D.Ivashchuk and V.N.Melnikov, *Grav. and Cosm.*, **5** (1999) 313.
[11] V.D.Ivashchuk and V.N.Melnikov, *Grav. and Cosm.*, **6** (2000) 27.
[12] V.D.Ivashchuk and V.N.Melnikov, *Class. Quantum Grav.* **17** (2000) 2073.
[13] V.D. Ivashchuk, V.S. Manko and V.N. Melnikov, *Grav, and Cosm.* **6** (2000) 23.

EFFECTIVE FOUR–DIMENSIONAL DILATON GRAVITY FROM FIVE–DIMENSIONAL CHERN–SIMONS GRAVITY

Alfredo Macias*
Departamento de Física, Universidad Autónoma Metropolitana–Iztapalapa, Apartado Postal 55-534, C.P. 09340, México, D.F., México.

Alberto Garcia†
Departamento de Física, CINVESTAV-IPN, Apartado Postal 14–740, C.P. 07000, México, D.F., México.

Abstract In this paper we consider the five–dimensional theory of gravity proposed recently by Chamseddine, we apply to this action the five–dimensional principal fiber bundle structure and the toroidal dimensional reduction process in order to study what modifications are obtained to the standard Einstein-Maxwell–dilaton theories. We obtain an effective four dimensional gravity theory endowed with a cosmological term, a scalar dilaton field, an Einstein term, and a non–linear electrodynamics.

Keywords: Chern–Simons, dilaton gravity, non–linear electrodynamics.

1. INTRODUCTION

The study of Chern–Simons (CS) theories in $(2 + 1)$–dimensions [1] has recently been of much interest due to a number of reasons. These range from the very mathematical applications related to knot theories [2] to the more physical implications in the description of the quantum Hall effect through the idea of anyons, which correspond to particles with fractional statistic [3]. Another very interesting property of the

*E–mail: amac@xanum.uam.mx
†E–mail: aagarcia@fis.cinvestav.mx

$(2+1)$ Chern–Simons theories is their exact solubility in terms of a finite number of degrees of freedom, which make them an adequate laboratory to test many ideas related to the quantization of gauge and gravitational theories [4]. In $(2+1)$–dimensions, the standard Einstein theory of gravity together with the de Sitter gravity [4], conformal gravity [5], and supergravity [6] turn out to be equivalent to Chern–Simons theories.

Since Chern–Simons theories are of topological nature, their description is most naturally based on the holonomies of the connection, thus providing a convenient realization to study loop quantization methods in gauge theories.

Chern–Simons theories have been constructed in any number of odd dimensions [7] and for dimensions higher then three, some dimensional reduction process allows the possibility of studying their implications in our $(3+1)$–dimensional world.

A simple possibility that arises is to consider a $4+1$ dimensional Chern–Simons theory of the group $SO(1,5)$ or $SO(2,4)$, which naturally incorporates Einstein gravity through the use of an extended connection that unifies the standard five–dimensional Ricci rotation coefficients with the fünfbein into the *six*–dimensional connection corresponding to the above mentioned groups. We further apply to this action the five–dimensional principal fiber bundle structure and the toroidal dimensional reduction process in order to show that the standard Einstein–Maxwell–dilaton theories are contained in it and to study what modifications are obtained to them. Since the $(4+1)$ Chern–Simons action is quadratic in the six–dimensional curvature one expects to produce terms of the same type in the four–dimensional Riemann curvature and also in the electromagnetic field strength two–form. The former will arise in the Gauss–Bonnet combination, while the later produce non–linear additions to the standard Maxwell action. Besides, the action contains a cosmological constant term plus extra electromagnetic–gravitational interactions.

The plan of the paper is as follows: In Sec. II a Chern–Simons theory in five dimensions is revisited. In Sec. III we present the five dimensional principle fiber boundle structure; in Sec. IV the effective action for the theory is constructed; and in Sec. V the we discuss the results.

2. CHERN–SIMONS ACTION IN FIVE DIMENSIONS

Resently, Chamseddine [7] proposed a topological gauge theory of gravity in five dimensions, being the 5–dimensional de Sitter groups $SO(1,5)$ or $SO(2,4)$ the gauge group. The action is based on the Chern–

Simons five–form defined from the trilinear $SO(1,5)$ or $SO(2,4)$ group invariant

$$< J_{\widetilde{A}_1\widetilde{A}_2} J_{\widetilde{A}_3\widetilde{A}_4} J_{\widetilde{A}_5\widetilde{A}_6} >= \epsilon_{\widetilde{A}_1\widetilde{A}_2\widetilde{A}_3\widetilde{A}_4\widetilde{A}_5\widetilde{A}_6} . \tag{1}$$

where $J_{\widetilde{A}\widetilde{B}}$ is a generator of the Lie algebra. The Chern–Simons five–form is given as follows

$$\omega_5 =< A(dA)^2 + \frac{3}{2}A^3 dA + \frac{3}{5}A^5 >, \tag{2}$$

where A is a Lie algebra valued gauge field (connection)

$$A = A_{\widetilde{\mu}}^{\widetilde{A}\widetilde{B}} J_{\widetilde{A}\widetilde{B}} dx^{\widetilde{\mu}} \tag{3}$$

and is a one–form on an arbitrary five–dimensional spacetime manifold. Here $\widetilde{\mu} = (\mu, 5)$ with $\mu = 0,1,2,3$, $\widetilde{B} = (B,6)$, $B = (b,5)$ with $b = 0,1,2,3$. If A is given over a six dimensional spacetime it will satisfy

$$d\omega_5 = \epsilon_{\widetilde{A}_1 \ldots \widetilde{A}_6} F^{\widetilde{A}_1\widetilde{A}_2} \wedge F^{\widetilde{A}_3\widetilde{A}_4} \wedge F^{\widetilde{A}_5\widetilde{A}_6}, \tag{4}$$

where $F = dA + A^2$.

The action to be considered is the following

$$I_5 = k \int_{M_5} \omega_5 , \tag{5}$$

here k is a dimensionless coupling constant since the gauge field A has dimension one. The field equations for $A^{\widetilde{A}\widetilde{B}}$ in (5) are given by

$$\epsilon_{\widetilde{A}\widetilde{B}\widetilde{C}\widetilde{D}\widetilde{E}} F^{\widetilde{A}\widetilde{B}} \wedge F^{\widetilde{A}\widetilde{D}} = 0, \tag{6}$$

where $F^{\widetilde{A}\widetilde{B}} = dA^{\widetilde{A}\widetilde{B}} + A^{\widetilde{A}\widetilde{C}} \wedge A_{\widetilde{C}}^{\;\widetilde{B}}$.

On the other hand, as it is well known, under a gauge transformation $g(x)$, the gauge field A transforms into $A_g = g^{-1}Ag + g^{-1}dg$, and

$$\omega_5(A_g) = \omega_5(A) + d\alpha_4 + \frac{1}{10} < (g^{-1}dg)^5 >, \tag{7}$$

where

$$\begin{aligned} \alpha_4 &= < -\frac{1}{2} dgg^{-1} \left(AdA + dAA + A^3\right) + \frac{1}{4} dgg^{-1} Adgg^{-1} A \\ &+ \frac{1}{2} \left(dgg^{-1}\right)^3 A > . \end{aligned} \tag{8}$$

For compact manifolds, the term $\int_{M_5} d\alpha_4$ vanishes and the last term in (8) is also a boundary term related to the fifth cohomology groups

$H^5(M_5, \pi_5(SO(1,5)))$ or $H^5(M_5, \pi_5(SO(1,5)))$. In what follows we assume that M_5 is compact so that the action (5) is gauge invariant.

We identify the components of the six dimensional gauge field as

$$A^{AB} = \widetilde{\omega}^{AB}, \qquad A^{A5} = \widetilde{e}^{A}, \qquad A = 0,1,2,3,5, \tag{9}$$

with $\widetilde{\omega}^{AB}$ the five dimensional connection and $\widetilde{e}^{A}$ the five dimensional coframe, they are related by the first Cartan equation $d\widetilde{e}^{B} + \widetilde{\omega}_{C}{}^{B} \wedge e^{C} = 0$, where the corresponding indices are lowered or raised by the local Lorentz metric. Then, we write the action (5) in terms of the new variables, ignoring boundary terms as follows

$$\begin{aligned} I_5 &= 3k \int_{M_5} \epsilon_{ABCDE} \left(\widetilde{e}^{A} \wedge \widetilde{R}^{BC} \wedge \widetilde{R}^{DE} + \frac{2}{3} \lambda \widetilde{e}^{A} \wedge \widetilde{e}^{B} \wedge \widetilde{e}^{C} \wedge \widetilde{R}^{DE} \right. \\ &+ \left. \frac{1}{5} \lambda^2 \widetilde{e}^{A} \wedge \widetilde{e}^{B} \wedge \widetilde{e}^{C} \wedge \widetilde{e}^{D} \wedge \widetilde{e}^{E} \right), \end{aligned} \tag{10}$$

where

$$\widetilde{R}^{AB} = d\widetilde{\omega}^{AB} + \widetilde{\omega}^{AC} \wedge \widetilde{\omega}_{C}{}^{B}, \tag{11}$$

is the field strength of the $SO(1,4)$–valued connection and λ is related to the signature of the fifth group index $\lambda = 1$ for $SO(2,4)$ and $\lambda = -1$ for $SO(1,5)$.

The first term in (10) is a Gauss–Bonnet one, the second term is the Einstein term, and the last one is a cosmological constant term. Decomposing the field equations (6) in terms of the component field $\widetilde{\omega}^{AB}$ and $\widetilde{e}^{A}$ we obtain

$$\begin{aligned} \epsilon_{ABCDE} \left(\widetilde{R}^{AB} + \lambda \widetilde{e}^{A} \wedge \widetilde{e}^{B} \right) \wedge \left(\widetilde{R}^{CD} + \lambda \widetilde{e}^{C} \wedge \widetilde{e}^{D} \right) &= 0 \qquad (12) \\ \epsilon_{ABCDE} \left(\widetilde{R}^{AB} + \lambda \widetilde{e}^{A} \wedge \widetilde{e}^{B} \right) \wedge \widetilde{T}^{C} &= 0, \qquad (13) \end{aligned}$$

where

$$T^{A} = d\widetilde{e}^{A} + \widetilde{\omega}^{A}{}_{C} \wedge \widetilde{e}^{C}. \tag{14}$$

In what follows we will restrict ourselves to the case $T^{A} = 0$.

3. FIVE DIMENSIONAL PRINCIPAL FIBER BOUNDLE

The topology of the vacuum is assumed to be $M^4 \times S^1$ and thus $0 \leq \theta < 2\pi$ or equivalently $0 \leq x^5 \leq 2\pi r$. The five-dimensional coordinates are denoted by $C^M(x^\mu, \theta)$ and the metric $g_{MN}(C)(M, N, L, \ldots = 0,1,2,3,5)$ has signature $(+,-,-,-,-)$. To account for the fact that below Plank energy the physical quantities appear only to change with respect to the coordinates x^μ of the usual space–time, we introduce as usual the cylinder condition $\tilde{g}_{AB,5} = 0$.

Then, the metric tensor in a non–diagonal basis is usually written as [8, 9]

$$ds^2 = \sigma^{-1/3}(x)\left[g_{\mu\nu}(x)dx^\mu dx^\nu - \sigma(x)\left(dx^5 + \kappa A_\mu(x)dx^\mu\right)^2\right], \quad (15)$$

where $g_{\mu\nu}(x)$ is the four–dimensional metric tensor, $A_\mu(x)$ the electromagnetic potentials and $\sigma(x)$ is the dilaton field, r is the compactification radius [8] of the fifth dimension ($x^5 = \theta r$). It is much practical to work in a horizontal lift basis (HLB), were the vector potential does not appear explicitly in the metric

$$\theta^\mu = dx^\mu\,, \quad \theta^5 = dx^5 + A_\mu dx^\mu \quad (16)$$

the θ^μ is a 1–form basis. The basis vectors e_μ which are dual to the θ_μ are given by

$$e_\mu = \partial_\mu - A_\mu\partial_5 \qquad \tilde{e}^5 = \partial_5\,. \quad (17)$$

It is important to notice that the HLB is an anholonomic basis, which means that some of the commutators among the vectors of the dual basis Eq. (17) are different from zero. It is easy to show that

$$[e_\mu, e_\nu] = -F_{\mu\nu}\partial_5 \qquad [e_\mu, e_5] = 0\,, \quad (18)$$

where $F_{\mu\nu} = \partial_\mu A_\nu - \partial_\nu A_\mu$. Given a set of basis vectors $\{e_M\}$, the commutation coefficients $C^L{}_{MN}$ are defined as follows

$$[e_M, e_N] = C_{MN}{}^L e_L\,. \quad (19)$$

The only non–vanishing commutators are

$$C_{\mu\nu}{}^5 = -F_{\mu\nu}\,. \quad (20)$$

In the next section we compute the action (10) for the ansatz (15) and compactify it to four dimensions.

4. EFFECTIVE ACTION

The 5–dimensional connection reads

$$\tilde{\omega}^a{}_b = \omega^a{}_b + \frac{\kappa}{2}\sigma^{1/2}F^a{}_b\tilde{e}^5\,, \quad (21)$$

$$\tilde{\omega}^5{}_b = \frac{\kappa}{2}\sigma^{1/2}F_{bc}e^c + \frac{1}{2}\sigma^{-1}\sigma_{,b}\tilde{e}^5\,, \quad (22)$$

where the 4–dimensional connection $\omega^a{}_b$ satisfies $de^a + \omega^a{}_b e^b = 0$, and F_{ab} is the zero–form $F_{ab} = e_a{}^\mu e_b{}^\nu(\partial_\mu A_\nu - \partial_\nu A_n u)$.

Next we calculate the components of the curvature 2–form $\widetilde{R}^A{}_B$ obtaining

$$\begin{aligned}
\widetilde{R}^a{}_b &= R^a{}_b + \frac{\kappa}{2}\sigma^{1/2}\nabla F^a{}_b \wedge \widetilde{e}^5 + \frac{\kappa^2}{4}\sigma\left(F^a{}_c F_{bd} + F^a{}_b F_{cd}\right) e^c \wedge e^d \\
&+ \frac{\kappa}{4}\sigma^{-1/2}\left(\sigma^{,a} F_{bc} - \sigma_{,b}F^a{}_c + \sigma_{,c}F^a{}_b\right) e^c \wedge \widetilde{e}^5\,, \qquad (23)\\
\widetilde{R}^a{}_5 &= \frac{\kappa}{2}\sigma^{1/2}\nabla F^a{}_b \wedge e^b + \frac{1}{4}\left(\kappa^2\sigma F^a{}_c F^c{}_b - \sigma^{-2}\sigma^{,a}\sigma_{,b}\right)\widetilde{e}^5 \wedge e^b \\
&+ \frac{\kappa}{4}\sigma^{-1/2}\sigma_{,c}F^a{}_b e^c \wedge e^b\,. \qquad (24)
\end{aligned}$$

Here $R^a{}_b$ is the 4–dimensional curvature $R^a{}_b = d\omega^a{}_b + \omega^a{}_c \wedge \omega^c{}_b$ and $\nabla F^a{}_b$ is the covariant derivative of the zero–form $F^a{}_b$ in the 4–dimensional manifold, which is given by $\nabla F^a{}_b = dF^a{}_b + \omega^a{}_c F^c{}_b - F^a{}_c\omega^c{}_b$.

Now we calculate the contribution S_1 to the action (10) which is quadratic in R^{AB}. Separating the a and 5 parts, this term splits into two pieces, namely $\epsilon_{abcd}\widetilde{e}^5 \wedge \widetilde{R}^{ab} \wedge \widetilde{R}^{cd} - 4\epsilon_{5bcd}\widetilde{e}^a \wedge \widetilde{R}^{5b} \wedge \widetilde{R}^{cd}$, where we subsequently substitute the equations (23), (24) for the Riemann 2–forms. The second expression involves the contribution of two covariant derivatives which are transformed into surface terms plus non–derivative terms using an integration by parts together with the identity $\nabla\nabla F^{ab} = R^a{}_c F^{cb} - F^{ac} R^b{}_c$. We obtain

$$\begin{aligned}
S_1 &= 3k\int\Bigg\{\epsilon_{5bcde}\,\widetilde{e}^5 \wedge \Bigg[R^{bc} \wedge R^{de} + \frac{\kappa^2}{4}\sigma\left(F^d{}_g F^e{}_r + F^{de}F_{gr}\right) \\
&\times\; R^{bc} \wedge e^g \wedge e^r \\
&+\; \frac{\kappa^2}{16}\sigma^2\left(F^b{}_g F^c{}_r + F^{bc}F_{gr}\right)\left(F^d{}_p F^e{}_q + F^{de}F_{pq}\right) \\
&\times\; e^g \wedge e^r \wedge e^p \wedge e^q + \frac{\kappa^2}{4}\sigma\left(F^b{}_g F^c{}_r + F^{bc}F_{gr}\right) e^g \wedge e^r \wedge R^{de}\Bigg] \\
&+\; 4\epsilon_{a5cde}\,e^a \wedge \Bigg[\kappa^2\left(\sigma F^c{}_m F^m{}_b - \sigma^{-2}\sigma^{,c}\sigma_{,b}\right)\widetilde{e}^5 \wedge e^b \wedge R^{de} \\
&+\; \frac{\kappa}{4}\left(\sigma^{-1/2}\sigma_{,m}F^c{}_b\, e^m + 2\sigma^{1/2}\nabla F^c{}_b\right) \wedge e^b \wedge R^{de} \\
&-\; \frac{\kappa^2}{16}\sigma^{-1}\sigma_{,m}F^c{}_b\left(\sigma^{,d}F^e{}_p - \sigma^{,e}F^d{}_p + \sigma^{,p}F^{de}\right)\widetilde{e}^5 \wedge e^m \wedge e^b \wedge^p \\
&+\; \frac{\kappa^3}{16}\sigma\left(F^d{}_p F^e{}_q + F^{de}F_{pq}\right)\left(\sigma F^c{}_m F^m{}_b - \sigma^{-2}\sigma^{,c}\sigma_{,b}\right) \\
&\times\; \widetilde{e}^5 \wedge e^b \wedge e^p \wedge e^q
\end{aligned}$$

$$+ \quad \frac{\kappa^3}{16}\sigma^{-1/2}\sigma_{,m}F^c{}_b\left(F^d{}_pF^e{}_q + F^{de}F_{pq}\right)e^m\wedge e^b\wedge e^p\wedge e^q$$
$$+ \quad \frac{\kappa^3}{8}\left(\sigma^{,d}F^e{}_m - \sigma^{,e}F^d{}_m + \sigma^{,m}F^{de}\right)\nabla F^c{}_b\wedge e^b\wedge e^m\wedge e^5$$
$$+ \quad \frac{\kappa^3}{8}\sigma^{3/2}\left(F^d{}_pF^e{}_q + F^{de}F_{pq}\right)\nabla F^c{}_b\wedge e^b\wedge e^p\wedge e^q\Big]\Big\}\,. \tag{25}$$

The next term in the action (10), which is linear in the R^{AB}, is the standard one in the Kaluza–Klein approach and leads to

$$S_2 = 3k\int\Big\{2\lambda\epsilon_{5bcde}\,\tilde{e}^5\wedge e^b\wedge e^c\wedge\Big[R^{de}$$
$$+ \quad \frac{\kappa^2}{4}\sigma\left(F^d{}_gF^e{}_a + F^{de}F_{ga}\right)e^g\wedge e^a\Big]$$
$$+ \quad \frac{4}{3}\lambda\epsilon_{abc5e}\,e^a\wedge e^b\wedge e^c\wedge\Big[\frac{\kappa}{2}\sigma^{1/2}\nabla F^e{}_g\,e^g$$
$$+ \quad \frac{\kappa^2}{4}\left(\sigma F^e{}_hF^h{}_g - \sigma^{-2}\sigma^{,e}\sigma_{,g}\right)e^g\wedge\tilde{e}^5 + \frac{\kappa}{4}\sigma^{1/2}\sigma_{,c}F^e{}_ge^c\wedge e^g\Big]\Big\} \tag{26}$$

The final piece of the action (10) corresponds to a cosmological constant term and is given by

$$S_3 = 3k\int\lambda^2e^a\wedge e^b\wedge e^c\wedge e^d\wedge\tilde{e}^5\,. \tag{27}$$

Thus, the complete action (10) can be written as

$$S = S_o + S_1 + S_2\,. \tag{28}$$

In order to understand better the type of interaction produced by our model it is more convenient to rewrite the above action (28) in terms of components. The final reduced action reads, neglecting surface terms,

$$S = 3kr\int d^4x\sqrt{-g}\Big\{-8\lambda R + \kappa^2\lambda\sigma F^{ab}F_{ab} - \sigma^{-2}\sigma_{,a}\sigma^{,a}$$
$$+ \quad \kappa^2\left(\sigma F^{ab}F_{ab} - \sigma^{-2}\sigma_{,a}\sigma^{,a}\right)R$$
$$- \quad \kappa^2\sigma^{-1}\left(\sigma_{,a}\sigma^{,a}F^{bc}F_{bc} + \sigma_{,a}\sigma^{,c}F^{ab}F_{bc}\right)$$
$$- \quad \kappa^2\sigma\left(3F^{dr}F_{rs}R^{bs}{}_{bd} - \frac{1}{2}F^d{}_bF^e{}_cR^{bc}{}_{de} - 2F^{de}F_{rs}R^{rs}{}_{de}\right)$$
$$+ \quad \frac{\kappa^3}{8}\left(\sigma_{,d}F^d{}_m\nabla_bF^m{}_b + \sigma^{,e}F^d{}_m\nabla_eF^d{}_m + \sigma^{,m}F^{de}\nabla_dF^m{}_e\right)$$

$$+ \quad \frac{\kappa^2}{16}\sigma^2 \left[38F^{gc}F_{ce}F^{er}F_{rg} - 11F^{gc}F_{gc}F^{de}F_{de}\right] - 24\lambda^2 \Bigg\} . \qquad (29)$$

Here we have neglected the term quadratic in R^{ab} arising from S_1 which is of Gauss–Bonnet type and thus reduces to a surface term in four dimensions. Also we have introduced the radius r of the fifith dimension which we assume to be compactified.

5. DISCUSSION

We showed that the five–dimensional Chamseddine theory of gravity under principal fiber bundle structure and toroidal dimensional reduction process contains the standard Einstein–Maxwell–dilaton theories with nonlinear modifications

We obtain an effective four dimensional gravity theory endowed with a cosmological term, a scalar dilaton field, an Einstein term, and a non–linear electrodynamics.

Acknowledgments

We dedicate this work to Heinz Dehnen and to Dietrich Kramer on occasion of their 65th and 60th birthday, respectively. Moreover, we thank Heinz Dehnen and Claus Laemmerzahl for hospitality at the University of Konstanz during the completion of this work. This research was supported by CONACyT Grants: 28339E, 32138E, by a FOMES Grant: P/FOMES 98–35–15, and by the joint German–Mexican project DLR–Conacyt MXI 010/98 OTH — E130–1148.

References

[1] D. Birmingham, M. Blau, M. Rakowski, and G. Thompson, *Phys. Rep.* **209** (1991) 129.

[2] E. Witten, *Commun. Math. Phys.* **121**)1989) 351. *Nucl. Phys.* **B322** (1989) 629.

[3] R. Iengo and K. Lechner, *Phys. Rep.* **213** (1992) 1.

[4] E. Witten, *Nucl. Phys.* **B311** (1988/89) 46.

[5] J.H. Horne and E. Witten *Phys. Rev. Lett.* **62** (1989) 501.

[6] K. Koehler, F. Mansouri, C. Vaz, and L. Witten, *Mod. Phys. Lett.* **A5** (1990) 935. *Nucl. Phys.* **B341** (1990) 167. *Nucl. Phys.* **B348** (1992) 373.

[7] A.H. Chamseddine, *Phys. Lett.* **B233** (1989) 291.

[8] Th. Kaluza, *Sitzungsber. Preuss. Akad. Wiss.* Berlin, Math. Phys. **K1** (1921) 966. O.Klein, Z. Phys. **37** (1926) 895; Nature **118** (1927) 516. P. Bergmann, *Introduction to the theory of Relativity* (Dover, New York, 1976).

[9] C. F. Chyba, *Am. J. Phys.* **53** (1985) 863.

A PLANE–FRONTED WAVE SOLUTION IN METRIC–AFFINE GRAVITY

Dirk Puetzfeld*

Institute for Theoretical Physics, University of Cologne, D–50923 Köln, Germany,

Abstract We study plane–fronted electrovacuum waves in metric–affine gravity (MAG) with cosmological constant in the triplet ansatz sector of the theory. Their field strengths are, on the gravitational side, curvature $R_\alpha{}^\beta$, nonmetricity $Q_{\alpha\beta}$, torsion T^α and, on the matter side, the electromagnetic field strength F. Here we basically present, after a short introduction into MAG and its triplet subcase, the results of earlier joint work with García, Macías, and Socorro [1]. Our solution is based on an exact solution of Ozsváth, Robinson, and Rózga describing type N gravitational fields in general relativity as coupled to electromagnetic null–fields.

Keywords: MAG, exact solutions, plane waves.

1. INTRODUCTION

Metric–affine gravity (MAG) represents a gauge theoretical formulation of a theory of gravity which, in contrast to general relativity theory (GR), is no longer confined to a pseudo–Riemannian spacetime structure (cf. [2]). There are new geometric quantities emerging in this theory, namely torsion and nonmetricity, which act as additional field strengths comparable to curvature in the general relativistic case. Due to this general ansatz, several alternative gravity theories are included in MAG, the Einstein-Cartan theory e. g., in which the nonmetricity vanishes and the only surviving post–Riemannian quantity is given by the torsion. One expects that the MAG provides the correct description for early stages of the universe, i. e. at high energies at which the general relativistic description is expected to break down. In case of vanishing post–Riemannian quantities, MAG proves to be compatible with GR. In

*E–mail: dp@thp.uni-koeln.de

Exact Solutions and Scalar Fields in Gravity: Recent Developments
Edited by Macias *et al.*, Kluwer Academic/Plenum Publishers, New York, 2001

contrast to GR, there are presently only a few exact solutions known in MAG (cf. [3]), what could be ascribed to the complexity of this theory.

In the following we will give a short overview of the field equations of MAG and the geometric quantities featuring therein. Especially, we will present the results of the work of Obukhov et al. [4] who found that a special case of MAG, the so called triplet ansatz, is effectively equivalent to an Einstein–Proca theory. Within the framework of this ansatz, we will show how one is able to construct a solution of the MAG field equations on the basis of a plane–fronted wave solution of GR which was originally presented by Ozsváth et al. in [5].

2. MAG IN GENERAL

In MAG we have the metric $g_{\alpha\beta}$, the coframe ϑ^α, and the connection 1-form $\Gamma_\alpha{}^\beta$ [with values in the Lie algebra of the four-dimensional linear group $GL(4,R)$] as new independent field variables. Here $\alpha, \beta, \ldots = 0,1,2,3$ denote (anholonomic) frame indices. Spacetime is described by a metric–affine geometry with the gravitational field strengths nonmetricity $Q_{\alpha\beta} := -Dg_{\alpha\beta}$, torsion $T^\alpha := D\vartheta^\alpha$, and curvature $R_\alpha{}^\beta := d\Gamma_\alpha{}^\beta - \Gamma_\alpha{}^\gamma \wedge \Gamma_\gamma{}^\beta$. A Lagrangian formalism for a matter field Ψ minimally coupled to the gravitational potentials $g_{\alpha\beta}$, ϑ^α, $\Gamma_\alpha{}^\beta$ has been set up in [2]. The dynamics of this theory is specified by a total Lagrangian

$$L = V_{\rm MAG}(g_{\alpha\beta}, \vartheta^\alpha, Q_{\alpha\beta}, T^\alpha, R_\alpha{}^\beta) + L_{\rm mat}(g_{\alpha\beta}, \vartheta^\alpha, \Psi, D\Psi). \tag{1}$$

The variation of the action with respect to the independent matter field and the gauge potentials leads to the field equations:

$$\frac{\delta L_{\rm mat}}{\delta \Psi} = 0, \tag{2}$$

$$DM^{\alpha\beta} - m^{\alpha\beta} = \sigma^{\alpha\beta}, \tag{3}$$

$$DH_\alpha - E_\alpha = \Sigma_\alpha, \tag{4}$$

$$DH^\alpha{}_\beta - E^\alpha{}_\beta = \Delta^\alpha{}_\beta. \tag{5}$$

Equation (4) represents the generalized Einstein equation with the energy–momentum 3-form Σ_α as its source whereas (3) and (5) are additional field equations which take into account other aspects of matter, such as spin, shear, and dilation currents represented collectively by the hypermomentum $\Delta^\alpha{}_\beta$. We made use of the definitions of the gauge field excitations,

$$M^{\alpha\beta} := -2\frac{\partial V_{\rm MAG}}{\partial Q_{\alpha\beta}}, \quad H_\alpha := -\frac{\partial V_{\rm MAG}}{\partial T^\alpha}, \quad H^\alpha{}_\beta := -\frac{\partial V_{\rm MAG}}{\partial R_\alpha{}^\beta}, \tag{6}$$

and of the canonical energy–momentum, the metric stress–energy, and the hypermomentum current of the gauge fields,

$$m^{\alpha\beta} := 2\frac{\partial V_{\rm MAG}}{\partial g_{\alpha\beta}}, \quad E_\alpha := \frac{\partial V_{\rm MAG}}{\partial \vartheta^\alpha}, \quad E^\alpha{}_\beta := -\vartheta^\alpha \wedge H_\beta - g_{\beta\gamma} M^{\alpha\gamma}. \quad (7)$$

Moreover, we introduced the canonical energy–momentum, the metric stress–energy, and the hypermomentum currents of the matter fields, respectively,

$$\sigma^{\alpha\beta} := 2\frac{\delta L_{\rm mat}}{\delta g_{\alpha\beta}}, \quad \Sigma_\alpha := \frac{\delta L_{\rm mat}}{\delta \vartheta^\alpha}, \quad \Delta^\alpha{}_\beta := \frac{\delta L_{\rm mat}}{\delta \Gamma_\alpha{}^\beta}. \quad (8)$$

Provided the matter equation (2) is fulfilled, the following Noether identities hold:

$$D\Sigma_\alpha = \left(e_\alpha \rfloor T^\beta\right) \wedge \Sigma_\beta - \frac{1}{2}\left(e_\alpha \rfloor Q_{\beta\gamma}\right) \sigma^{\beta\gamma} + \left(e_\alpha \rfloor R_\beta{}^\gamma\right) \wedge \Delta^\beta{}_\gamma, \quad (9)$$

$$D\Delta^\alpha{}_\beta = g_{\beta\gamma}\sigma^{\alpha\gamma} - \vartheta^\alpha \wedge \Sigma_\beta. \quad (10)$$

They show that the field equation (3) is redundant. Thus we only need to take into account (4) and (5). As suggested in [2], the most general parity conserving Lagrangian expressed in terms of the irreducible pieces (cf. [2]) of nonmetricity $Q_{\alpha\beta}$, torsion T^α , and curvature $R_{\alpha\beta}$ reads

$$\begin{aligned} V_{\rm MAG} = {} & \frac{1}{2\kappa}\Bigg[-a_0\, R^{\alpha\beta} \wedge \eta_{\alpha\beta} - 2\lambda\eta + T^\alpha \wedge {}^*\left(\sum_{I=1}^{3} a_I\, {}^{(I)}T_\alpha\right) \\ & + Q_{\alpha\beta} \wedge {}^*\left(\sum_{I=1}^{4} b_I\, {}^{(I)}Q^{\alpha\beta}\right) + b_5\left({}^{(3)}Q_{\alpha\gamma} \wedge \vartheta^\alpha\right) \wedge {}^*\left({}^{(4)}Q^{\beta\gamma} \wedge \vartheta_\beta\right) \\ & + 2\left(\sum_{I=2}^{4} c_I\, {}^{(I)}Q_{\alpha\beta}\right) \wedge \vartheta^\alpha \wedge {}^*T^\beta\Bigg] \\ & - \frac{1}{2\rho} R^{\alpha\beta} \wedge {}^*\Bigg[\sum_{I=1}^{6} w_I\, {}^{(I)}W_{\alpha\beta} + \sum_{I=1}^{5} z_I\, {}^{(I)}Z_{\alpha\beta} + w_7\, \vartheta_\alpha \wedge \left(e_\gamma \rfloor {}^{(5)}W^\gamma{}_\beta\right) \\ & + z_6\, \vartheta_\gamma \wedge \left(e_\alpha \rfloor {}^{(2)}Z^\gamma{}_\beta\right) + \sum_{I=7}^{9} z_I\, \vartheta_\alpha \wedge \left(e_\gamma \rfloor {}^{(I-4)}Z^\gamma{}_\beta\right)\Bigg]. \end{aligned} \quad (11)$$

Note that we decompose the curvature 2–form $R_\alpha{}^\beta$ into its antisymmetric and symmetric parts, i.e. $R_{\alpha\beta} = W_{\alpha\beta} + Z_{\alpha\beta} = R_{[\alpha\beta]} + R_{(\alpha\beta)} \sim$ rotational $\oplus$ strain curvature. The constants entering eq. (11) are the cosmological constant λ, the weak and strong coupling constant κ and ρ, respectively, and the 28 dimensionless parameters

$$a_0, \ldots, a_3, b_1, \ldots, b_5, c_2, \ldots, c_4, w_1, \ldots, w_7, z_1, \ldots, z_9. \quad (12)$$

We have here the following dimensions: $[\lambda]$ =length^{-2}, $[\kappa]$ =length^2, $[\rho] = [\hbar] = [c] = 1$. The Lagrangian (11) and the presently known exact solutions in MAG have been reviewed in [3].

3. THE TRIPLET ANSATZ

In the following we will briefly review the results of Obukhov et al. [4]. Starting from the most general gauge Lagrangian $V_{\rm MAG}$ in (11), we now investigate the special case with

$$w_1, \ldots, w_7 = 0, \quad z_1, \ldots, z_3, z_5, \ldots, z_9 = 0, \quad z_4 \neq 0. \tag{13}$$

Thus we consider a general weak part, i. e., we do not impose that one of the weak coupling constants vanishes right from the beginning. However, the strong gravity part of (11) is truncated for simplicity. Its only surviving piece is given by the square of the dilation part of the segmental curvature ${}^{(4)}Z_{\alpha\beta} := \frac{1}{4} g_{\alpha\beta} Z_\gamma{}^\gamma$. In this case, the result of Obukhov et al. [4] reads as follows: Effectively, the curvature $R_{\alpha\beta}$ may be considered as Riemannian, torsion and nonmetricity may be represented by a 1-form ω,

$$\begin{aligned} Q &= k_0\,\omega, \qquad \Lambda = k_1\,\omega, \qquad T = k_2\,\omega, && (14)\\ T^\alpha &= {}^{(2)}T^\alpha = \frac{1}{3}\vartheta^\alpha \wedge T, && (15)\\ Q_{\alpha\beta} &= {}^{(3)}Q_{\alpha\beta} + {}^{(4)}Q_{\alpha\beta} = \frac{4}{9}\left(\vartheta_{(\alpha} e_{\beta)} \rfloor \Lambda - \frac{1}{4} g_{\alpha\beta}\Lambda\right) + g_{\alpha\beta} Q. && (16) \end{aligned}$$

With the aid of the Riemannian curvature $\tilde{R}_{\alpha\beta}$, we denote Riemannian quantities by a tilde, the field equation (4) looks like the Einstein equation with an energy-momentum source that depends on torsion and nonmetricity. Therefore, the field equation (5) becomes a system of differential equations for torsion and nonmetricity alone. In the vacuum case (i. e. $\Sigma_\alpha = 0$ and $\Delta_\alpha{}^\beta = 0$), these differential equations reduce to

$$\begin{aligned} \frac{a_0}{2}\,\eta_{\alpha\beta\gamma} \wedge \tilde{R}^{\beta\gamma} + \lambda\eta_\alpha &= \kappa \Sigma_\alpha^{(\omega)}, && (17)\\ d^\star d\omega + m^2\,{}^\star\omega &= 0. && (18) \end{aligned}$$

The four constants m, k_0, k_1, and k_2, which appear in (18) and (14), depend uniquely on the parameters of the MAG Lagrangian (13):

$$\begin{aligned} k_0 &= 4\,(a_2 - 2a_0)\left(b_3 + \frac{a_0}{8}\right) - 3\,(c_3 + a_0)^2,\\ k_1 &= \frac{9}{2}\,(a_2 - 2a_0)\,(b_5 - a_0) - 9\,(c_3 + a_0)\,(c_4 + a_0), \end{aligned}$$

$$\begin{aligned} k_2 &= 12\left(b_3 + \frac{a_0}{8}\right)(c_4 + a_0) - \frac{9}{2}(b_5 - a_0)(c_3 + a_0)\,, \\ m^2 &= \frac{1}{z_4 \kappa}\left(-4b_4 + \frac{3}{2}a_0 + \frac{k_1}{2k_0}(b_5 - a_0) + \frac{k_2}{k_0}(c_4 + a_0)\right). \end{aligned} \tag{19}$$

The energy–momentum source of torsion and nonmetricity $\Sigma_\alpha^{(\omega)}$, which appears in the effective Einstein equation (17), reads

$$\begin{aligned} \Sigma_\alpha^{(\omega)} &= \frac{z_4 k_0^2}{2\rho}\Big\{(e_\alpha \rfloor d\omega) \wedge {}^\star d\omega - (e_\alpha \rfloor {}^\star d\omega) \wedge d\omega \\ &\quad + m^2\left[(e_\alpha \rfloor \omega) \wedge {}^\star \omega + (e_\alpha \rfloor {}^\star \omega) \wedge \omega\right]\Big\}. \end{aligned} \tag{20}$$

This energy–momentum is exactly that of a Proca 1–form field. The parameter m in (18) has the meaning of the mass parameter ($[m]$ =length^{-1}). If m vanishes, the constrained MAG theory looks similar to the Einstein–Maxwell theory, as can be seen immediately by comparing (20) with the energy–momentum current of the Maxwell theory

$$\Sigma_\alpha^{\rm Max} = \frac{1}{2}\Big\{(e_\alpha \rfloor dA) \wedge {}^\star dA - (e_\alpha \rfloor {}^\star dA) \wedge dA\Big\}, \tag{21}$$

where A denotes the electromagnetic potential 1–form ($F = dA$). Note that $m = 0$ leads to an additional constraint among the coupling constants (cf. eq. (19)).

4. PLANE–FRONTED WAVES IN GR

Ozsváth, Robinson and Rózga [5] dealt with a solution of the Einstein–Maxwell equations. Here we sketch their procedure in order to show how to generalize it to the triplet subcase of MAG. Since we perform our calculations with arbitrary gravitational coupling constant κ, the results presented here differ by some factor of κ from the original ones in [5]. Using the coordinates $(\rho, \sigma, \zeta, \bar\zeta)$, we start with the line element

$$ds^2 = 2(\vartheta^{\hat 0} \otimes \vartheta^{\hat 1} + \vartheta^{\hat 2} \otimes \vartheta^{\hat 3}), \tag{22}$$

and the coframe (the bar denotes complex conjugation)

$$\vartheta^{\hat 0} = \frac{1}{p}\,d\zeta, \quad \vartheta^{\hat 1} = \frac{1}{p}\,d\bar\zeta, \quad \vartheta^{\hat 2} = -d\sigma, \quad \vartheta^{\hat 3} = \left(\frac{q}{p}\right)^2 (s\,d\sigma + d\rho)\,. \tag{23}$$

In order to write down the coframe in a compact form, we made use of the abbreviations p, q, and s which are defined in the following way:

$$p(\zeta, \bar\zeta) = 1 + \frac{\lambda}{6}\,\zeta\bar\zeta, \tag{24}$$

$$q(\sigma,\zeta,\bar{\zeta}) = \left(1-\frac{\lambda}{6}\,\zeta\bar{\zeta}\right)\alpha(\sigma)+\zeta\bar{\beta}(\sigma)+\bar{\zeta}\beta(\sigma), \tag{25}$$

$$\begin{aligned} s(\rho,\sigma,\zeta,\bar{\zeta}) &= -\frac{\rho^2\lambda}{6}\,\alpha^2(\sigma)-\rho^2\beta(\sigma)\,\bar{\beta}(\sigma)+\rho\,\partial_\sigma\left(\ln|q|\right) \\ &\quad +\frac{p}{2q}\,H(\sigma,\zeta,\bar{\zeta}). \end{aligned} \tag{26}$$

Here $\alpha(\sigma)$, $\beta(\sigma)$, and $H(\sigma,\zeta,\bar{\zeta})$ are arbitrary functions of the coordinates and λ is the cosmological constant. Additionally, we introduce the notion of the so–called propagation 1–form $k := k_\mu\vartheta^\mu$ which inherits the properties of the geodesic, shear–free, expansion–free and twistless null vector field k^μ representing the propagation vector of a plane–fronted wave. We proceed by imposing some restrictions on the electromagnetic 2–form F and a 2-form $S_{\alpha\beta}$ defined in terms of the irreducible decomposition of the Riemannian curvature 2–form $\tilde{R}_{\alpha\beta}$ in the following way:

$$\begin{aligned} S_{\alpha\beta} &:= \tilde{R}_{\alpha\beta}-{}^{(6)}\tilde{R}_{\alpha\beta} = {}^{(1)}\tilde{R}_{\alpha\beta}+{}^{(4)}\tilde{R}_{\alpha\beta} \\ &= \tilde{R}_{\alpha\beta}+\frac{1}{12}(e_\nu\rfloor e_\mu\rfloor\tilde{R}^{\nu\mu})\,\vartheta_\alpha\wedge\vartheta_\beta \overset{\text{in vacuum}}{=} {}^{(1)}\tilde{R}_{\alpha\beta} =: C_{\alpha\beta} \end{aligned} \tag{27}$$

They shall obey the so–called radiation conditions

$$S_{\alpha\beta}\wedge k=0, \quad \text{and} \quad (e_\alpha\rfloor k)\,S^\alpha{}_\beta=0, \tag{28}$$

and

$$F\wedge k=0, \quad \text{and} \quad \frac{1}{2}(e^\alpha\rfloor k)\,e_\alpha\rfloor F=0. \tag{29}$$

If one imposes the conditions (28) and (29), the formerly arbitrary functions $\alpha(\sigma)$ and $\beta(\sigma)$ in (24)–(26) become restricted, namely $\alpha(\sigma)$ to the real and $\beta(\sigma)$ to the complex domain, see [5]. In the next step we insert the ansatz for the coframe (23) into the Einstein–Maxwell field equations with cosmological constant

$$\eta_{\alpha\beta\gamma}\wedge\tilde{R}^{\beta\gamma}+2\,\lambda\,\eta_\alpha = 2\kappa\Sigma_\alpha^{\text{Max}}, \tag{30}$$

$$dF=0, \quad d\,{}^\star F = 0, \tag{31}$$

where $\Sigma_\alpha^{\text{Max}}$ is the energy–momentum current of the electromagnetic field (cf. eq. (21)). Let us consider the vacuum field equations first before switching on the electromagnetic field, i. e. there are only gravitational waves and we only have to take into account the left hand side of (30). As shown in [5], this equation, after inserting the coframe into it (cf. eq. (4.34) of [5]), turns into a homogeneous PDE for the unknown function $H(\sigma,\zeta,\bar{\zeta})$:

$$H_{,\zeta\bar{\zeta}}+\frac{\lambda}{3p^2}H=0. \tag{32}$$

Equation (4.39) of [5] supplies us with the solution for $H(\sigma,\zeta,\bar\zeta)$ in terms of an arbitrary holomorphic function $\phi(\sigma,\zeta)$

$$H(\sigma,\zeta,\bar\zeta) = \phi_{,\zeta} - \frac{\lambda\,\bar\zeta}{3\,p}\phi + \bar\phi_{,\bar\zeta} - \frac{\lambda\,\zeta}{3\,p}\bar\phi. \tag{33}$$

Observe that H in (33) is a real quantity. We are now going to switch on the electromagnetic field. We make the following ansatz for the electromagnetic 2–form F in terms of an arbitrary complex function $f(\sigma,\zeta)$:

$$F = dA = -d\left[\left(\int^{\zeta} d\zeta'\, f(\zeta',\sigma) + \int^{\bar\zeta} d\bar\zeta'\, \bar f(\bar\zeta',\sigma)\right)\vartheta^{\hat 2}\right]. \tag{34}$$

In compliance with [5], this ansatz for F leads to $\Sigma_\alpha^{\rm Max} = -2\,\delta_\alpha^{\hat 2}\,p^2 f\bar f\,\vartheta^{\hat 0}\wedge\vartheta^{\hat 1}\wedge\vartheta^{\hat 2}$. Now the field equations (30)–(31) turn into an inhomogeneous PDE for $H(\sigma,\zeta,\bar\zeta)$ (cf. eq. (4.35) of [5]):

$$H_{,\zeta\bar\zeta} + \frac{\lambda}{3p^2}H = \frac{2\kappa p}{q}\, f\,\bar f. \tag{35}$$

The homogeneous solution $H_{\rm h}(\sigma,\zeta,\bar\zeta)$ of this equation is again given by (33). The particular solution $H_{\rm p}(\sigma,\zeta,\bar\zeta)$ of the inhomogeneous equation can be written in a similar form,

$$H_{\rm p}(\sigma,\zeta,\bar\zeta) = \mu_{,\zeta} - \frac{\lambda\,\bar\zeta}{3\,p}\mu + \bar\mu_{,\bar\zeta} - \frac{\lambda\,\zeta}{3\,p}\bar\mu, \tag{36}$$

where the function $\mu(\sigma,\zeta,\bar\zeta)$ can be expressed in the following integral form:

$$\mu(\sigma,\zeta,\bar\zeta) = \kappa\int^{\bar\zeta} d\bar\zeta p^2 \int^{\zeta} d\zeta'\frac{1}{p^2}\int^{\zeta'} d\zeta''\frac{p\,f\,\bar f}{q}. \tag{37}$$

Of course, one is only able to derive $H_{\rm p}$ explicitly after choosing the arbitrary functions $\alpha(\sigma)$ and $\beta(\sigma)$ in (24)-(26). They enter the coframe and, as a consequence, the function $\mu(\sigma,\zeta,\bar\zeta)$ in (37). The general solution reads [1]

$$H(\sigma,\zeta,\bar\zeta) = H_{\rm h}(\sigma,\zeta,\bar\zeta) + H_{\rm p}(\sigma,\zeta,\bar\zeta). \tag{38}$$

We proceed with a particular choice for the functions entering the coframe and the ansatz for the electromagnetic potential

$$\alpha = 1, \qquad \beta = 0, \qquad f = f_0\,\zeta^n \quad \text{where} \quad {\rm n} = 0, \pm 1, \pm 2, \ldots\,, \tag{39}$$

and $[f_0]$ =length^{-2-n}. The electromagnetic potential is now given by

$$A = -f_0\left(\int^{\zeta} d\zeta'\,\zeta'^n + \int^{\bar\zeta} d\bar\zeta'\,\bar\zeta'^n\right)\vartheta^{\hat 2}. \tag{40}$$

The only unknown function is $H_\mathrm{p}(\sigma,\zeta,\bar\zeta)$. Thus we have to carry out the integration (cf. (7.7) of [5]) in eq. (37), after substituting the function $f = f_0\,\zeta^n$ which originates from our ansatz (39). The solution of this integration for different choices of n is given in (7.10)–(7.13) of [5]. We will present this solution in a more compact form as (i) $n<-1$

$$\begin{aligned} H_\mathrm{p} &= \frac{2\kappa p f_0^2}{q}\left(\frac{(\zeta\bar\zeta)^{1+n}}{(1+n)^2}+4\left(\frac{\lambda}{6}\right)^{-n-1}\ln|q|-4\left(\frac{\lambda}{6}\right)^{-n-1}\ln|p-1|\right. \\ &\left. +4\sum_{r=1}^{-n-1}\frac{\left(\frac{\lambda}{6}\right)^{-n-r-1}}{r\,(\zeta\bar\zeta)^r}\right)+\frac{8\kappa f_0^2\,(\zeta\bar\zeta)^{n+1}}{(1+n)\,p}, \end{aligned} \tag{41}$$

(ii) $n=-1$

$$H_\mathrm{p} = \frac{2\kappa f_0^2}{p}\left(4q\ln|q|+\frac{2\lambda\zeta\bar\zeta}{3}\ln\left(f_0^2\zeta\bar\zeta\right)+\frac{q}{2}\ln^2\left(f_0^2\zeta\bar\zeta\right)\right), \tag{42}$$

(iii) $n>-1$

$$\begin{aligned} H_\mathrm{p} &= \frac{8\kappa q f_0^2}{p}\left(\frac{\lambda}{6}\right)^{-n-1}\left(\ln|q|+\sum_{r=1}^{n}\frac{\binom{n}{r}}{r}\left((p-2)^r-(-1)^r\right)\right) \\ &+\frac{2\kappa f_0^2(\zeta\bar\zeta)^{n+1}}{p\,(n+1)^2}(4(n+1)+q). \end{aligned} \tag{43}$$

Note that the original solution in [5] is not completely correct, as admitted by Ozsváth [6]. Needless to say that we displayed in (41)–(43) the correct expressions for H_p. We characterize the solutions obtained above by means of the selfdual part of the conformal curvature 2-form ${}^{+}C_{\alpha\beta}$, the trace–free Ricci 1-form $\tilde{\mathcal{R}}_\alpha$ and the Ricci scalar $\tilde R$. Here we list only the results for the general ansatz (24)–(26), i.e. for arbitrary $\alpha(\sigma)$, $\beta(\sigma)$, and $H(\sigma,\zeta,\bar\zeta)$. For the sake of brevity we make use of the structure functions p and q as defined in (24) and (25):

$$\begin{aligned} {}^{+}C_{\hat2\hat0} &= -{}^{+}C_{\hat0\hat2}=\frac{1-i}{4}\,\partial_\zeta\left[q^2\partial_\zeta\left(\frac{p}{q}H\right)\right]\vartheta^{\hat0}\wedge\vartheta^{\hat2}, & (44)\\ {}^{+}C_{\hat2\hat1} &= -{}^{+}C_{\hat1\hat2}=\frac{1+i}{4}\,\partial_{\bar\zeta}\left[q^2\partial_{\bar\zeta}\left(\frac{p}{q}H\right)\right]\vartheta^{\hat1}\wedge\vartheta^{\hat2}, & (45)\\ \tilde{\mathcal{R}}_{\hat2} &= \frac{2pq\kappa p}{q}\,f\,\bar f\,\vartheta^{\hat2}=2\kappa p^2\,f\,\bar f\,\vartheta^{\hat2}. & (46) \end{aligned}$$

5. PLANE-FRONTED WAVES IN MAG

We now turn to the triplet subcase of MAG. Thus we are concerned with the triplet of 1–forms in eq. (14). As shown in section 3, the triplet ansatz reduces the electrovacuum MAG field equations (4) and (5) to an effective Einstein–Proca–Maxwell system:

$$a_0\, \eta_{\alpha\beta\gamma} \wedge \tilde{R}^{\beta\gamma} + 2\lambda\eta_\alpha = 2\kappa \left[\Sigma^{(\omega)}_\alpha + \Sigma^{\rm Max}_\alpha\right], \tag{47}$$
$$d^\star d\omega + m^2\, {}^\star\omega = 0, \tag{48}$$
$$dF = 0, \quad d^\star F = 0. \tag{49}$$

From here on we will presuppose that $m^2 = 0$, reducing eq. (48) to $d^\star d\omega = 0$ and the energy–momentum $\Sigma^{(\omega)}_\alpha$ of the triplet field to the first line of eq. (20). As one realizes immediately, the system (47)-(49) now becomes very similar to the one investigated in the Einstein–Maxwell case in (30)-(31). Let us start with the same ansatz for the line element, coframe (22)–(26), and electromagnetic 2-form (34). The only thing missing up to now is a suitable ansatz for the 1–form ω which governs the non–Riemannian parts of the system and enters eqs. (47)–(48):

$$\omega = -\left[\int^\zeta d\zeta'\, g(\sigma,\zeta') + \int^{\bar\zeta} d\bar\zeta'\, \bar g(\sigma,\bar\zeta')\right] \vartheta^{\hat 2}. \tag{50}$$

Here $g(\sigma,\zeta)$ represents an arbitrary complex function of the coordinates. Since the first field equation (47) in the MAG case differs from the Einsteinian one only by the emergence of $\Sigma^{(\omega)}_\alpha$. Accordingly, we expect only a linear change in the PDE (35). Thus, in case of switching on the electromagnetic and the triplet field, the solution for $\mathcal{H}(\sigma,\zeta,\bar\zeta)$ entering the coframe, is determined by

$$\mathcal{H}_{,\zeta\bar\zeta} + \frac{\lambda}{3p^2}\mathcal{H} = \frac{2\kappa p}{q}\left(f\bar f + g\bar g\right). \tag{51}$$

We use calligraphic letters for quantities that belong to the MAG solution. Consequently, the homogeneous solution $\mathcal{H}_{\rm h}$, corresponding to $f = g = 0$, is given again by (33). In order to solve the inhomogeneous equation (51), we modify the ansatz for $H_{\rm p}$ made in (36) and (37). For clarity, we distinguish between the Einstein-Maxwell and the MAG case by changing the name of $\mu(\sigma,\zeta,\bar\zeta)$ in (36) into $M(\sigma,\zeta,\bar\zeta)$ which leads to the following form of $\mathcal{H}_{\rm p}$:

$$\mathcal{H}_{\rm p}(\sigma,\zeta,\bar\zeta) = M_{,\zeta} - \frac{\lambda}{3}\frac{\bar\zeta}{p}M + \bar M_{,\bar\zeta} - \frac{\lambda}{3}\frac{\zeta}{p}\bar M, \tag{52}$$

$$\text{where} \quad M = \kappa \int^{\bar\zeta} d\bar\zeta p^2 \int^\zeta d\zeta' \frac{1}{p^2} \int^{\zeta'} d\zeta'' \frac{p}{q}\left(f\bar f + g\bar g\right). \tag{53}$$

Thus, the general solution of (51) is given by

$$\mathcal{H}(\sigma,\zeta,\bar{\zeta}) = \mathcal{H}_{\rm h}(\sigma,\zeta,\bar{\zeta}) + \mathcal{H}_{\rm p}(\sigma,\zeta,\bar{\zeta}). \tag{54}$$

Substitution of this ansatz into the field equations yields the following constraint for the coupling constants of the constrained MAG Lagrangian:

$$a_0 = 1, \quad z_4 = \frac{\rho}{2k_0}. \tag{55}$$

Here, we made use of the definition of k_0 mentioned in (19). We will now look for a particular solution $\mathcal{H}_{\rm p}$ of (51). As in the Riemannian case, we choose $\alpha = 1$, $\beta = 0$ and make a polynomial ansatz for the functions f and g which govern the Maxwell and triplet regime of the system,

$$f = f_0\zeta^n \quad n = 0, \pm 1, \pm 2, \dots, \qquad g = g_0\zeta^l \quad l = 0, \pm 1, \pm 2, \dots, \tag{56}$$

with $[g_0]$ =length^{-2-l}. Now we have to perform the integration in (53) which yields the solution for $\mathcal{H}_{\rm p}$ via eq. (52). At this point we remember that the solution for $\mathcal{H}_{\rm p}$, in case of excitations corresponding to f, is already known from eqs. (41)–(43). Let us introduce a new name for $H_{\rm p}$ as displayed in (41)–(43), namely $\mathcal{H}^f_{\rm p}$. Furthermore, we introduce the quantity $\mathcal{H}^g_{\rm p}$, defined in the same way as $\mathcal{H}^f_{\rm p}$ but with the ansatz for g from (56). Thus, one has to perform the substitutions $n \to l$ and $f_0 \to g_0$ in eqs. (41)-(43) in order to obtain $\mathcal{H}^g_{\rm p}$. Due to the linearity of our ansatz in (53), we can infer that the particular solution $\mathcal{H}_{\rm p}$ in the MAG case is given by the sum of the appropriate branches for $\mathcal{H}^f_{\rm p}$ and $\mathcal{H}^g_{\rm p}$, i.e. the general solution form eq. (54) now reads

$$\mathcal{H}(\sigma,\zeta,\bar{\zeta}) = \mathcal{H}_{\rm h}(\sigma,\zeta,\bar{\zeta}) + \mathcal{H}^f_{\rm p}(\sigma,\zeta,\bar{\zeta}) + \mathcal{H}^g_{\rm p}(\sigma,\zeta,\bar{\zeta}). \tag{57}$$

Note that we have to impose the same additional constraints among the coupling constants as in the general case (cf. eq. (55)). In contrast to the general relativistic case, there are two new geometric quantities entering our description, namely the torsion T_α and the nonmetricity $Q_{\alpha\beta}$ given by

$$\begin{aligned} Q_{\alpha\beta} &= -\frac{4k_1}{9}\vartheta_{(\alpha}e_{\beta)}\rfloor\left[\int^\zeta d\zeta' g(\sigma,\zeta') + \int^{\bar{\zeta}} d\bar{\zeta}' \bar{g}(\sigma,\bar{\zeta}')\right]\vartheta^{\hat{2}} \\ &\quad + g_{\alpha\beta}\left(\frac{k_1}{9} - k_0\right)\left[\int^\zeta d\zeta' g(\sigma,\zeta') + \int^{\bar{\zeta}} d\bar{\zeta}' \bar{g}(\sigma,\bar{\zeta}')\right]\vartheta^{\hat{2}}, \end{aligned} \tag{58}$$

$$T^\alpha = -\frac{k_2}{3}\left[\int^\zeta d\zeta' g(\sigma,\zeta') + \int^{\bar{\zeta}} d\bar{\zeta}' \bar{g}(\sigma,\bar{\zeta}')\right]\vartheta^\alpha \wedge \vartheta^{\hat{2}}. \tag{59}$$

6. SUMMARY

We investigated plane–fronted electrovacuum waves in MAG with cosmological constant in the triplet ansatz sector of the theory. The spacetime under consideration carries curvature, nonmetricity, torsion, and an electromagnetic field. Apart from the cosmological constant, the solution contains several arbitrary functions, namely $\alpha(\sigma)$, $\beta(\sigma)$, $f(\sigma,\zeta)$, $g(\sigma,\zeta)$, and $\phi(\sigma,\zeta)$. One may address these functions by the generic term *wave parameters* since they control the different sectors of the solution like the electromagnetic or the non–Riemannian regime. In this way, we generalized the class of solutions obtained by Ozsváth, Robinson, and Rózga [5].

Acknowledgments

The author is grateful to Prof. F.W. Hehl, C. Heinicke, and G. Rubilar for their help. The support of our German–Mexican collaboration by CONACYT–DFG (E130–655—444 MEX 100) is gratefully acknowledged.

Notes

1. We changed the name of the vacuum solution from $H(\sigma,\zeta,\bar{\zeta})$, mentioned in (33), into $H_{\rm h}(\sigma,\zeta,\bar{\zeta})$.

References

[1] A. García, A. Macías, D. Puetzfeld, and J. Socorro, *Phys. Rev.* **D62** (2000) 044021.

[2] F.W. Hehl, J.D. McCrea, E.W. Mielke, and Y. Ne'eman, *Phys. Rep.* **258** (1995) 1.

[3] F.W. Hehl, A. Macías, *Int. J. Mod. Phys.* **D8** (1999) 399.

[4] Y.N. Obukhov, E.J. Vlachynsky, W. Esser, F.W. Hehl, *Phys. Rev.* **D56** (1997) 7769.

[5] I. Ozsváth, I. Robinson, K. Rózga, *J. Math. Phys.* **26** (1985) 1755.

[6] I. Ozsváth. Private communication of February 2000.

III

COSMOLOGY AND INFLATION

NEW SOLUTIONS OF BIANCHI MODELS IN SCALAR-TENSOR THEORIES

Jorge L. Cervantes–Cota*, M. A. Rodriguez–Meza†, Marcos Nahmad
Departamento de Física,
Instituto Nacional de Investigaciones Nucleares (ININ)
P.O. Box 18–1027, México D.F. 11801, México.

Abstract We use scaled variables to display a set coupled equations valid for isotropic and anisotropic Bianchi type I, V, IX models in scalar-tensor theories, including Dehnen's induced gravity theory. In particular, for the Brans–Dicke theory the equations decouple and one is able to integrate the system for FRW and Bianchi I and V models. Additionally, we analyze the possibility that anisotropic models asymptotically isotropize, and/or possess inflationary properties.

Keywords: Scalar–tensor theories, exact solutions, cosmology.

1. INTRODUCTION

The CMBR measured by the COBE satellite [1], as well as current cosmological x–ray surveys, imply that the universe is nowadays at big scales homogeneous and isotropic, and must also has been having these properties since, at least, the era of nucleosynthesis [2]. In order to explain the isotropy of the universe from theoretical anisotropic models, many authors have considered the Bianchi models that can in principle evolve to a Friedmann–Robertson–Walker (FRW) cosmology. It has been shown that some Bianchi models in General Relativity (GR) tend to their isotropic solutions, up some extent [3, 4], and even that they can explain the level of anisotropy measured by COBE [5]. Motived by these facts, we have been working in Brans-Dicke (BD) theory [6] to investigate if Bianchi universes are able to isotropize as the universe evolves, and if its evolution can be inflationary. In previous investigations we have shown that anisotropic, Bianchi type I, V, and IX models

*E–mail:jorge@nuclear.inin.mx
†On leave from Instituto de Física, Universidad Autónoma de Puebla.

tend to isotropize as models evolve [7, 8]. However, this may happen for some restrictive values of ω in the cases of Bianchi type I and IX, and only Bianchi type V model can accomplish an isotropization mechanism within BD current constraints [9] on ω. It has been also shown that the isotropization mechanism in the Bianchi type V model can be inflationary, without the presence of any cosmological constant, when small values for the coupling constant ω are considered [9], as in the case of some induced gravity (IG) models [10, 11]. We have recently shown, however, that the isotropization mechanism can be attained with sufficient amount of e–folds of inflation, only for negative values of ω [12]. In the present report we present the set of differential equations that have be solve to find solutions of scalar-tensor theories, and particular results are presented for BD theory.

This paper is organized as follows. In section 2 FRW and anisotropic equations in scalar–tensor theories are presented in convenient variables. In section 3 we specialize our formalism to the BD theory and review the main results on these cosmological models. Finally, conclusions are in section 4.

2. EQUATIONS FOR FRW AND BIANCHI MODELS

In previous investigations [7, 9, 12, 13] we have used scaled variables, in terms of which our solutions have been given, therefore following we use them: the scaled field $\psi \equiv \phi a^{3(1-\nu)}$, a new cosmic time parameter $d\eta \equiv a^{-3\nu}dt$, $()' \equiv \frac{d}{d\eta}$, the 'volume' $a^3 \equiv a_1 a_2 a_3$, and the Hubble parameters $H_i \equiv a_i'/a_i$ corresponding to the scale factors $a_i = a_i(\eta)$ for $i = 1, 2, 3$. We assume comoving coordinates and a perfect fluid with barotropic equation of state, $p = \nu\rho$, where ν is a constant. Using these definitions one obtains the cosmological equations for Bianchi type I, V and IX models (in units with $G = c = 1$ and $i = 1, 2, 3$):

$$(\psi H_i)' - \psi a^{6\nu} R_{ij} = \frac{8\pi\, a^{3(1+\nu)}}{3+2\omega} \left[[1 + (1-\nu)\omega]\, \rho + (3+2\omega)V + \delta V \right] \tag{1}$$

$$H_1H_2 + H_1H_3 + H_2H_3 + [1 + (1-\nu)\omega]\, (H_1 + H_2 + H_3)\, \frac{\psi'}{\psi}$$

$$-(1-\nu)[1 + \omega(1-\nu)/2](H_1 + H_2 + H_3)^2 - \frac{\omega}{2}\left(\frac{\psi'}{\psi}\right)^2 - \frac{R_j}{2} a^{6\nu}$$

$$= 8\pi \frac{(\rho+V)a^{3(1+\nu)}}{\psi} , \tag{2}$$

$$\begin{aligned}\psi'' + (\nu-1)a^{6\nu}R_j\psi &= \frac{8\pi\, a^{3(1+\nu)}}{3+2\omega}\Big[\left[2(2-3\nu)+3(1-\nu)^2\omega\right]\rho \\ &+ 3(1-\nu)(3+2\omega)V + (1-3\nu)\delta V\Big]\end{aligned} \tag{3}$$

where $\delta V = (\frac{\partial V}{\partial\psi})(\frac{\partial\psi}{\partial\phi})$ and a column sum is given by $R_j \equiv \Sigma_i R_{ij}$, where j =I, V or IX and

$$R_{ij} = \begin{array}{ccc} I & V & IX \\ 0 & 2/a_1^2 & [a_1^4 - a_2^4 - a_3^4 + 2a_2^2a_3^2]/(-2a^6) \\ 0 & 2/a_1^2 & [a_2^4 - a_3^4 - a_1^4 + 2a_1^2a_3^2]/(-2a^6) \\ 0 & 2/a_1^2 & [a_3^4 - a_1^4 - a_2^4 + 2a_1^2a_2^2]/(-2a^6) \end{array} \tag{4}$$

For the Bianchi type V model one has the additional constraint: $H_2 + H_3 = 2H_1$, implying that a_2 and a_3 are inverse proportional functions, $a_2a_3 = a_1^2$; note that the mean Hubble parameter, $H \equiv \frac{1}{3}(H_1+H_2+H_3)$, is for this Bianchi type model $H = H_1$. FRW models are obtained if $a_1 = a_2 = a_3$, whereby the flat model corresponds to the Bianchi I curvature term, the open to Bianchi V, and closed model to Bianchi IX.

Additionally, the continuity equation yields: $\rho a^{3(1+\nu)} = \text{const.} \equiv M_\nu$, M_ν being a dimensional constant depending on the fluid present. The vacuum case is attained when $M_\nu = 0$.

The system of ordinary differential equations, Eqs. (1-12), can be recast in the following set:

$$\begin{aligned}&\psi\psi'' + (1-\nu)\omega\psi'^2 + [2+(1-\nu)(1+3\nu)\omega]m_\nu\psi + (1-\nu)(h_1^2+h_2^2+h_3^2) \\ &+\frac{2(1-\nu)}{3}\left[2-3\nu+\frac{3}{2}(1-\nu)^2\omega\right][(H_1+H_2+H_3)\psi]^2 \\ &-2(1-\nu)[1+(1-\nu)\omega](H_1+H_2+H_3)\psi\psi' \\ &= 8\pi(1-\nu)\psi a^{3(1+\nu)}\left[V + \frac{1-3\nu}{1-\nu}\frac{1}{3+2\omega}\delta V\right]\end{aligned} \tag{5}$$

and

$$\psi'' + (\nu-1)[(H_1+H_2+H_3)\psi]' - (1-3\nu)m_\nu + \frac{16\pi}{3+2\omega}\delta V a^{3(1+\nu)} = 0 , \tag{6}$$

where $m_\nu \equiv \frac{8\pi M_\nu}{3+2\omega}$, and we have written the Hubble rates as follows (similar to the Bianchi type I model deduced in Ref. [14]):

$$H_i = \frac{1}{3}(H_1+H_2+H_3) + \frac{h_i}{\psi} , \tag{7}$$

where the h_i's are some unknown functions of η that determine the anisotropic character of the solutions. Furthermore, Bianchi models obey the condition

$$h_1 + h_2 + h_3 = 0 \tag{8}$$

to demand consistency with Eq. (7). For the Bianchi type V model one has additionally that $h_1 = 0$, since $H_1 = H$ as mentioned above. For FRW models one has that $h_1 = h_2 = h_3 = 0$.

In order to analyze the anisotropic character of the solutions, we consider the anisotropic shear, $\sigma \equiv -(H_1 - H_2)^2 - (H_2 - H_3)^2 - (H_3 - H_1)^2$. $\sigma = 0$ is a necessary condition to obtain a FRW cosmology since it implies $H_1 = H_2 = H_3$, cf. Ref. [7, 15]. If the sum of the squared differences of the Hubble expansion rates tends to zero, it would mean that the anisotropic scale factors tend to a single function of time which is, certainly, the scale factor of a FRW solution.

The dimensionless, anisotropic shear parameter (σ/H^2) [16] becomes, using Eqs. (7) and (8):

$$\frac{\sigma}{H^2} = -\frac{3(h_1^2 + h_2^2 + h_3^2)}{(\psi H)^2} \ . \tag{9}$$

If the above equations admit solutions such that $\sigma/H^2 \to 0$ as $\eta \to \infty$ $(t \to \infty)$, then one has time asymptotic isotropization solutions, similar to solutions found for the Bianchi models in GR [4]. In fact, one does not need to impose an asymptotic, infinity condition, but just that $\eta \gg \eta_*$, where η_* is yet some arbitrary value to warrant that σ/H^2 can be bounded from above.

3. SOLUTIONS AND ASYMPTOTIC BEHAVIOR IN BD THEORY

The problem to find solutions of Bianchi and FRW models in the scalar-tensor theories is, firstly, to solve Eqs. (4) and (6), and, secondly, to solve for each scale factor through Eq. (1). This task, is very complicated to achieve analytically, therefore this work have to be done numerically. However, we observe from Eq. (6) that if $\delta V = 0$, then this equation canbe once integrated to get that

$$H = \frac{\psi' - (1 - 3\nu) m_\nu \eta - \delta}{3(1 - \nu)\psi} \, , \tag{10}$$

where δ is an integration constant. This equation permit us to know the average Hubble rate in terms of ψ, and substituting it into Eq. (7), gives us each of the Hubble rates. The point that is left is to have solutions for ψ. Let us substitute Eq. (10) into Eq. (4), then we get that

$$\begin{aligned}
&\psi\,\psi'' - \frac{2}{3(1-\nu)}\psi'^2 - \frac{2(1-3\nu)}{3(1-\nu)}\left[m_\nu(1-3\nu)\eta + \delta\right]\psi' \\
+ \quad & [2+(1-\nu)(1+3\nu)\omega]m_\nu\,\psi + \frac{2}{3(1-\nu)}\left[2-3\nu+\frac{3}{2}(1-\nu)^2\omega\right] \times \\
& [m_\nu(1-3\nu)\,\eta+\delta]^2 + (1-\nu)(h_1^2+h_2^2+h_3^2) \\
= \quad & 8\pi(1-\nu)\,a^{3(1+\nu)}\,\psi\,V \qquad (11)
\end{aligned}$$

where $V=$ const, since we have assumed $\delta V = 0$. Still Eq. (10) is too complicated, and it is coupled to the other equations through the h_i's and the source term. But, if V is identically zero, that is, for the BD theory, we can make some further progress. Then, from here and on we will assume a vanishing potential. In this case, FRW models, for which $h_i = 0$, imply that Eq. (10) decouples, and the problem is reduced to solve this single, second order differential equation. For Bianchi type I and V models it turns out that the only possible solutions imply that the h_i's are constants (hence, the equations decouple again), whereas for type IX they are unknown functions of η. An explanation of this fact resides in the property that Bianchi type I and V models have curvature terms of FRW type, whereas type IX has a very much complicated form, see Eq. (4). So the things, it seems that the most general solution of Eq. (10) with h_i's constants would give general solutions for Bianchi models I and V, as well as for the FRW models.

Let us present in the following subsections the known and new solutions.

3.1. FRW SOLUTIONS

It was found the following quadratic–polynomial solution [14] $\psi = A\,\eta^2 + B\,\eta + C$, where A, B and C are constants that depend on ω and ν. This is the general solution for the flat FRW model, and a particular solution for the non–flat FRW models [7]. Other particular solutions, possibly of non-polynomial nature, are expected to be found for Eq. (10) that will reveal new aspects of non–flat FRW models.

We have found [13] a new solution of Eq. (10) that is valid for $k \neq 0$ FRW cosmologies. The new solution is:

$$\psi = m_\nu\left(\frac{1-3\nu}{1+3\nu}\right)^2 \left[\kappa_1 + (1-3\nu)\,\eta\right]\left[\kappa_1 + 2\eta + \kappa_2\left[\kappa_1+(1-3\nu)\,\eta\right]^{\frac{2}{1-3\nu}}\right], \qquad (12)$$

where κ_1, κ_2 are arbitrary integration constants. This is the general solution of curved FRW cosmologies when the following relationships are valid: $\delta = 0$ and $\omega = -18\nu/(1+3\nu)^2$. Though the latter relationship constrains the range of possible values of ω and ν, one can find values of physical interest, e.g. $\omega = -1$ that has some interest in string effective theories, see for instance Ref. [17]. Moreover, when $\nu \to -1/3$ one obtains the GR limit $\omega \to \infty$. Finally, one gets the dust model ($\nu_{\rm dust} = 0$) in the limit when $\omega \to 0$, like in Dehnen's IG theory [11].

The Hubble parameter is given by:

$$H = \frac{\kappa_1 + \frac{1-9\nu}{1-3\nu}\eta + \kappa_2\left[\kappa_1 + (1-3\nu)\eta\right]^{\frac{2}{1-3\nu}}}{\left[\kappa_1 + (1-3\nu)\eta\right]\left[\kappa_1 + 2\eta + \kappa_2\left[\kappa_1 + (1-3\nu)\eta\right]^{\frac{2}{1-3\nu}}\right]}, \qquad (13)$$

from which one can find the scale factor:

$$a = \left[\frac{-k}{\kappa_2}\right]^{\frac{1}{2(1-3\nu)}} \frac{\left[\kappa_1 + 2\eta + \kappa_2\left[\kappa_1 + (1-3\nu)\eta\right]^{\frac{2}{1-3\nu}}\right]^{\frac{1}{2(1-3\nu)}}}{\left[\kappa_1 + (1-3\nu)\eta\right]^{\frac{3\nu}{(1-3\nu)^2}}}. \qquad (14)$$

The BD field ($\phi = \psi a^{-3(1-\nu)}$) is obtained through Eqs. (12) and (14):

$$\begin{aligned} \phi &= m_\nu \left(\frac{1-3\nu}{1+3\nu}\right)^2 \left(\frac{-\kappa_2}{k}\right)^{\frac{3(1-\nu)}{2(1-3\nu)}} \times \\ & \frac{\left[\kappa_1 + (1-3\nu)\eta\right]^{\frac{1+3\nu}{(1-3\nu)^2}}}{\left[\kappa_1 + 2\eta + \kappa_2\left[\kappa_1 + (1-3\nu)\eta\right]^{\frac{2}{1-3\nu}}\right]^{\frac{1+3\nu}{2(1-3\nu)}}}. \end{aligned} \qquad (15)$$

Eqs. (14) and (14) imply that the sign of κ_2 is equal to the sign of $-k$ for most values of ν. For open ($k=-1$) models this implies that κ_2 must be positive, and for closed ($k=+1$) models κ_2 must be negative which allows the solutions to (re)collapse: The value of κ_2 determines the time of maximum expansion, so it is very related to the mass (m_ν) of the model. On the other hand, κ_1 represents a η-time shift.

Because of the mathematical form of Eq. (13) it is not possible to have an inflationary era that lasts for a sufficient time period to solve the horizon and flatness problems. Although the solutions differ for small η-times from other known solutions, one recovers for large times standard behaviors, that is, $a \sim t$, $\phi = $ const. for the case that $2/(1-3\nu) > 1$, and $a \sim t^{\alpha_1}$, $\phi \sim t^{\alpha_2}$ for $2/(1-3\nu) < 1$, where α_i are constants that depend on ν, see Ref. [13].

3.2. BIANCHI TYPE I MODEL

For this Bianchi model the known, the general solution is found in which the h_i's are constants, and the solution is $\psi = A_I\eta^2 + B_I\eta +$

C_I, where the constants A_I, B_I, C_I are reported elsewhere [7, 14]. This model can be solved in a general way since the curvature sum column $R_I = 0$, then Eq. (12), with $V = 0$, can be directly integrated. This solution represents a particular solution of master Eq. (10). Direct substitution of ψ into Eq. (7) gives the Hubble rates, and into Eq. (9) shows that solutions isotropize as time evolves, that is, $\sigma/H^2 \to 0$ as $\eta \to \infty$, see Ref. [7]. However, the isotropization mechanism is only possible for solutions such that $\Delta_I \equiv B_I^2 - 4A_I C_I$

$$\begin{aligned}\Delta_I &= \frac{2(2-3\nu)+3(1-\nu)^2\omega}{3(1-\nu)^2(3+2\omega)}\left[\frac{(1-\nu)^2 B_I^2}{2(2-3\nu)+3(1-\nu)^2\omega}\right.\\ &- 2(1-3\nu)\delta\, B + [2(2-3\nu)+3(1-\nu)^2\omega]\delta^2\\ &+ \left. 3(1-\nu)^2(h_1^2+h_2^2+h_3^2)\right] \qquad (16)\end{aligned}$$

is negative [9]. Then, some restrictions on ν and ω apply. For instance, in Dehnen's IG theory [11] $\omega \ll 1$, then the isotropization mechanism is not possible within this Bianchi model [9].

3.3. BIANCHI TYPE V MODEL

This Bianchi model is more complicated because curvature terms are different than zero. Still, it is possible to find particular solutions of Eq. (10), since for this model the h_i's are constants too, and this equation is decoupled as is the case of Bianchi type I model. The general solution of this model should be obtained through the general solution of Eq. (10), yet unknown. We have found a particular solution that is again of the form $\psi = A_V\,\eta^2 + B_V\,\eta + C_V$, where the constants A_V, B_V, C_V are reported in Refs. [7, 12]. This solution is such that

$$\Delta_V = \frac{-8(1-3\nu)^2}{18\nu+(1+3\nu)^2\omega}h_2^2 \qquad (17)$$

is negative for $\omega > -18\nu/(1+3\nu)^2$, so the solution tends to isotropic solution within BD theory constraints [18], $\omega \geq 500$, that is $\sigma/H^2 \to 0$ as $\eta \to \infty$. For this Bianchi type model, Dehnen's IG theory [11] can achieve an isotropization mechanism [9]. An inflationary behavior may be observed in type V models, but to get enough e-foldings of inflation ($N \sim 68$) one must demand that $\omega \leq -\frac{3}{2}$ [12] in consistency with previous results [19].

3.4. BIANCHI TYPE IX MODEL

Bianchi type IX model is the most complicated to solve, since curvature terms involve quartic polynomials of the scale factors, see Eq.

(4). For this Bianchi model it implies, by imposing the condition that h_i's are constants, severe algebraic constraints on the scale factors, so it seems more likely that h_i's are functions. This explains why no totally anisotropic ($H_1 \neq H_2 \neq H_3$) solution has been found yet. In Ref. [7] we have analyzed the case when the polynomial solution for ψ is valid. In this case, unfortunately, we could not found explicitly the values of the constants A_{IX}, B_{IX}, C_{IX}. If this solution is valid, however, one has that $h_1^2 + h_2^2 + h_3^2 = D\eta^2 + F\eta + G$, where D, F, G are constants. Accordingly, Eq. (9) indicates that the solution must tend, as time evolves, to the positive curvature FRW solution, i.e. one has again that $\sigma/H^2 \to 0$ as $\eta \to \infty$. However, a definitive answer will arrive by obtaining explicitly the values of A_{IX}, B_{IX}, C_{IX}.

4. CONCLUSIONS

We have presented a set of differential equations written in rescaled variables that are valid for Bianchi type I, V and IX models, as well as for FRW models within general scalar–tensor theories, including Dehnen's IG theory. It seems to be very difficult to solve analytically this system given by Eqs. (1), (4) and (6). However, for the special case in which the potential vanishes, or when matter terms dominate over the potential [9], that is for the BD case, one is able to find solutions. Accordingly, one can integrate Eq. (10) for the cases of FRW models, and Bianchi type I and V models, because in these cases the anisotropic parameters (h_i's) are constants.

The solutions discussed here are:

(i) A particular solution for non-flat FRW models. We have found a new solution of Eq. (10) valid for curved ($k \neq 0$) FRW cosmologies, that is, a solution with $h_i = 0$. This is the general solution subject to some constraint that permits one to have two important physical cases: the case when $\omega = -1$, having some interest in string cosmology [17], that implies an equation of state of a quasi dust model ($\nu \gtrsim 0$), and the case when $\omega = 500$, consistent with current BD local experimental constraints [18], implying that $\nu \sim -1/3$. The new solution is non-inflationary and for asymptotic times is of power-law type.

(ii) The general solution of Bianchi type I model and a particular solution of type V; both solutions are quadratic polynomials. These solutions let the models isotropize as time evolves, however, this can happens only for some parameter (ν, ω) range. The polynomial solution may also be valid for Bianchi type IX, but it is not proved yet. If it were, isotropization would be also guaranteed.

Further solutions of Eq. (10) are in order, which can be either within FRW cosmologies or Bianchi type V or IX models. In particular, the general solution of this equation with h_i's constants should provide the general solution of both curved FRW cosmologies and Bianchi type V model.

Acknowledgments

This work has been financially supported by Conacyt, grant number 33278-E.

References

[1] C.L. Bennett *et al.* (1996). *Astrophys. J.* **464** (1996) L1; K.M. Górski *et al.*, *ibid.* L11; G. Hinshaw, *ibid.* L17.

[2] S.W. Hawking and R.J. Taylor, *Nature* (London) **209** (1966) 1278; J.D. Barrow, *Mon. Not. R. Astron. Soc.* **175** (1976) 359.

[3] C.B. Collins and S.W. Hawking, *Astrophys. J.* **180** (1973) 317.

[4] J.D. Barrow and D.H. Sonoda, *Phys. Rep.* **139** (1986) 1.

[5] J.D. Barrow, *Phys. Rev.* **D51** (1995) 3113.

[6] C. Brans and R. Dicke, *Phys. Rev.* **124** (1961) 925.

[7] P. Chauvet and J.L. Cervantes–Cota, *Phys. Rev.* **D52** (1995) 3416.

[8] J.P. Mimoso, and D. Wands, *Phys. Rev.* **D52** (1995) 5612.

[9] J.L. Cervantes–Cota and P.A. Chauvet, *Phys. Rev.* **D59** (1999) 043501.

[10] R. Fakir and W. G. Unruh, *Phys. Rev.* **D41** (1990) 1783; *ibid.* 1792.

[11] H. Dehnen, H. Frommert, and F. Ghaboussi, *Int. J. Theo. Phys.* **31** (1992) 109; H. Dehnen, and H. Frommert, *Int. J. Theo. Phys.* **32** (1993) 135; J.L. Cervantes–Cota and H. Dehnen, *Phys. Rev.* **D51** (1995) 395; *ibid. Nucl. Phys.* **B442** (1995) 391.

[12] J.L. Cervantes–Cota, *Class. Quant. Grav.* **16** (1999) 3903.

[13] J.L. Cervantes-Cota and M. Nahmad, to appear in *Gen. Rel. Grav.* May issue (2001) gr-qc/0005032.

[14] V. A. Ruban and A.M. Finkelstein, *Gen. Rel. Grav.* **6** (1975) 601.

[15] P. Chauvet, J. Cervantes–Cota, and H.N. Núñez-Yépez, in: *Proceedings of the 7th Latin American Symposium on General Relativity and Gravitation, SILARG VII, 1991*, D'Olivo, J. C., et al., eds. (World Scientific, Singapore) pp. 487.

[16] J. Wainwright and G.F.R. Ellis, in: *Dynamical Systems in Cosmology* (Cambridge University Press, Cambridge, 1997).

[17] R. Dick, *Gen. Rel. Grav.* **30** (1998) 435.

[18] C. M. Will, *Theory and experiment in gravitational physics* (Cambridge University Press, Cambridge, 1993) chapter 8.

[19] J.J. Levin, *Phys. Rev.* **D51** (1995) 462; J.J. Levin and K. Freese, *ibid.* **D47** (1993) 4282; J.J. Levin and Freese, *Nucl. Phys.* **B421** (1994) 635.

SCALAR FIELD DARK MATTER

Tonatiuh Matos, F. Siddhartha Guzmán, L. Arturo Ureña–López
Departamento de Física, Centro de Investigación y de Estudios Avanzados del IPN, AP 14–740, 07000 México D.F., México.

Dario Núñez
Instituto de Ciencias Nucleares,
Universidad Nacional Autónoma de México,
A.P. 70–543, 04510 México D.F., México.

Abstract The main goal of this work is to put the last results of the Scalar Field Dark Matter model of the Universe at cosmological and at galactic level in a work together. We present the complete solution to the 95% scalar field cosmological model in which the dark matter is modeled by a scalar field Φ with the scalar potential $V(\Phi) = V_o\left[\cosh\left(\lambda\sqrt{\kappa_o}\Phi\right) - 1\right]$ and the dark energy is modeled by a scalar field Ψ, endowed with the scalar potential $\tilde{V}(\Psi) = \tilde{V}_o\left[\sinh\left(\alpha\sqrt{\kappa_o}\Psi\right)\right]^{\beta}$. This model has only two free parameters, λ and the equation of state ω_{Ψ}. The results of the model are: 1) the fine tuning and the cosmic coincidence problems are ameliorated for both dark matter and dark energy and the models agrees with astronomical observations. 2) The model predicts a suppression of the Mass Power Spectrum for small scales having a wave number $k > k_{min,\Phi}$, where $k_{min,\Phi} \simeq 4.5\,h\,\mathrm{Mpc}^{-1}$ for $\lambda \simeq 20.3$. This last fact could help to explain the dearth of dwarf galaxies and the smoothness of galaxy core halos. 3) From this, all parameters of the scalar dark matter potential are completely determined. 4) The dark matter consists of an ultra-light particle, whose mass is $m_{\Phi} \simeq 1.1 \times 10^{-23}\,\mathrm{eV}$ and all the success of the standard cold dark matter model is recovered. 5) If the scale of renormalization of the model is of order of the Planck Mass, then the scalar field Φ can be a reliable model for dark matter in galaxies. 6) The predicted scattering cross section fits the value required for self-interacting dark matter. 7) Studying a spherically symmetric fluctuation of the scalar field Φ in cosmos we show that it could be the halo dark matter in galaxies. 8) The local space–time of the fluctuation of the scalar field Φ contains a three dimensional space–like hypersurface with surplus of angle. 9) We also present a model for the dark matter in the halos of spiral galaxies, we obtain that 10) the effective energy density goes like $1/(r^2 + K^2)$ and 11) the resulting circular velocity profile

Exact Solutions and Scalar Fields in Gravity: Recent Developments
Edited by Macias *et al.*, Kluwer Academic/Plenum Publishers, New York, 2001

of tests particles is in good agreement with the observed one in spiral galaxies. This implies that a scalar field could also be a good candidate as the dark matter of the Universe.

Keywords: Scalar fields, dark matter, cosmology.

1. INTRODUCTION

There is no doubt that we are living exciting times in Cosmology. We are now convinced that the last results of the observations of the Universe using new technologies and instruments will conduce to new physics and a qualitative new knowledge of Nature. One question which seems not to have an answer using the well-known physics is the question of the nature of the matter of the Universe. It is amazing that after so much effort dedicated to such question, what is the Universe composed of?, it has not been possible to give a conclusive answer. From the latest observations, we do know that about 95% of matter in the Universe is of non baryonic nature. The old belief that matter in Cosmos is made of quarks, leptons and gauge bosons is being abandoned due to the recent observations and the inconsistences which spring out of this assumption [1]. Now we are convinced on the existence of an exotic non baryonic sort of matter which dominates the structure of the Universe, but its nature is until now a puzzle.

In this work we pretend to summarize the results of one proposal for the nature of the matter in the Universe, namely, the Scalar Field Dark Matter model (SFDM). This model has had relative success at cosmological level as well as at galactic level. In this work we pretend to put both models together and explain which is the possible connection between them. Most of the material contained in this work has been separately published elsewhere. The main goal of this work is to write all together and to explain which is the possible connection between galactic and cosmological levels. First, let us give a breve introduction of the Dark Matter problem.

The existence of dark matter in the Universe has been firmly established by astronomical observations at very different length–scales, ranging from single galaxies, to clusters of galaxies, up to cosmological scale (see for example [2]). A large fraction of the mass needed to produce the observed dynamical effects in all these very different systems is not seen. At the galactic scale, the problem is clearly posed: The measurements of rotation curves (tangential velocities of objects) in spiral galaxies show that the coplanar orbital motion of gas in the outer parts of these galaxies keeps a more or less constant velocity up to several luminous radii [3], forming a radii independent curve in the outer parts of the rotational curves profile; a motion which does not correspond to the one due to the observed matter distribution, hence there must be present some type of dark matter causing the observed motion. The flat profile of the rotational curves is maybe the main feature observed in many galaxies. It is believed that the dark matter in galaxies has an almost spherical distribution which decays like $1/r^2$. With this distribution of some kind of matter it is possible to fit the rotational curves of galaxies quite well [4]. Nevertheless, the main question of the dark matter problem remains; which is the nature of the dark matter in galaxies? The problem is not easy to solve, it is not sufficient to find out an exotic particle which could exist in galaxies in the low energy regime of some theory. It is necessary to show as well, that this particle (baryonic or exotic) distributes in a very similar manner in all these galaxies, and finally, to give some reason for its existence in galaxies.

Recent observations of the luminosity–redshift relation of Ia Supernovae suggest that distant galaxies are moving slower than predicted by Hubble's law, implying an accelerated expansion of the Universe [5]. These observations open the possibility to the existence of an energy component in the Universe with a negative equation of state, $\omega < 0$, being $p = \omega\rho$, called dark energy. This result is very important because it separates the components of the Universe into gravitational (also called the "matter component" of the Universe) and antigravitational (repulsive, the dark energy). The dark energy would be the currently dominant component in the Universe and its ratio relative to the whole energy would be $\Omega_{DE} \sim 70\%$. The most simple model for this dark energy is cosmological constant (Λ), in which $\omega = -1$.

Observations in galaxy clusters and dynamical measurements of the mass in galaxies indicate that the matter component of the Universe is $\Omega_M \sim 30\%$. But this component of the Universe decomposes itself in baryons, neutrinos, etc. and cold dark matter which is responsible of the formation of structure in the Universe. Observations indicate that stars and dust (baryons) represent something like 0.3% of the whole matter of the Universe. The new measurements of the neutrino mass indicate that neutrinos contribute as much with a same quantity as dust. In other words, say $\Omega_M = \Omega_m + \Omega_{DM} = \Omega_b + \Omega_\nu + \cdots + \Omega_{DM} \sim 0.05 + \Omega_{DM}$, where Ω_{CDM} represents the dark matter part of the matter contributions which has a value of $\Omega_{CDM} \sim 0.25$. The value of the amount of baryonic matter ($\sim 5\%$) is in concordance with the limits imposed by nucleosynthesis (see for example [1]). Everything seems to agree. Then, this model considers a flat Universe ($\Omega_\Lambda + \Omega_M \approx 1$) full with 95% of unknown matter but which is of great importance at cosmological level. Moreover, it seems to be the most succesful model fitting current cosmological observations [6].

The SFDM model [7, 8, 9, 10] consist in to suppose that the dark matter and the dark energy are of scalar field nature. A particular model we have considered in the last years is the following. For the dark energy we adopt a quintessence field Ψ with a sinh like scalar potential. In [11], it was showed that the potential

$$\begin{aligned} \tilde{V}(\Psi) &= \tilde{V}_o\left[\sinh\left(\alpha\sqrt{\kappa_o}\Psi\right)\right]^{\beta} \qquad (1) \\ &= \begin{cases} \tilde{V}_o\left(\alpha\sqrt{\kappa_o}\Psi\right)^{\beta} & |\alpha\sqrt{\kappa_o}\Psi| \ll 1 \\ \left(\tilde{V}_o/2^{\beta}\right)\exp\left(\alpha\beta\sqrt{\kappa_o}\Psi\right) & |\alpha\sqrt{\kappa_o}\Psi| \gg 1 \end{cases}; \end{aligned}$$

is a reliable model for the dark energy, because of its asymptotic behaviors. The main point of the SFDM model is to suppose that the dark matter is a scalar field Φ endowed with the potential [7, 10]

$$\begin{aligned} V(\Phi) &= V_o\left[\cosh\left(\lambda\sqrt{\kappa_o}\Phi\right) - 1\right] \qquad (2) \\ &= \begin{cases} \frac{1}{2}m_\Phi^2\Phi^2 + \frac{1}{24}\lambda^2 m_\Phi^2\kappa_0\Phi^4 & |\lambda\sqrt{\kappa_o}\Phi| \ll 1 \\ (V_o/2)\exp\left(\lambda\sqrt{\kappa_o}\Phi\right) & |\lambda\sqrt{\kappa_o}\Phi| \gg 1 \end{cases}. \end{aligned}$$

The mass of the scalar field Φ is defined as $m_\Phi^2 = V''|_{\Phi=0} = \lambda^2\kappa_o V_o$. In this case, we will deal with a massive scalar field.

2. COSMOLOGICAL SCALAR FIELD SOLUTIONS

In this section, we give all the solutions to the model at the cosmological scale and focus our attention in the scalar dark matter. Since current observations of CMBR anisotropy by BOOMERANG and MAXIMA [12] suggest a flat Universe, we use as ansatz the flat Friedmann-Robertson-Walker (FRW) metric

$$ds^2 = -dt^2 + a^2(t)\left[dr^2 + r^2 d\Omega^2\right], \qquad (3)$$

where a is the scale factor ($a = 1$ today) and we have set $c = 1$. The components of the Universe are baryons, radiation, three species of light neutrinos, etc., and two minimally coupled and homogeneous scalar fields Φ and Ψ, which represent the dark matter and the dark energy, respectively. The evolution equations for this Universe reads

$$H^2 \equiv \left(\frac{\dot{a}}{a}\right)^2 = \frac{\kappa_o}{3}\left(\rho + \rho_\Phi + \rho_\Psi\right) \qquad (4)$$

$$\ddot{\Phi} + 3H\dot{\Phi} + \frac{dV(\Phi)}{d\Phi} = 0 \quad (5)$$

$$\ddot{\Psi} + 3H\dot{\Psi} + \frac{d\tilde{V}(\Psi)}{d\Psi} = 0 \quad (6)$$

$$\dot{\rho} + 3H(\rho + p) = 0, \quad (7)$$

being $\kappa_o \equiv 8\pi G$ and ρ (p) is the energy density (pressure) of radiation, plus baryons, plus neutrinos, etc. The scalar energy densities (pressures) are $\rho_\Phi = \frac{1}{2}\dot{\Phi}^2 + V(\Phi)$ ($p_\Phi = \frac{1}{2}\dot{\Phi}^2 - V(\Phi)$) and $\rho_\Psi = \frac{1}{2}\dot{\Psi}^2 + \tilde{V}(\Psi)$ ($p_\Psi = \frac{1}{2}\dot{\Psi}^2 - \tilde{V}(\Psi)$). Here overdots denote derivative with respect to the cosmological time t.

2.1. RADIATION DOMINATED ERA (RD)

We start the evolution of the Universe at the end of inflation, *i.e.* in the radiation dominated (RD) era. The initial conditions are set such that $(\rho_{i\Phi}, \rho_{i\Psi}) \leq \rho_{i\gamma}$. Let us begin with the dark energy. For the potential (1) an exact solution in the presence of nonrelativistic matter can be found [11, 13] and the parameters of the potential are given by

$$\alpha = \frac{-3\omega_\Psi}{2\sqrt{3(1+\omega_\Psi)}},$$
$$\beta = \frac{2(1+\omega_\Psi)}{\omega_\Psi}, \quad (8)$$
$$\kappa_0 \tilde{V}_0 = \frac{3(1-\omega_\Psi)}{2}\left(\frac{\Omega_{oM}}{\Omega_{o\Psi}}\right)^{\frac{1+\omega_\Psi}{\omega_\Psi}} \Omega_{o\Psi} H_0^2$$

with $\Omega_{o\Psi}$ and Ω_{oCDM} the current values of dark energy and dark matter, respectively; and $-0.6 \geq \omega_\Psi \geq -0.9$ the range for the current equation of state. With these values for the parameters (α, $\beta < 0$) the solution for the dark energy (Ψ) becomes a tracker one, is only reached until a matter dominated epoch and the scalar field Ψ would begin to dominate the expansion of the Universe after matter domination. Before this, at the radiation dominated epoch, the scalar energy density ρ_Ψ is frozen, strongly subdominant and of the same order than today [11]. Then the dark energy contribution can be neglected during this epoch.

Now we study the behavior of the dark matter. For the potential (2) we begin the evolution with large and negative values of Φ, when the potential behaves as an exponential one. It is found that the exponential potential makes the scalar field Φ mimic the dominant energy density, that is, $\rho_\Phi = \rho_{i\Phi} a^{-4}$. The ratio of ρ_Φ to the total energy density is [13, 14]

$$\frac{\rho_\Phi}{\rho_\gamma + \rho_\Phi} = \frac{4}{\lambda^2}. \tag{9}$$

This solution is self-adjusting and it helps to avoid the fine tuning problem of matter, too. Here appears one restriction due to nucleosynthesis [14] acting on the parameter $\lambda > \sqrt{24}$. Once the potential (2) reaches its polynomial behavior, Φ oscillates so fast around the minimum of the potential that the Universe is only able to feel the average values of the energy density and pressure in a scalar oscillation. Both Φ and $< \omega_\Phi >$ go down to zero and $< \rho_\Phi >$ scales as non-relativistic matter [15]. If we would like the scalar field Φ to act as cold dark matter, in order to recover all the successful features of the standard model, we need first derive a relation between the parameters (V_0, λ). The required relation reads [10]

$$\kappa_0 V_0 \simeq \frac{1.7}{3}\left(\lambda^2 - 4\right)^3 \left(\frac{\Omega_{0CDM}}{\Omega_{o\gamma}}\right)^3 \Omega_{0CDM} H_o^2. \tag{10}$$

Notice that V_0 depends on both current amounts of dark matter and radiation (including light neutrinos) and that we can choose λ to be the only free parameter of potential (2). Since now, we can be sure that $\rho_\Phi = \rho_{CDM}$ and that we will recover the standard cold dark matter evolution.

2.2. MATTER DOMINATED ERA (MD) AND SCALAR FIELD Ψ DOMINATED ERA (ΨD)

During this time, the scalar field Φ continues oscillating and behaving as nonrelativistic matter and there is a matter dominated era just like that of the standard model. A short after matter completely dominates the evolution of the Universe, the scalar field Ψ reaches its tracker solution [11] and it begins to be an important component. Lately, the scalar field Ψ becomes the dominant component of the Universe and the scalar potential (1) is effectively an exponential one [9, 11]. Thus, the scalar field Ψ drives the Universe into a power-law inflationary stage ($a \sim t^p$, $p > 1$). This solution is distinguishable from a cosmological constant one.

A complete numerical solution for the dimensionless density parameters Ω's are shown in fig. 1 until today. The results agree with the solutions found in this section. It can be seen that eq. (10) makes the scalar field Φ behave quite similar to the standard cold dark matter model once the scalar oscillations begin and the required contributions

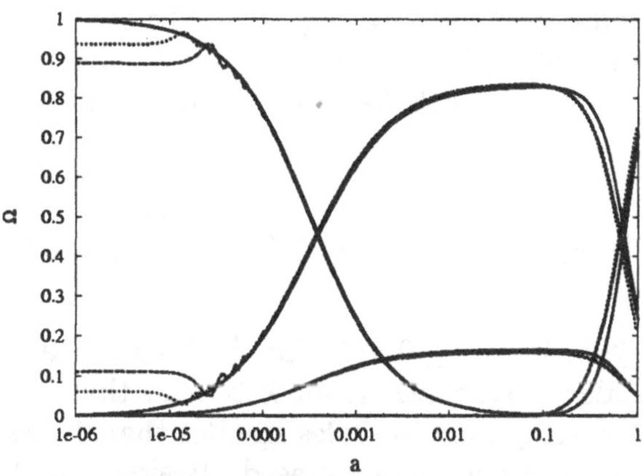

Figure 1. Evolution of the dimensionless density parameters *vs.* the scale factor a with $\Omega_{oM} = 0.30$: ΛCDM (solid-curves) and $\Psi\Phi DM$ for two values of $\lambda = 6$ (dashed–curves), $\lambda = 8$ (dotted-curves). The equation of state for the dark energy is $\omega_{\Psi} = -0.8$.

of dark matter and dark energy are the observed ones [10, 11, 16]. We recovered the standard cosmological evolution and then we can see that potentials (1,2) are reliable models of dark energy and dark matter in the Universe.

2.3. SCALAR POWER SPECTRUM FOR Φ DARK MATTER

In this section, we analyze the perturbations of the space due to the presence of the scalar fields Φ, Ψ. First, we consider a linear perturbation of the space given by h_{ij}. We will work in the synchronous gauge formalism, where the line element is $ds^2 = a^2[-d\tau^2 + (\delta_{ij} + h_{ij})dx^i dx^j]$. We must add the perturbed equations for the scalar fields $\Phi(\tau) \rightarrow \Phi(\tau) + \phi(k,\tau)$ and $\Psi(\tau) \rightarrow \Psi(\tau) + \psi(k,\tau)$ [14, 17]

$$\ddot{\phi} + 2\mathcal{H}\dot{\phi} + k^2\phi + a^2V''\phi + \frac{1}{2}\dot{\Phi}\dot{h} = 0 \quad (11)$$

$$\ddot{\psi} + 2\mathcal{H}\dot{\psi} + k^2\psi + a^2\tilde{V}''\psi + \frac{1}{2}\dot{\Psi}\dot{h} = 0. \quad (12)$$

to the linearly perturbed Einstein equations in k–space ($\vec{k} = k\hat{k}$) (see [18]). Here, overdots are derivatives with respect to the conformal time τ and primes are derivatives with respect of the unperturbed scalar fields Φ and Ψ, respectively.

It is known that scalar perturbations can only grow if the k^2–term in eqs. (11,12) is subdominant with respect to the second derivative of the scalar potential, that is, if $k < a\sqrt{V''}$ [19]. According to the solution given above for potential (2), $k_\Phi = a\sqrt{V''}$ has a minimum value given by [16]

$$k_{min,\Phi} = m_\Phi a^* \simeq 1.3\,\lambda\sqrt{\lambda^2 - 4}\frac{\Omega_{0CDM}}{\sqrt{\Omega_{o\gamma}}}H_o, \tag{13}$$

and then for $\lambda \geq 5$, $k_{min,\Phi} \geq 0.375\,Mpc^{-1}$. Then, it can be assured that there are no scalar perturbations for $k > m_\Phi$, that is, bigger than k_Φ today. These k corresponds to scales smaller than $1.2\,kpc$ (here $\lambda = 5$). They must have been completely erased. Besides, modes which $m_\Phi > k > k_{min,\Phi}$ must have been damped during certain periods of time. From this, we conclude that the scalar power spectrum of Φ will be damped for $k > k_{min,\Phi}$ with respect to the standard case. Therefore, the Jeans length must be [16]

$$L_J(a) = 2\pi\, k^{-1}_{min,\Phi}, \tag{14}$$

and it is a universal constant because it is completely determined by the mass of the scalar field particle. L_J is not only proportional to the quantity $(\kappa_0 V_0)^{-1/2}$, but also the time when scalar oscillations start (represented by a^*) is important.

On the other hand, the wave number for the dark energy $k_\Psi = a\sqrt{\tilde{V}''}$ is always out of the Hubble horizon, then only structure at larger scales than H^{-1} can be formed by the scalar fluctuations ψ. Instead of a minimum, there is a maximum $k_{max,\Psi} \sim 10^{-3}\,Mpc^{-1}$. All scalar perturbations of the dark energy which $k > 10^{-3}\,Mpc^{-1}$ must have been completely erased. Perturbations with $k \leq 10^{-3}\,Mpc^{-1}$ have started to grow only recently. For a more detailed analysis of the dark energy fluctuations, see [19, 20].

In fig. 2, a numerical evolution of $\delta \equiv \delta\rho/\rho$ is shown compared with the standard CDM case [10]. The numerical evolution for the density contrasts was done using an amended version of CMBFAST [21]. Due to its oscillations around the minimum, the scalar field Φ changes to a complete standard CDM and so do its perturbations. All the standard

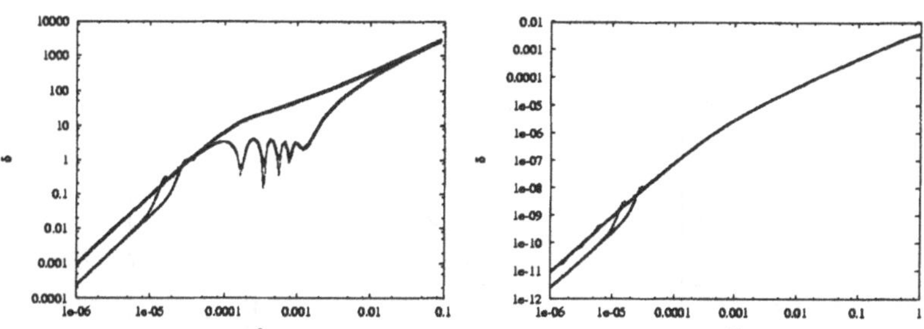

Figure 2. Evolution of the density contrasts for baryons δ_b, standard cold dark matter δ_{CDM} and scalar dark matter δ_Φ *vs.* the scale factor a taking $\Omega_{oM} = 0.30$ for the models given in fig. (1). The modes shown are $k = 1.0 \times 10^{-5}\, Mpc^{-1}$ (left) and $k = 0.1\, Mpc^{-1}$ (right).

growing behavior for modes $k < k_{min,\Phi}$ is recovered and preserved until today by potential (2).

In fig. 3 [16] we can see δ^2 at a redshift $z = 50$ from a complete numerical evolution using the amended version of CMBFAST. We also observe a sharp cut–off in the processed power spectrum at small scales when compared to the standard case, as it was argued above. This suppression could explain the smooth cores of dark halos in galaxies and a less number of dwarf galaxies [22].

The mass power spectrum is related to the CDM case by the semi–analytical relation (see [23])

$$P_\Phi(k) \simeq \left(\frac{\cos x^3}{1 + x^8} \right)^2 P_{CDM}(k), \tag{15}$$

but using $x = (k/k_{min,\Phi})$ with $k_{min,\Phi}$ being the wave number associated to the Jeans length (14). If we take a cut–off of the mass power spectrum at $k = 4.5\, h\, \mathrm{Mpc}^{-1}$ [22], we can fix the value of parameter λ. Using eq. (13), we find that [16]

$$\begin{aligned} \lambda &\simeq 20.3, \\ V_0 &\simeq \left(3.0 \times 10^{-27}\, M_{Pl} \simeq 36.5\, \mathrm{eV}\right)^4, \\ m_\Phi &\simeq 9.1 \times 10^{-52}\, M_{Pl} \simeq 1.1 \times 10^{-23}\, \mathrm{eV}. \end{aligned} \tag{16}$$

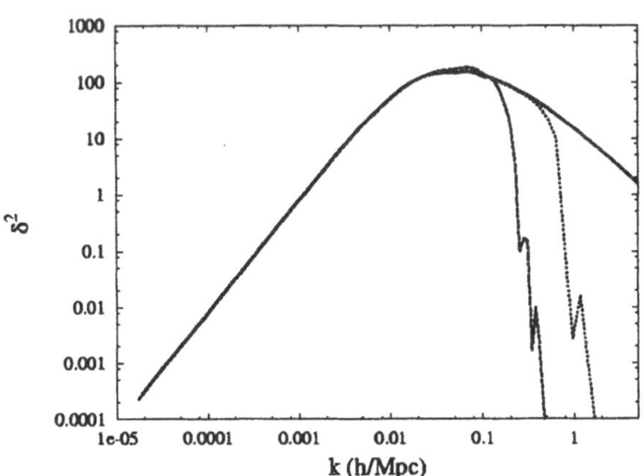

Figure 9. Power spectrum at a redshift $z = 50$: ΛCDM (solid-curve), and ΦCDM with $\lambda = 5$ (dashed-curve) and $\lambda = 10$ (dotted-curve). The normalization is arbitrary.

where $M_{Pl} = 1.22 \times 10^{19}\,\mathrm{GeV}$ is the Plank mass. All parameters of potential (2) are now completely determined and we have the right cut–off in the mass power spectrum.

3. SCALAR DARK MATTER AND PLANCK SCALE PHYSICS

At galactic scale, numerical simulations show some discrepancies between dark–matter predictions and observations [24]. Dark matter simulations show cuspy halos of galaxies with an excess of small scale structure, while observations suggest a constant halo core density [25] and a small number of subgalactic objects [26]. There have been some proposals for resolving the dark matter crisis (see for example [24, 26]). A promising model is that of a self–interacting dark matter [27]. This proposal considers that dark matter particles have an interaction characterized by a scattering cross section by mass of the particles given by [25, 27]

$$\frac{\sigma_{2\to 2}}{m} = 10^{-25} - 10^{-23} cm^2\, GeV^{-1}. \tag{17}$$

This self–interaction provides shallow cores of galaxies and a minimum scale of structure formation, that it must also be noticed as a cut–off in the Mass Power Spectrum [22, 26].

Because of the presence of the scalar field potential (2), there must be an important self–interaction among the scalar particles. That means that this scalar field dark matter model belongs to the so called group of self-interacting dark matter models mentioned above, and it should be characterized by the scattering cross section (17).

Let us calculate $\sigma_{2\to 2}/m_\Phi$ for the scalar potential (2). The scalar potential (2) can be written as a series of even powers in Φ, $V = \Sigma_{n=1}^{\infty} V_0 \lambda^{2n} \kappa_0^n \Phi^{2n}$. Working on 4 dimensions, it is commonly believed that only Φ^4 and lower order theories are renormalizable. But, following the important work [28], if we consider that there is only one intrinsic scale Λ in the theory, we conclude that there is a momentum cut–off and that we can have an effective Φ^4 theory which depends upon all couplings in the theory. In addition, it was recently demonstrated [29] that scalar exponential–like potentials of the form (we have used the notation of potential (2) and for example, parameter μ in eqs. 7–8 in Ref. [29] is $\mu^{-1} = \lambda\sqrt{\kappa_0}$)) [31]

$$U_\Lambda(\Phi) = M^4 \exp\left(-\frac{\lambda^2 \kappa_0 \Lambda^2}{32\pi^2}\right) \exp\left(\pm\lambda\sqrt{\kappa_0}\Phi\right) \tag{18}$$

are non–perturbative solutions of the exact renormalization group equation in the Local Potential Approximation (LPA). Here M and λ are free parameters of the potential and Λ is the scale of renormalization. Being our potential for scalar dark matter a cosh–like potential (non-polynomial), it is then a solution to the renormalization group equations in the (LPA), too. Comparing eqs. (2, 18), we can identify $V_0 = M^4 \exp\left[-\left(\lambda^2\kappa_0\Lambda^2\right)/(32\pi^2)\right]$. We see that an additional free parameter appears, the scale of renormalization Λ. From this, we can assume that potential (2) is renormalizable with only one intrinsic scale: Λ.

Following the procedure shown in [30] for a potential with even powers of the dimensionless scalar field ϕ, $U(\phi) = \Sigma_{n=1}^{\infty} u_{2n}\phi^{2n}$, the cross section for $2 \to 2$ scattering in the center-of-mass frame is [31]

$$\sigma_{2\to 2} = \frac{|A_4|^2}{16\pi E^2} = \frac{M^8 \kappa_0^4 \lambda^8}{16\pi E^2}, \tag{19}$$

where E is the total energy. Thus, we arrive to a real effective ϕ^4 theory with a coupling $g = M^4 \kappa_0^2 \lambda^4$, leading to a different cross section for the self-interacting scalar field than the previously discussed polynomial models. Near the threshold $E \simeq 4m^2$, and taking the values of λ, m_Φ and the expected result (17), we find $M = (6.7 \pm 1.9) \times 10^2$ TeV and $\Lambda \simeq (1.93 \pm 0.01) M_{Pl} \simeq 2.3 \times 10^{19}$ GeV. The scale of renormalization is of order of the Planck Mass. All parameters are completely fixed now [31].

4. SPHERICAL SCALAR FIELD FLUCTUATIONS AS GALACTIC HALOS

In this section we explore whether a scalar field can fluctuate along the history of the Universe and thus forming concentrations of scalar field density. If, for example, the scalar field evolves with a scalar field potential $V(\Phi) \sim \Phi^2$, the evolution of this scalar field will be similar to the evolution of a perfect fluid with equation of state $p = 0$, *i.e.*, it would evolve as cold dark matter [32]. However, it is not clear whether a scalar field fluctuation can serve as dark matter in galaxies. In this section we show that this could be the case. We assume that the halo of a galaxy is a spherical fluctuation of cosmological scalar dark matter and study the consequences for the space-time background at this scale, in order to restrict the state equation corresponding to the dark matter inside the fluctuation.

The region of the galaxy we are interested in is that which goes from the limits of the luminous matter (including the stars far away of the center of the galaxy) over the limits of the halo. In this region measurements indicate that the stars and the hydrogen constituting the halo, present a very curious behavior, their circular velocity is almost independent of the radii at which they are located within the equatorial plane [33].

Observational data show that the galaxies are composed by almost 90% of dark matter. Nevertheless the halo contains a larger amount of dark matter, because otherwise the observed dynamics of particles in the halo is not consistent with the predictions of Newtonian theory, which explains well the dynamics of the luminous sector of the galaxy. So we can suppose that luminous matter does not contribute in a very important way to the total energy density of the halo of the galaxy at least in the mentioned region, instead the scalar matter will be the main contributor to it. As a first approximation we thus can neglect the baryonic matter contribution to the total energy density of the galactic halo. On the other hand, the exact symmetry of the halo is stills unknown, but in order to provide the stability of the disk it is necessary to be spheroidal and we will consider it to be spherical since such symmetry is the first approximation is a scenario of structure formation. Furthermore, the rotation of the galaxy does not affect the motion of test particles around the galaxy, dragging effects in the halo of the galaxy should be too small to affect the tests particles (stars and dust) traveling around the galaxy. Hence, in the region of interest we can suppose the space–time to be static, given that the circular velocity of stars (like the sun) of about 230 Km/s seems not to be affected by the rotation of the galaxy and we can consider a time reversal symmetry of the space-time.

We start from the general spherically symmetric line element and find out the conditions on the metric in order that the test particles in the galaxy possess a flat rotation curve in the region where the scalar field (the dark matter) dominates.

Assuming that the halo has spherical symmetry and that in such regions dragging effects on stars and dust are inappreciable, i.e. the space–time is static, the following line element is the appropriate

$$ds^2 = -B(r)dt^2 + A(r)dr^2 + r^2 d\theta^2 + r^2 \sin^2\theta d\varphi^2 \tag{20}$$

where A and B are arbitrary functions of the coordinate r.

The dynamics of test particles on this space-time can be derived from the Lagrangian

$$2\mathcal{L} = -B(r)\dot{t}^2 + A(r)\dot{r}^2 + r^2\,(\dot{\theta}^2 + \sin^2\theta\,\dot{\varphi}^2). \tag{21}$$

We have two conserved quantities, the energy $E = B(r)\dot{t}$, the φ-momentum $L_\varphi = r^2\sin^2\theta\dot{\varphi}$, and the total angular momentum, $L^2 = {L_\theta}^2 + (\frac{L_\varphi}{\sin\theta})^2$, with $L_\theta = r^2\dot{\theta}$. The radial motion equation can thus be written as:

$$\dot{r}^2 + V(r) = 0, \tag{22}$$

with the expression for the potential $V(r)$

$$V(r) = -\frac{1}{A(r)}(\frac{E^2}{B(r)} - \frac{L^2}{r^2} - 1). \tag{23}$$

Notice that, due to the spherical symmetry, we do not need to restrict the study to equatorial orbits, this last radial motion is valid for any angle θ. For circular stable orbits, we have the conditions, $\dot{r} = 0, V_{,r} = 0$, and $V_{,rr} > 0$, which imply the following expression for the energy and total momentum of the particles in such orbits:

$$E = \frac{2B(r)^2}{2B(r) - rB(r)_{,r}}, \tag{24}$$

$$L = \frac{r^3 B(r)_{,r}}{2B(r) - rB(r)_{,r}}, \tag{25}$$

and for the second derivative of the potential evaluated at the extrema

$$V(r)_{,rr}|_{extr} = 2\frac{\frac{rB(r)_{,rr}}{B} + \frac{B(r)_{,r}}{B}(3 - \frac{2rB(r)_{,r}}{B})}{rA(r)(2 - \frac{rB(r)_{,r}}{B})}. \tag{26}$$

Combining the expressions above it is found that the circular (tangential) velocity, $(v_c)^2 = \frac{r^2}{B(r)}((\frac{d\theta}{dt})^2 + \sin^2\theta(\frac{d\varphi}{dt})^2)$, for particles in stable circular orbits is given by:

$$(v_c)^2 = \frac{rB(r)_{,r}}{2B(r)}. \tag{27}$$

Thus, imposing the observed condition that this circular velocity is constant for the radius, this last equation can be integrated for the metric coefficient g_{tt}:

$$B(r) = B_0\, r^{2\,(v_c)^2}, \tag{28}$$

with B_0 an integration constant.

In this way, we again arrive to a theorem, stating that for a static spherically symmetric spacetime, v_c, the circular velocity of particles moving in circular stable orbits is radii independent if and only if the g_{tt} metric coefficient has the form $g_{tt} = B_0\, r^{2\,(v_c)^2}$.

Notice that in this case, one of the metric coefficients was completely integrated and the other one, $A(r)$ remains arbitrary. Also, as mentioned, the analysis made no suppositions on the plane of motion, so the result is valid for any circular stable trajectory. This result is not surprising. Remember that the Newtonian potential ψ is defined as $g_{00} = -exp(2\psi) = -1 - 2\psi - \cdots$. On the other side, the observed rotational curve profile in the dark matter dominated region is such that the rotational velocity v^φ of the stars is constant, the force is then given by $F = -(v^\varphi)^2/r$, which respective Newtonian potential is $\psi = (v^\varphi)^2 \ln(r)$. If we now read the Newtonian potential from the metric, we just obtain the same result. Metric is then the one of the general relativistic version of a matter distribution, which test particles move in constant rotational curves. Function A will be determined by the kind of substance we are supposing the dark matter is made of.

We show now that a spherical fluctuation of the scalar field could be the dark matter in galaxies. Assuming thus that the dark matter is scalar, we start with the energy momentum tensor $T_{\mu\nu} = \Phi_{,\mu}\Phi_{,\nu} - 1/2 g_{\mu\nu}\Phi^{,\sigma}\Phi_{,\sigma} - g_{\mu\nu}V(\Phi)$, being Φ the scalar field and $V(\Phi)$ the scalar potential. The Klein–Gordon and Einstein equations respectively are:

$$\Phi^{;\mu}_{;\mu} - \frac{dV}{d\Phi} = 0$$

$$R_{\mu\nu} = \kappa_0[\Phi_{,\mu}\Phi_{,\nu} + g_{\mu\nu}V(\Phi)],$$

where $R_{\mu\nu}$ is the Ricci tensor, $\sqrt{-g}$ the determinant of the metric, $\kappa_0 = 8\pi G$ and a semicolon stands for covariant derivative according to the background space-time; $\mu, \nu = 0, 1, 2, 3$.

Assuming the *flat curve condition* in the scalar dark matter hypothesis, we are in the position to write down the set of field equations. Using the metric, the Klein Gordon equation reads

$$\Phi'' + \frac{1}{2r}\left[l + 4 - \frac{A'}{A}r\right]\Phi' - \frac{1}{4}A\frac{dV(\Phi)}{d\Phi} = 0 \tag{29}$$

and the Einstein equations are

$$\frac{A - (l+1)}{r^2} = -\kappa_0\left[\frac{1}{2}\Phi'^2 - AV(\Phi)\right] \tag{30}$$

$$\frac{1}{4r^2}\left[l^2 - \frac{A'}{A}r\,(l+2)\right] = -\kappa_0\left[\frac{1}{2}\Phi'^2 + AV(\Phi)\right] \tag{31}$$

$$\frac{1}{r^2}\left[1 - A - \frac{A'}{A}r\right] = -\kappa_0\left[\frac{1}{2}\Phi'^2 + AV(\Phi)\right] \tag{32}$$

In order to solve equations (30–32), observe that the combination of the previous equations $[(2 - -l)$(eq. 30)$- - 4$ (eq. 31)$+(2 + l)$(eq. 32)$]$ implies

$$V = -\frac{l}{\kappa_0(2-l)}\frac{1}{r^2} \tag{33}$$

This is a very important result, namely the scalar potential goes always as $1/r^2$ for a spherically symmetric metric with the *flat curve condition.* It is remarkable that this behavior of the stress tensor coincides with the expected behavior of the energy density of the dark matter in a galaxy. We can go further and solve the field equations, the general solution of equations (30-32) is

$$A(r) = \left(4 - l^2\right) / \left(4 + C\left(4 - l^2\right)r^{-(l+2)}\right),$$

being C an integration constant and we can thus integrate the function Φ. Nevertheless, in this letter we consider the most simple solution of the field equations with $C = 0$. Observe that for this particular solution the stress tensor goes like $1/r^2$. The energy momentum tensor is made essentially of two parts. One is the scalar potential and the other one contains products of the derivatives of the scalar field, both going as $1/r^2$

Furthermore, as $(\Phi_{,r})^2 \sim 1/r^2$, this means that $\Phi \sim \ln(r)$, implying that the scalar potential is exponential $V \sim \exp(2\alpha\Phi)$ such as has been found useful for structure formation scenarios [14] and scaling solutions with a primordial scalar field in the cosmological context [14, 34] including quintessential scenarios [35]. Thus, the particular solution for the system (29 -32) that we are considering is

$$A = \frac{4-l^2}{4}, \tag{34}$$

$$\Phi = \sqrt{\frac{l}{\kappa_0}}\ln(r) + \Phi_0, \tag{35}$$

$$V(\Phi) = -\frac{l}{2-l}\exp[-2\sqrt{\frac{\kappa_0}{l}}(\Phi - \Phi_0)]. \tag{36}$$

where (34) and (35) approach asymptotically ($r \to \infty$) the case with $m = 2, n = 2l/(2-l)$ in the general study of the global properties of spherically symmetric solutions in dimensionally reduced apace-times [36]. Function A corresponds to an exact solution of the Einstein equations of a spherically symmetric space-time, in which the matter contents is a scalar field with an exponential potential. Let us perform the rescaling $r^2 \to 4r^2/\left(4-l^2\right)$. In this case the three dimensional space corresponds to a *surplus of angle* (analogous to the deficit of angle) one; the metric reads

$$ds^2 = -B_0 r^l dt^2 + dr^2 + \frac{4}{4-l^2} r^2 \left[d\theta^2 + \sin^2\theta d\varphi^2\right] \tag{37}$$

for which the two dimensional hypersurface area is $4\pi r^2 \times 4/(4-l^2) = 4\pi r^2/(1-(v^\varphi)^4)$. Observe that if the rotational velocity of the test particles were the speed of light $v^\varphi \to 1$, this area would grow very fast. Nevertheless, for a typical galaxy, the rotational velocities are $v^\varphi \sim 10^{-3}$ $(300km/s)$, in this case the rate of the difference of this hypersurface area and a flat one is $(v^\varphi)^4/(1-(v^\varphi)^4) \sim 10^{-12}$, which is too small to be measured, but sufficient to give the right behavior of the motion of stars in a galaxy.

Let us consider the components of the scalar field as those of a perfect fluid, it is found that the components of the stress–energy tensor have the following form

$$-\rho = T^0{}_0 = -\frac{l^2}{(4-l^2)}\frac{1}{\kappa_0 r^2} \tag{38}$$

$$P = T^r{}_r = -\frac{l(l+4)}{(4-l^2)}\frac{1}{\kappa_0 r^2} \tag{39}$$

while the angular pressures are $P_\theta = P_\varphi = -\rho$. The analysis of an axially symmetric perfect fluid in general is given in [37], where a similar result was found (see also [8]).

The effective density (38) depends on the velocities of the stars in the galaxy, $\rho = (v^\varphi)^4/(1-(v^\varphi)^4) \times 1/(\kappa_0 r^2)$ which for the typical velocities in a galaxy is $\rho \sim 10^{-12} \times 1/(\kappa_0 r^2)$, while the effective radial pressure is $|P| = (v^\varphi)^2((v^\varphi)^2+2)/(1-(v^\varphi)^2) \times 1/(\kappa_0 r^2) \sim 10^{-6} \times 1/(\kappa_0 r^2)$, *i.e.*, six orders of magnitude greater than the scalar field density. This is the reason why it is not possible to understand a galaxy with Newtonian dynamics. Newton theory is the limit of the Einstein theory for weak fields, small velocities but also for small pressures (in comparison with densities). A galaxy fulfills the first two conditions, but it has pressures six orders of magnitude bigger than the dark matter density, which is the dominating density in a galaxy. This effective pressure is the responsible for the behavior of the flat rotation curves in the dark matter dominated part of the galaxies.

Metric (37) is not asymptotically flat, it could not be so. An asymptotically flat metric behaves necessarily like a Newtonian potential provoking that the velocity profile somewhere decays, which is not the observed case in galaxies. Nevertheless, the energy density in the halo of the galaxy decays as

$$\rho \sim \frac{10^{-12}}{\kappa_0 r^2} = \frac{10^{-12} H_0^{-2}}{3r^2} \rho_{crit} \tag{40}$$

where $H_0^{-1} = \sqrt{3}/h \ 10^6 Kpc$ is the Hubble parameter and ρ_{crit} is the critical density of the Universe. This means that after a relative small distance $r_{crit} \sim \sqrt{3/h^2} \approx 3Kpc$ the effective density of the halo is similar as the critical density of the Universe. One expects, of course, that the matter density around a galaxy is smaller than the critical density, say $\rho_{around} \sim 0.06\rho_{crit}$, then $r_{crit} \approx 14Kpc$. Observe also that metric (37) has an almost flat three dimensional space-like hypersurface. The difference between a flat three dimensional hypersurface area and the three dimensional hypersurface area of metric (37) is $\sim 10^{-12}$, this is the reason why the space-time of a galaxies seems to be so flat. We think that these results show that it is possible that the scalar field could be the missing matter (the dark matter) of galaxies and maybe of the Universe.

5. THE GALAXY CENTER

The study of the center of the galaxy is much more complicated because we do not have any direct observation of it. If we follow the hypothesis of the scalar dark matter, we cannot expect that the center of a galaxy is made of "ordinary matter" , we expect that it contains baryons, self-interacting scalar fields, etc. There the density contrast of baryonic matter and scalar dark matter is the same (see [10]) and their states are in extreme conditions. At the other hand, we think that the center of the galaxy is not static at all. Observations suggest this for instance in active nuclei. A regular solution that explains with a great accuracy the galactic nucleus is the assumption that there lies a boson star [38], which on the other hand would be an excellent source for the scalar dark matter [39]. It is possible to consider also the galactic nucleus as the current stage of an evolving collapse, providing a boson star in the galactic center made of a complex scalar field or a regular structure made of a time dependent real scalar field (oscillaton), both being stable systems [40].

The energy conditions are no longer valid in nature, as it is shown by cosmological observations on the dark energy. Why should the energy conditions be valid in a so extreme state of matter like it is assumed to exist in the center of galaxies? If dark matter is of scalar nature, why should it fulfill the energy conditions in such extreme situation? In any case, the center of the galaxy still remains as a place for speculations.

6. CONCLUSIONS

We have developed most of the interesting features of a 95% scalar-nature cosmological model. The interesting implications of such a model are direct consequences of the scalar potentials (1,2).

The most interesting features appear in the scalar dark matter model. As we have shown in this work, the solutions found alleviate the fine tuning problem for dark matter. Once the scalar field Φ begins to oscillate around the minimum of its potential (2), we can recover the evolution of a standard cold dark matter model because the dark matter density contrast is also recovered in the required amount. Also, we find a Jeans length for this model. This provokes the suppression in the power spectrum for small scales, that could explain the smooth core density of galaxies and the dearth of dwarf galaxies. Up to this point, the model has only one free parameter, λ. However, if we suppose that the scale of suppression is $k = 4.5h\,\mathrm{Mpc}$, then $\lambda \approx 20.3$, and then all parameters are completely fixed. From this, we found that $V_0 \simeq (36.5\,\mathrm{eV})^4$ and the mass of the ultra-light scalar particle is $m_\Phi \simeq 1.1 \times 10^{-23}\,\mathrm{eV}$. Considering that potential (2) is renormalizable, we calculated the scattering cross section $\sigma_{2\rightarrow 2}$. From this, we find that the renormalization scale Λ is of order of the Planck mass if we take the value required for self-interacting dark matter, as our model is also self-interacting.

The hypothesis of the scalar dark matter is well justified at galactic level (see also [41]) and at the cosmological level too [10]. But at this moment it is only that, a hypothesis which is worth to be investigated.

Summarizing, a model for the Universe where the dark matter and energy are of scalar nature can be realistic and could explain most of the observed structures and features of the Universe.

Acknowledgments

We would like to thank Vladimir Avila–Reese, Michael Reisenberger, Ulises Nucamendi and Hugo Villegas Brena for helpful discussions. L.A.U. would like to thank Rodrigo Pelayo and Juan Carlos Arteaga Velázquez for many helpful discussions. F.S.G. wants to thank the criticism provided by Paolo Salucci related to the galactic phenomenology and the ICTP for its kind hospitality during the beginning of this discussion. We also want to express our acknowledgment to the relativity group in Jena for its kind hospitality. This work is partly supported by CONACyT México under grant 34407–E, and by grants 94890 (F.S.G.), 119259 (L.A.U.), DGAPA–UNAM IN121298 (D.N.) and by a cooperation grant DFG–CONACyT.

References

[1] D.N. Schramm, in: *Nuclear and Particle Astrophysics*, ed. J.G. Hirsch and D. Page, Cambridge Contemporary Astrophysics, 1998; X. Shi, D.N. Schramm, and D. Dearborn, *Phys. Rev. D* **50** (1995) 2414.

[2] P.J.E. Peebles, in: *Principles of Physical Cosmology*, Princeton University Press, 1993.

[3] M. Persic, P. Salucci, and F. Stel, *MNRAS* **281** (1996) 27.

[4] K.G. Begeman, A.H. Broeils, and R.H. Sanders, *MNRAS* **249** (1991) 523.

[5] S. Perlmutter *et al. Ap. J.* **517** (1999) 565; A.G. Riess *et al.*, *A. J.* **116** (1998) 1009.

[6] N.A. Bahcall, J.P. Ostriker, S. Perlmutter, and P.J. Steinhardt, *Science* **284** 1481; V. Sahni, A. Starobinsky, *Int. J. Mod. Phys. D* **9** (2000) 373.

[7] V. Sahni and L. Wang, *Phys. Rev. D* **62** (2000) 103517.

[8] F.S. Guzmán, and T. Matos, *Class. Quantum Grav.* **17** (2000) L9; T. Matos and F. S. Guzmán, *Ann. Phys. (Leipzig)* **9** (2000) 133; T. Matos, F.S. Guzmán, and D. Núñez, *Phys. Rev. D* **62** (2000) 061301.

[9] T. Matos, F.S. Guzmán, and L. A. Ureña-López, *Class. Quantum Grav.* **17** (2000) 1707.

[10] T. Matos and L.A. Ureña-López, *Class. Quantum Grav.* **17** (2000) L71.

[11] L.A. Ureña-López and T. Matos, *Phys. Rev. D* **62** (2000) 081302.

[12] P. de Bernardis *et al.*, *Nature* **404** (2000) 955; S. Hanany *et al.* , *preprint* astro-ph/0005123.

[13] L.P. Chimento, A.S. Jakubi, *Int. J. Mod. Phys. D* **5** (1996) 71.

[14] Pedro G. Ferreira and Michael Joyce, Phys. Rev. **D 58** (1998) 023503.

[15] M.S. Turner, *Phys. Rev. D* **28** (1983) 1253; L.H. Ford, *Phys. Rev. D* **35** (1987) 2955.

[16] T. Matos and L. A. Ureña-López, *Phys. Rev. D* in press. *preprint* astro-ph/0006024.

[17] R.R. Caldwell, R. Dave, and P.J. Steinhardt, *Phys. Rev. Lett.* **80** (1998) 1582; I. Zlatev, L. Wang, and P.J. Steinhardt, *Phys. Rev. Lett.* **82** (1999) 896; P.J. Steinhardt, L. Wang, and I. Zlatev, *Phys. Rev. D* **59** (1999) 123504.

[18] C.-P. Ma and E. Berthshinger, *Ap. J.* **455** (1995) 7.

[19] C.-Pei Ma *et al*, *A. J.* **521** (1999) L1.

[20] P. Brax, J. Martin, and A. Riazuelo, *preprint* astro-ph/0005428.

[21] U. Seljak and M. Zaldarriaga, *Ap. J.* **469** (1996) 437.

[22] M. Kamionkowski and A.R. Liddle, *Phys. Rev. Lett.* **84** (2000) 4525.

[23] W. Hu, R. Barkana, and A. Gruzinov, *Phys. Rev. Lett.* **85** (2000) 1158.

[24] B.D. Wandelt, R. Dave, G.R. Farrar, P.C. McGuire, D.N. Spergel, and P.J. Steinhardt, *preprint* astro-ph/0006344.

[25] C. Firmani, E. D'Onghia, G. Chincarini, X. Hernández, and V. Ávila-Reese, *preprint* astro-ph/0005010; V. Ávila-Reese, C. Firmani, E. D'Onghia, and X. Hernández, *preprint* astro-ph/0007120.

[26] P. Colín, V. Ávila-Reese, and O. Valenzuela, *preprint* astro-ph/0004115.

[27] D.N. Spergel and P.J. Steinhardt, *Phys. Rev. Lett.* **84** (2000) 3760.

[28] K. Halpern and K. Huang, *Phys. Rev. Lett.* **74** (1995) 3526; K. Halpern and K. Huang, *Phys. Rev. D* **53** (1996) 3252.

[29] V. Branchina, *preprint* hep-ph/0002013.

[30] K. Halpern, *Phys. Rev. D* **57** (1998) 6337.

[31] T. Matos and L.A. Ureña-López, *preprint* astro-ph/0010226.

[32] M.S. Turner, *Phys. Rev. D* **28** (1983) 1243; L.H. Ford, *Phys. Rev. D* **35** (1987) 2955.

[33] V.C. Rubin, W.K. Ford, and N. Thonnard, *A. J.* **238** (1980) 471.

[34] E.J. Copeland, A.R. Liddle, and D. Wands, *Phys. Rev. D* **57** (1998) 4686.

[35] T. Barreiro, E. J. Copeland, and N. J. Nunes, *Phys. Rev. D* **61** (2000) 127301.

[36] S. Mignemi and D. L. Wiltshire, *Class. Quantum Grav.* **6** (1989) 987.

[37] T. Matos, D. Nuñez, F.S. Guzmán, and E. Ramírez, *preprint* astro-ph/0003105; T. Matos, D. Nuñez, F.S. Guzmán, and E. Ramírez, "Conditions on the Flat Behavior of the Rotational Curves in Galaxies", *to be published.*

[38] D.F. Torres, S. Capozziello and G. Lambiase, *Phys. Rev. D* **62** (2000) 104012.

[39] T. Matos, L. A. Ureña-López, and F.S. Guzmán, *to be published.*

[40] E. Seidel and W. Suen, *Phys. Rev. Lett.* **72** (1994) 2516.

[41] T. Matos, F.S. Guzmán, and D. Núñez, *Phys. Rev. D* **62** (2000) 061301(R).

INFLATION WITH A BLUE EIGENVALUE SPECTRUM

Eckehard W. Mielke*
Departamento de Física,
Universidad Autónoma Metropolitana–Iztapalapa,
Apartado Postal 55-534, C.P. 09340, México, D.F., Mexico

Franz E. Schunck†
Institut für Theoretische Physik, Universität zu Köln,
50923 Köln, Germany

Abstract We reconsider the second order slow–roll approximation of Stewart and Lyth. For the resulting nonlinear *Abel equation*, we find a new eigenvalue spectrum of the spectral index in the 'blue' regime $n_s > 1$. Some of the *discrete* values have consistent fits to the cumulative COBE data as well as to recent ground-base CMB experiments.

Keywords: Inflation, COBE data, blue spectrum.

1. INTRODUCTION

For a wide range of inflationary models [1], a single scalar field, *the inflaton*, is presumed to be rolling in some underlying potential $U(\phi)$. This standard scenario is generically referred to as *chaotic inflation.* A variance of it is the recently proposed *quintessence.*

Our rather ambitious aim of reconstruction [2] is to employ observational data to deduce the complete functional form $U(\phi)$ of the inflaton self–interaction potential over the range corresponding to large scale structures, allowing a so–called graceful exit to the Friedmann cosmos [3]. The generation of density perturbations and gravitational waves has been extensively investigated. The usual strategy is an expansion in

*E–mail: ekke@xanum.uam.mx
†E–mail: fs@thp.uni-koeln.de

the deviation from scale invariance, formally expressed as the slow–roll expansion [4].

In general, *exact* inflationary solutions, after an elegant coordinate transformation [5], depend on the Hubble expansion rate H as a new "inverse time", and the regime of inflationary potentials allowing a graceful exit has already been classified [6]. In our more phenomenological approach, the inflationary dynamics is not prescribed by one's theoretical prejudice. On the contrary, in this solvable framework, the 'graceful exit function' $g(H)$, which determines the *inflaton potential* $U(\phi)$ and the exact Friedmann type solution, is *reconstructed* in order to fit the data. Even more, the transparent description of inflationary phase within the H–formalism was the foundation for deriving a new mechanism of inflation, called *assisted inflation* [7].

Observations of the cosmic microwave background (CMB) confirm that the Universe expands rather homogeneously on the large scale. From cumulative four years CMB observations, the spectral index is now measured by the satellite COBE as $n_s = 1.2 \pm 0.3$, including the quadrupole ($l = 2$) anisotropy [8]. This experiment has been complemented by ground-based CMB experiments: The preliminary data [9] from the Cambridge Cosmic Anisotropy Telescope (CAT) e.g., are consistent with COBE, albeit a *slightly higher* spectral index of $n_s = 1.3 \pm 0.4$. Moreover, balloon experiments like BOOMERANG [10] confirm for a small solid angle of the sky a spatially *flat* Universe. This allows [11] to constrain cosmological parameters, such as matter density $\Omega_{\rm m} = 0.32 \pm 0.08$ and the Hubble constant to $H_0 = 47 \pm 6$ km/(s Mpc). A more recent analysis [12] of the CMB anisotropy strongly constrain the spatial curvature of the Universe to near zero, i.e. $k \cong 0$. All data imply that a total density $\Omega = 1$ (or $k = 0$) is 2×10^7 more probable then $\Omega \cong 0.4$. Nowaday's best-fit to *all* CMB data is a Hubble constant of $H_0 = 65$ km/(s Mpc) and requires *dark energy* with $\Omega_\Lambda = 0.69$ and a spectral index $n_s = 1.12$.

On the theoretical side, Stewart and Lyth [13] proceeded from the exact power–law inflation in order to analytically compute the second order slow–roll correction to the standard formula [14]. At present, this is the most accurate approximation available. This remarkable accuracy has been confirmed [15] by numerical perturbations of the exact analytical result for the power–law inflation.

By applying the H–formalism to this accurate second order perturbation formalism, we can transform the nonlinear equation of Stewart and Lyth into an *Abel equation.* In extension of our earlier work [16], we determine a *discrete eigenvalue spectrum* of this nonlinear equation. In particular, our new *blue* discrete eigenvalue spectrum has the scale invariant Harrison–Zel'dovich spectrum with index $n_s = 1$ as a *limiting point.* Thus some specific eigenvalues $1 < n_s < 1.5$ of the family of solutions can be related rather accurately to existing as well as future observations.

2. SPATIALLY FLAT INFLATIONARY UNIVERSE

In the Einstein frame, a rather general class of inflationary models can be modeled by the Lagrangian density:

$$\mathcal{L}_{\rm E} = \frac{1}{2\kappa}\sqrt{|\,g\,|}\left(R + \kappa\left[g^{\mu\nu}(\partial_\mu\phi)(\partial_\nu\phi) - 2U(\phi)\right]\right)\,, \tag{1}$$

where R is the curvature scalar, ϕ the scalar field, $U(\phi)$ *the self–interaction potential* and $\kappa = 8\pi G/c^4$ the gravitational coupling constant. Natural units with $c = \hbar = 1$ are used and our signature for the metric is $(+1, -1, -1, -1)$ as it is common in particle physics.

For the *flat* ($k = 0$ or $\Omega = 1$) Robertson-Walker metric favored by recent observations, the evolution of the generic inflationary model (1) is determined by the autonomous first order equations

$$\dot{H} = g(\phi, H)\,, \qquad \dot{\phi} = \pm\sqrt{2/\kappa}\sqrt{-g(\phi, H)}\,, \tag{2}$$

where $g(\phi, H) := \kappa U(\phi) - 3H^2$. For this system of equation, de Sitter inflation with $\dot{H} = 0$ is a *singular*, but well–known and trivial subcase, cf. Ref. [17]. Otherwise, for $g(\phi, H) \neq 0$ we can simply *reparametrize* the inflationary potential $U(\phi) \rightarrow \widetilde{U} = \widetilde{U}(H) := U(\phi(t(H)))$ in terms of the Hubble expansion rate $H := \dot{a}(t)/a(t)$ as the *new* "inverse time" *coordinate.* For this non-singular coordinate transformation, the term $g(\phi, H)$ reduces to the graceful exit function $g(H) = \kappa\widetilde{U}(H) - 3H^2 < 0$ and the following general metric and scalar field solution of (2) emerges [5, 6]:

$$ds^2 = \frac{dH^2}{g^2(H)} - {a_0}^2 \exp\left(2\int\frac{H dH}{g(H)}\right)\left[dr^2 + r^2\left(d\theta^2 + \sin^2\theta d\varphi^2\right)\right] \tag{3}$$

$$\phi = \phi(H) = \mp\sqrt{\frac{2}{\kappa}}\int\frac{dH}{\sqrt{-g(H)}}\,. \tag{4}$$

Once we know $g(H)$ from some experimental input, we can (formally) invert (4) in order to recover from the chain of substitutions

$$[g(H) + 3H^2]/\kappa = \widetilde{U}(H) = \widetilde{U}(H \circ \phi^{-1}(\phi)) = U(\phi) \tag{5}$$

the inflaton potential in a simple and rather elegant manner.

3. ABEL EQUATION FROM SECOND-ORDER SLOW–ROLL APPROXIMATION

In order to reconstruct the potential $U(\phi)$, we adopt here the equations of Stewart and Lyth [13] for the *second order* slow–roll approximation. In terms of the energy density $\epsilon = -g(H)/H^2$, this *nonlinear* second order differential equation involving the *scalar* spectral index n_s reads [16, 18]

$$2C\epsilon\,\ddot{\epsilon} - (2C+3)\epsilon\,\dot{\epsilon} - \dot{\epsilon} + \epsilon^2 + \epsilon + \Delta = 0\,, \tag{6}$$

where $\cdot \hat{=} d/d\ln H^2$, $C := -2 + \ln 2 + \gamma \simeq -0.73$, and

$$\Delta := (n_s - 1)/2 \tag{7}$$

denotes the deviation from the scale invariant Harrison–Zel'dovich spectrum. The corresponding second order equation for the spectral index n_g of *tensor* perturbations turns out to be

$$n_g = -2\epsilon\left[1 + \epsilon - 2(1+C)\,\dot{\epsilon}\right] \simeq n_s - 1\,, \tag{8}$$

where the approximation is valid in first order. Note that for the constant solution $\epsilon = A_0$, where

$$A_0^2 + A_0 + \Delta = 0\,, \quad \Rightarrow \quad A_0 = \frac{1}{2}\left(-1 \mp \sqrt{3 - 2n_s}\right)\,, \tag{9}$$

we recover from the second order equation (6) exactly the *first order consistency relation* $n_g = n_s - 1$. Moreover, we can infer already from (9) that a *real* eigenvalue spectrum is constrained by $n_s \leq 1.5$.

By introducing the flow of the energy density ϵ via $p(\epsilon) := \dot{\epsilon}$ with $\ddot{\epsilon} = p dp/d\epsilon$, the second order equation (6) can be brought to the following form:

$$\epsilon p \frac{dp}{d\epsilon} - (\beta + \alpha\epsilon)p + \beta(\epsilon^2 + \epsilon + \Delta) = 0\,, \tag{10}$$

where $\alpha = 1 + 3/(2C)$ and $\beta = 1/(2C)$. For invertible energy flow, it can be transformed via $u := 1/p$ into the canonical form of the *Abel equation* of the first kind, cf. Ref. [19],

$$2C\epsilon\frac{du}{d\epsilon} + u^2[1 + (2C+3)\epsilon] - u^3\left[\epsilon^2 + \epsilon + \Delta\right] = 0\,. \tag{11}$$

García et al. [20] have demonstrated that, within the class of *finite* polynomials $p = \sum_{i=0}^N b_i \epsilon^i$, the unique solution of (10) is given by a polynomial $p = \Delta + b_1\epsilon$ of *first* degree, where $b_1 = (\alpha \pm \sqrt{\alpha^2 - 4\beta})/2$ necessarily satisfies the additional constraint $b_1(1-b_1) = \Delta$ due to the *nonlinearity* of the Abel equation. Mathematically, this transformation improves the order of the polynomial by one in comparison with (3.1), where the solution is just a polynomial of zero degree. This first order polynomial corresponds to $\epsilon = A_0 + \epsilon_0 y^{1+A_0}$, where A_0 is again given by (9). Since Δ is determined due to the additional constraint mentioned above, cf. Ref. [21], the spectral index n_s is now *completely fixed* by the Euler constant γ as

$$\begin{aligned} n_s &= 1 - \frac{2C + 9 \mp 3\sqrt{4C^2 + 4C + 9}}{4C^2} \\ &= 1.49575 \quad (-6.57797)\,. \end{aligned} \tag{12}$$

In order to circumvent this 'no-go' theorem for polynomial solutions, we shall introduce a different transformation which leads us to new solutions in form of an *infinite* series.

3.1. SOLUTIONS WITH A BLUE SPECTRUM

If we assume that $\epsilon = \epsilon(y)$ with $y := H^2$, Eq. (6) can be rearranged to

$$2C\epsilon y^2\epsilon'' - (3\epsilon + 1)y\epsilon' + \epsilon^2 + \epsilon + \Delta = 0\,, \tag{13}$$

where $'\hat{=}d/dy$. If we try the ansatz $\epsilon = \sum_{i=0}^{\infty} A_i y^i$, it turns out that the odd powers have to vanish. This can be traced back to the fact that (13) is *invariant* under reflections $y \to -y$ only for *even* functions $\epsilon(-y) = +\epsilon(y)$. For even powers, this boils down to the following Taylor expansion:

$$\epsilon = \sum_{j=0}^{\infty} A_{2mj} y^{2mj}\,, \tag{14}$$

where m is a positive integer, which labels the first non-constant term in our expansion and thus distinguishes the *different subclasses* of solutions. Hence, the series for $m = 1$ starts as $\epsilon = A_0 + A_2 y^2 + A_4 y^4 + \ldots$, for the case $m = 2$ as $\epsilon = A_0 + A_4 y^4 + A_8 y^8 + \ldots$, and so on. In all cases, the coefficient A_0 of zeroth order (y^0) is again determined algebraically by a Δ via Eq. (9); or, as we shall show below, A_0 is determined by the next higher order, and the zeroth order establishes n_s.

In order $2m$, we find the following *discriminating relation*

$$\{[2m(2m-1)C - 3m + 1]A_0 - m + 1/2\}A_{2m} = 0\,, \tag{15}$$

due to the *nonlinearity* of the Abel type equation. This recursion relation implies $A_{2m} = 0$ (and $\epsilon = A_0 =$ *constant*) or the *new additional* constraint

$$A_0 = \frac{2m-1}{2[2m(2m-1)C-3m+1]} < 0\,. \tag{16}$$

Together with (9) this implies that the spectral index of our new class of solutions can only adopt the following *discrete eigenvalues*

$$n_s = 1 + \frac{(2m-1)[4m-1-4m(2m-1)C]}{2[2m(2m-1)C-3m+1]^2}\,. \tag{17}$$

For solutions with large m, we find asymptotically

$$A_0 \simeq \frac{1}{4mC}\,, \quad n_s \simeq 1 - \frac{1}{2mC}\,, \tag{18}$$

which reveals that the scale invariant Harrison–Zel'dovich solution with index $n_s = 1$ is the *limiting point* of our new class of solutions. Moreover, our new discrete spectrum approaches it from the *blue* side, which previously was considered rather difficult to achieve.

In order to correlate this with observational restrictions, let us display the highest eigenvalues for n_s. For $m = 1$, we have $(4CA_0-4A_0-1)A_2 = 0$ and, because of $A_2 \neq 0$, this implies $A_0 = 1/4(C-1) = -0.1445$. Then Eq. (9) delivers

$$n_s = 1.247 \tag{19}$$

for the value of the scalar spectral index. The integration constant A_2 remains arbitrary. The fourth order establishes the constant A_4 as an algebraic function of A_0, A_2 and similar for higher orders.

For $m = 2$, the fourth order constraint is $A_0 = 3/(24C-10) = -0.109$, while A_4 now plays the role of a free constant. Then, zeroth order vanishes if

$$n_s = 1.194\,, \tag{20}$$

which is very close to the experimental value $n_s = 1.2 \pm 0.3$ from COBE. Furthermore, it is revealed that $A_8 = f_1(A_4, A_0)$, $A_{12} = f_2(A_8, A_4, A_0)$, and so on (where f_1, f_2 are some algebraic relations). The higher order coefficients of the series can be derived recursively via MATHEMATICA, similarly as in the case of the Bartnik–McKinnon solution, cf. Ref. [22].

4. COMPLETE EIGENVALUE SPECTRUM

In cosmological applications, the energy density ϵ in our second order reconstruction has only a limited range. For the deceleration parameter $q(t) := -\ddot{a}\,a/\dot{a}^{\,2} = \epsilon - 1$ to become negative, as necessary for inflation, our solutions are constrained by $0 < \epsilon < 1$. In the case of the constant solution $\epsilon = A_0$, merely the spectral index is restricted to the *continuous* range $-3 < n_s < 1$. Since this corresponds to $g(h) = -A_0 H^2$, a simple inversion of the integral (4) and subsequent insertion into (5) leads to the already known potential

$$U(\phi) = \frac{3 - A_0}{\kappa} \tilde{C}^2 \exp\left(\pm\sqrt{2\kappa A_0}\,\phi\right), \tag{21}$$

of *power–law inflation*, where $\tilde{C}$ is an integration constant.

Similarly, for the extreme blue spectrum (12) the graceful exit function reads $g(H) = -H^2 \left[A_0 + \epsilon_0 H^{2(1+A_0)}\right]$ yielding via (4) and (5) the reconstructed potential

$$U(\phi) = \frac{1}{\kappa}\left[-\frac{A_0}{\epsilon_0}\left(1+\tan^2(P\phi)\right)\right]^{1/(1+A_0)}\left(3 + A_0 \tan^2(P\phi)\right), \tag{22}$$

where $P = \pm\sqrt{-\kappa A_0/2}\,(1 + A_0)$ is a constant. According to its shape, cf. Ref. [21], it interpolates between the power–law inflation in the limit $\epsilon_0 \to 0$ and intermediate inflation. Its form is similar to the exact solution of Easther [23] except that the latter yields the too large value $n_s = 3$ for the spectral index.

For some of the new *blue* spectrum (17) we have reconstructed the potential $U(\phi)$ numerically in Ref. [2]. Their form resembles those of Refs. [23, 16].

The range of ϕ for which inflation occurs depends also on the (so-far missing) information on the spectral index n_g of tensor perturbations, cf. Eq. (8). Since all our new solutions are 'deformations' of the constant solution $\epsilon = A_0$, we expect the first order consistency relation of power-law inflation to hold to some approximation, i.e. $n_g \simeq n_s - 1 \in (-0.4, 0.5]$ [24]. In second order, it is found [2] that for our solutions n_g is always negative.

Of particular importance [24] is also the squared ratio of the two further observables, the amplitude A_g of gravitational waves, versus its counter part A_s, the amplitude of primordial scalar density perturbations:

$$\left(A_g^2/A_s^2\right) = \epsilon\left[1 - 2C\,\dot{\epsilon}\right] \simeq \frac{1}{2}\left(\sqrt{3 - 2n_s} - 1\right), \tag{23}$$

where the approximation is valid in first order. The numerical results are displayed in Fig. 5 and 6 of [2]. From first order approximations, we expect the range of this ratio within $\sim [0, 0.2)$. For several well-known inflationary models (power-law and polynomial chaotic, e.g.) the predicted ratio can be found in Table 2 of [24]. Recently, assisted inflation was able to produce a scalar spectral index closer to scale-invariance (but for $n_s < 1$) because of the large number of non-inflationary real scalar fields present in this theory [7].

Table 1. Eigenvalues of the spectral index

Eigenvalue n_s	**Spectrum**	**Potential $U(\phi)$**
(-6.57797)	(unrealistic)	$\sim \tan^2(\sqrt{\kappa}\phi/4)$
$-3 < n_s < 1$	continuous	$\propto \exp(\pm\sqrt{2\kappa A_0}\,\phi)$
$\vdots$		$\vdots$
	discrete	
1.106	$(m = 5)$	
1.125	All CMB data	
1.153	$(m = 3)$	
1.194	COBE data	
1.247	$(m = 1)$	
1.49575		$\sim \tan^2(\sqrt{\kappa}\phi/4)$

5. DISCUSSION

Basically, we followed the idea of reconstructing the inflaton potential in second order, as exposed in Ref. [3]. However, instead of producing the potential just for a few values ϕ, we could construct the complete functional form of the potential for a series of constant spectral indices. As it is well-known, a full reconstruction can only be achieved if informations on the gravitational parts are included; scalar perturbations determine the potential up to an unknown constant. Since the differential equations are nonlinear, different constants yield different potentials but for the identical scalar spectrum. Any piece of knowledge concerning the tensor perturbations breaks this degeneracy; so we use our solutions to calculate both the tensor index n_g and the ratio of the amplitudes A_g^2/A_s^2 necessary in second order to define the inflaton potential uniquely. We recognize that both parameters are depending on the wave vector, in contrast to the scalar spectral index, which, in our approximation, is regarded as a constant. We have checked numerically several consistency equations in second order and found good confirmation within the inflationary regime. Furthermore, we analyzed the behavior of the slow-roll parameters ϵ and $\eta = -dg/dH^2$, cf. the definition in [3, 6]. At the beginning of inflation, we derive that $\eta > 2\epsilon$ which is valid for the $(m = 1)$-solution up to $\epsilon \simeq 0.32$ and for the $(m = 2)$-solution up to $\epsilon \simeq 0.28$. This feature resembles that of *hybrid inflation* [25], cf. Table 1 of [3], where the influence of gravitational waves is negligible in accordance with the small values of both n_g and A_g^2/A_s^2 as can be recognized within Fig. 5 and 6 of [2] close to $\phi \simeq 1$.

Recently, the combined approximate system of the Abel equation (6) and the tensor equation (8) has been qualitatively treated as an "inverse problem" for nonlinear equations [26]. However, in the case of inflation the bifurcation into regimes of stability or instability depends in the first place on the behavior of the *exact* autonomous system (2), which has already been analyzed [17, 27] under the more general aspect of *catastrophe theory*. Moreover, there have been recently raised some reservations [28] on the applicability and consistency of the second order slow–roll approximation, which need to be investigated further.

Summarizing, the nonlinear Abel equation (6) permits exact solutions for the discrete or continuous eigenvalues of Table 1. All positive discrete values are marginally consistent with the $n_s = 1.2 \pm 0.3$ from COBE as well as recent ground-based CMB experiments like BOOMERANG, our highest value for the blue spectrum is just below the $n_s \leq 1.5$ constraint in order to suppress distortions of the microwave spectrum and the formation of primordial black holes during the phase of reheating, cf. Ref. [3]. Since our *blue* discrete spectrum has the scale invariant Harrison–Zel'dovich solution as *limiting point* $n_s \to 1$, our expectations are high for the new more accurate constraints [29] the future PLANCK satellite will provide for inflation.

Acknowledgments

We would like to thank Ed Copeland, Andrew Liddle, Alfredo Macías, and Anupam Mazumdar for useful comments and hints to the literature. This work was partially supported by CONACyT, grant No. 28339–E, the grant P/FOMES 98-35-15, and the joint German–Mexican project DLR–Conacyt E130–1148 and MXI 010/98 OTH. One of us (E.W.M.) would like to thank his wife Noelia Méndez Córdova for encouragement and love.

References

[1] A.R. Liddle and D.H. Lyth, *Phys. Rep.* **231** (1993) 1.

[2] F.E. Schunck and E.W. Mielke, *Phys. Lett.* **B485** (2000) 231.

[3] J.E. Lidsey, A.R. Liddle, E.W. Kolb, E.J. Copeland, T. Barriero, and M. Abney, *Rev. Mod. Phys.* **69** (1997) 373.

[4] E.J. Copeland, E.W. Kolb, A.R. Liddle and J.E. Lidsey, *Phys. Rev.* **D49** (1994) 1840.

[5] F.E. Schunck and E.W. Mielke, *Phys. Rev.* **D50** (1994) 4794.

[6] E.W. Mielke and F.E. Schunck, *Phys. Rev.* **D52** (1995) 672.

[7] A.R. Liddle, A. Mazumdar, and F.E. Schunck, *Phys. Rev.* **D58** (1998) R061301.

[8] C.L. Bennett et al., *Astrophys. J.* **436** (1994) 423; *Astrophys. J. Lett.* **464** (1996) L1.

[9] P.F. Scott et al., *Astrophys. J. Lett.* **461** (1996) L1.

[10] P. de Bernadis et al., *Nature (London)* **404** (2000) 955.

[11] A.N. Lasenby, A.W. Jones, and Y. Dabrowski: *Review of ground-based* CMB *experiments*, in: Proc. Moriond Workshop, Les Arcs; France 1998, astro-ph/9810175.

[12] S. Dodelson and L. Knox, *Phys. Rev. Lett.* **84** (2000) 3523; L. Knox and L. Page, *Phys. Rev. Lett.* **85** (2000) 1366.

[13] E.D. Stewart and D.H. Lyth, *Phys. Lett.* **B302** (1993) 171.

[14] D.H. Lyth and E.D. Stewart, *Phys. Lett.* **B274** (1992) 168.

[15] I.J. Grivell and A.R. Liddle, *Phys. Rev.* **D54** (1996) 7191.

[16] E.W. Mielke, O. Obregón and A. Macías, *Phys. Lett.* **B391** (1997) 281.

[17] F.V. Kusmartsev, E.W. Mielke, Yu.N. Obukhov and F.E. Schunck, *Phys. Rev.* **D51** (1995) 924.

[18] E.W. Mielke, in: *Proc. of the Eighteenth Texas Symposium on Relativistic Astrophysics and Cosmology*, "Texas in Chicago", 15–20 Dec. 1996, eds. A.V. Olinto, J.A. Frieman, and D.N. Schramm (World Scientific, Singapore, 1998), pp. 688.

[19] J. Devlin, N.G. Lloyd and J.M. Pearson, *J. Diff. Eq.* **147** (1998) 435 .

[20] A. García, A. Macías and E.W. Mielke, *Phys. Lett.* **A229** (1997) 32.

[21] J. Benítez, A. Macías, E.W. Mielke, O. Obregón, and V.M. Villanueva, *Int. J. Mod. Phys.* **A12** (1997) 2835.

[22] F.E. Schunck, *Ann. Phys. (Leipzig)* **2** (1993) 647.

[23] R. Easther, *Class. Quantum Grav.* **13** (1996) 1775.

[24] E.W. Kolb: *Inflation in the postmodern era, Proceedings of the Erice School, Astrofundamental Physics*, astro-ph/9612138.

[25] A.D. Linde, *Phys. Rev.* **D49** (1994) 748; E.J. Copeland, A.R. Liddle, D.H. Lyth, E.D. Stewart, and D. Wands, *Phys. Rev.* **D49** (1994) 6410.

[26] E. Ayón-Beato, A. García, R. Mansilla, and C.A. Terrero-Escalante, *Phys. Rev.* **D62** (2000) 103513.

[27] A. García, A. Macías, and E.W. Mielke (1998): *Extended inflation and catastrophe theory*, in: *Recent Developments in Gravitation and Mathematical Physics*, Proc. of the 2nd Mexican School on Gravitation and Mathematical Physics, eds. A. García, C. Lämmerzahl, A. Macías, T. Matos, and D. Nunez, ISBN 3–9805735–0–8 (Science Network Publ., Konstanz, 1996).

[28] J. Martin and D. Schwarz, *Phys. Rev.* **D62** (2000) 103520.

[29] A.R. Liddle, *Contemp. Phys.* **39** (1998) 95.

CLASSICAL AND QUANTUM COSMOLOGY WITH AND WITHOUT SELF–INTERACTING SCALAR FIELD IN BERGMANN–WAGONER THEORY

Luis O. Pimentel*

Departamento de Física, Universidad Autónoma Metropolitana, Iztapalapa
P.O. Box 55–534, CP 09340, México D. F., México

Cesar Mora†

Departamento de Física, Universidad Autónoma Metropolitana, Iztapalapa
P.O. Box 55–534, CP 09340, México D. F., México
Departamento de Física, UPIBI-Instituto Politécnico Nacional,
Av. Acueducto s/n Col. Barrio La Laguna Ticomán,
CP 07340 México DF, México.

Abstract Classical and quantum cosmologies are considered in the context of generalized scalar–tensor theories of gravitation of Bergmann–Wagonner (BW)for Isotropic, and the Bianchi types I and II cosmologies with and without self interactions of the scalar field. The coupling function $\omega(\phi)$ is kept arbitrary and the self interacting potential is chosen to obtain exact solutions. Quantum wormholes and classical non–singular solutions are given.

Keywords: Quantum cosmology, scalar–tensor theories, Bergmann–Wagonner transformations.

1. INTRODUCTION

In this paper we consider classical and quantum cosmological models for isotropic, Bianchi type I and II models in the generalized vacuum scalar–tensor theories of gravity. By quantum cosmology we understand

*E-mail: lopr@xanum.uam.mx
†E-mail: ceml@xanum.uam.mx

the canonical quantization resulting in the Wheeler–DeWitt equation (WDW). More detailed calculations and some other aspects of these theories can be found in our previous work [1], here we review of part of it. Interest in these generalized scalar–tensor theories has been widespread in recent years in connection with inflation and string theories. They are defined by the action[2]

$$A = \int d^4x\sqrt{-g}\left[\phi R - \omega(\phi)\left(\nabla\phi\right)^2 - 2\phi\lambda(\phi)\right]. \tag{1}$$

The dilaton field ϕ plays the role of a time–varying gravitational constant with a potential $\lambda(\phi)$, $\omega(\phi)$ is a coupling function. Each scalar–tensor theory is characterized by specific functional forms of $\omega(\phi)$ and $\lambda(\phi)$. Some examples are: i) Jordan–Brans–Dicke theory, where $\omega(\phi) = \omega_0 = const.$ and $\lambda(\phi) = 0$[3], ii) low energy limit of string theory, $\omega(\phi) = -1$, iii) Kaluza–Klein theory in $D = 4 + n$ space time dimensions, $\omega(\phi) = (1 - n)/n$, iv) Barker's theory, $\omega(\phi) = (4 - 3\phi)/(2\phi - 2), \lambda = 0$. It is known that inflationary solutions exist in a wide class of scalar–tensor cosmologies [4], therefore they are relevant to the study of the very early Universe.

2. ISOTROPIC MODELS

We are interested to study an homogeneous and isotropic cosmological model, thus we use the FRW metric line element

$$ds^2 = -N^2(t)dt^2 + a^2(t)\left[\frac{dr^2}{1-kr^2} + r^2(d\theta^2 + \sin^2\theta\, d\Phi^2)\right], \tag{2}$$

N is the lapse function, a is the scale factor of the universe. When this metric is used, we obtain

$$A = \frac{1}{2}\int\left[-Nka\phi + \frac{a\phi}{N}\dot{a}^2 + \frac{a^2}{N}\dot{a}\dot{\phi} - \frac{N\omega(\phi)}{6\phi}a^3\dot{\phi}^2 + \frac{N}{3}a^3\phi\lambda(\phi)\right]dt, \tag{3}$$

dot are derivative with respect to t; now we introduce the new variables

$$x = a\phi^{\frac{1}{2}}\,, y = \int\left([2\omega(\phi)+3]/12\right)^{\frac{1}{2}} d\phi/\phi\,, \Lambda(y) = \lambda(\phi)/3\phi\,, d\tau = \phi^{\frac{1}{2}}dt, \tag{4}$$

then the BW action simplifies to

$$A_{BW} = \frac{1}{2}\int\left[\frac{x}{N}x'^2 - \frac{x^3}{N}y'^2 - Nkx + Nx^3\Lambda(y)\right]d\tau, \tag{5}$$

prime are derivatives with respect to τ. The Hamiltonian can be constructed and we expressed it as $H = N\mathcal{H}$. Since N is a Lagrange multiplier, we have the constraint $\mathcal{H} \approx 0$, and follow the Dirac quantization procedure $\mathcal{H}\psi = 0$, this equation is known as WDW equation. This equation is independent of N , however in the following we use the gauges $N = 1$ and $N = 1/x$, since we obtain simpler equations with those choices.

2.1. GAUGE N=1

With the choice $N = 1$, the action (5) becomes

$$A_{BW} = \frac{1}{2}\int \left[x x'^2 - x^3 y'^2 - kx + x^3\Lambda(y)\right] d\tau \equiv \int L\, d\tau, \tag{6}$$

hence the canonical conjugate momenta corresponding to x and $y$and the corresponding Hamiltonian can be calculated

$$\mathcal{H} = \frac{1}{2}\left[x^{-1}\pi_x^2 - x^{-3}\pi_y^2 + kx - x^3\Lambda(y)\right], \tag{7}$$

now, the canonical momenta in Eq. (7) are converted into operators in the standard way, $\pi_x^2 \to -x^{-p}\frac{\partial}{\partial x}\left(x^p\frac{\partial}{\partial x}\right)$ and $\pi_y^2 \to -\frac{\partial^2}{\partial y^2}$, the ambiguity of factor ordering is encoded in the parameter p. The resulting WDW equation is

$$\left[x^2\frac{\partial^2}{\partial x^2} + px\frac{\partial}{\partial x} - \frac{\partial^2}{\partial y^2} - kx^4 + x^6\Lambda(y)\right]\psi(x,y) = 0. \tag{8}$$

Next we present three simple solvable cases for different cosmological terms.

Case $\Lambda(y) = 0, k = 1$. The solution to WDW equation without a cosmological term and $k = 1$ is

$$\psi_s(x,y) = x^{\frac{1-p}{2}} y^{\frac{1}{2}}\left[C_1 I_m\left(\frac{x^2}{2}\right) + C_2 K_m\left(\frac{x^2}{2}\right)\right]\left[C_3 e^{isy} + C_4 e^{-isy}\right]. \tag{9}$$

where s is a separation constant, the Bessel function order is given by $m = \frac{1}{2}\sqrt{\left(\frac{1-p}{2}\right)^2 - s^2}$. By superposition of the above solutions we obtain

$$\psi_\mu(x,y) = e^{-\frac{x^2}{2}\cosh[2y+\mu]}. \tag{10}$$

This solution is a quantum wormhole, satisfying the Hawking–Page regularity boundary conditions, i.e., the wavefunction is exponentially damped for large spatial geometry, and is regular when the spatial geometry degenerates [5].

Case $\Lambda(y) = \Lambda_0$, $k = 0$. Now, we take $\Lambda(y) = \Lambda_0 = const.$, this case is equivalent to a cosmological term linear in the scalar field ϕ. The solution to WDW equation is,

$$\begin{aligned}\psi_s(x,y) &= x^{\frac{1-p}{2}} y^{\frac{1-q}{2}} \left[C_1 I_M \left(\frac{\sqrt{\Lambda_0}}{3} x^3 \right) \right. \\ &+ \left. C_2 I_{-M} \left(\frac{\sqrt{\Lambda_0}}{3} x^3 \right) \right] \left[C_3 e^{isy} + C_4 e^{-isy} \right], \qquad (11)\end{aligned}$$

where $I_{\pm M}$ is a modified Bessel function and $M = \sqrt{(1/2 - p/2)^2 - s^2/3}$. For $\Lambda_0 < 0$ and $p = 1$, a quantum wormhole solution is

$$\psi_\mu(x,y) = e^{-\sqrt{\Lambda_0} x^3 \cosh[3y-\mu]}. \qquad (12)$$

2.2. GAUGE N=1/X

With $N = 1/x$, and the new variables, $\alpha = x^2 \cosh(2y)$, $\beta = x^2 \sinh(2y)$, the action becomes

$$A = \frac{1}{2} \int \left[\frac{1}{4} \left(\alpha'^2 - \beta'^2 \right) + \Lambda_1 \alpha + \Lambda_2 \beta - k \right] d\tau, \qquad (13)$$

where we have chosen the cosmological term as $\Lambda(y) = \Lambda_1 \cosh(2y) + \Lambda_2 \sinh(2y)$, with Λ_i constants. The corresponding WDW equation is

$$\left[\frac{\partial^2}{\partial \alpha^2} - \frac{\partial^2}{\partial \beta^2} + \frac{1}{4} \left(\Lambda_1 \alpha + \Lambda_2 \beta - k \right) \right] \psi(\alpha,\beta) = 0. \qquad (14)$$

The universe wavefunction for this model is

$$\begin{aligned}\psi(\alpha,\beta) &= \left\{ C_1 Ai \left[(2\Lambda_1)^{-\frac{2}{3}} (k - 4s^2 - \Lambda_1 \alpha) \right] \right. \\ &+ \left. C_2 Bi \left[(2\Lambda_1)^{-\frac{2}{3}} (k - 4s^2 - \Lambda_1 \alpha) \right] \right\} \left\{ C_3 Ai \left[(2\Lambda_2)^{-\frac{2}{3}} (\Lambda_2 \beta - 4s^2) \right] \right. \\ &+ \left. C_4 Bi \left[(2\Lambda_2)^{-\frac{2}{3}} (\Lambda_2 \beta - 4s^2) \right] \right\}, \qquad (15)\end{aligned}$$

where s is the separation constant.

Classical solutions. The classical equations of motion for this model derived from the action (13) are

$$\alpha''(\tau) - 2\Lambda_1 = 0, \ \beta''(\tau) + 2\Lambda_2 = 0, \tag{16}$$

with the Hamiltonian constraint $\alpha'^2 - \beta'^2 - 4\Lambda_1\alpha - 4\Lambda_2\beta + 4k = 0$. The solution to system (16) is

$$\alpha(\tau) = \Lambda_1\tau^2 + C_1\tau + C_2, \ \beta(\tau) = -\Lambda_2\tau^2 + C_3\tau + C_4, \tag{17}$$

C_i are integration constants that satisfy $C_1^2 - C_3^2 - 4\Lambda_1 C_2 - 4\Lambda_2 C_4 + 4k = 0$.

Classical solutions for Brans–Dicke theory. Choosing $\omega(\phi) = \omega_0$, then $y = \frac{\rho}{2}\ln\phi$ and we obtain the following solution

$$\begin{aligned} a(\tau) &= \Big[(\Lambda_1 + \Lambda_2)\tau^2 + (C_1 - C_3)\tau + C_2 - C_4\Big]^m \\ &\quad \Big[(\Lambda_1 - \Lambda_2)\tau^2 + (C_1 + C_3)\tau + C_2 + C_4\Big]^{m-n}, \end{aligned} \tag{18}$$

$$\phi(\tau) = \left[\frac{(\Lambda_1 - \Lambda_2)\tau^2 + (C_1 + C_3)\tau + C_2 + C_4}{(\Lambda_1 + \Lambda_2)\tau^2 + (C_1 - C_3)\tau + C_2 - C_4}\right]^n, \tag{19}$$

where $m = (1+\rho)/4\rho$, $n = 1/2\rho$ and $\rho = \sqrt{(2\omega_0 + 3)/3}$. Next we take particular values of the parameters to have explicit solutions that are easier to analyze. We found singular and non singular solutions.

The non singular solution is obtained by setting $C_1 = C_3 = C_4 = 0, C_2 = k/\Lambda_1$ and $\Lambda_1 = \Lambda_2$ in the above equations, resulting in

$$a(\tau) = a_0\left[\frac{\tau^2}{\tau_0^2} + 1\right]^m, \ \phi(\tau) = \left[\frac{\tau^2}{\tau_0^2} + 1\right]^{-n}, \tag{20}$$

where $a_0 = (k^2/\Lambda_1^2)^{\frac{1}{4}}$ and $\tau_0^2 = k/2\Lambda_1^2$. We can verify that this solution is not singular by direct inspection of the metric invariants. The Hubble function and the deceleration parameter are,

$$H = \frac{2ma_0}{\tau_0^2}[1 + (\frac{\tau}{\tau_0})^2], \ q = -\frac{2\tau_0^2}{4m\tau^2}. \tag{21}$$

Another singular cosmological solution can be obtained by choosing $C_1 = C_3 = 0, \ C_2 = C_4 = k/2\Lambda_1$ and $\Lambda_1 = \Lambda_2$, thus (18)-(19) are reduced to

$$a(\tau) = a_0'\tau^{2m}, \ \phi(\tau) = \phi_0'\tau^{-2n}, \tag{22}$$

where $a_0' = 2^{\frac{1}{4}}\tau_0^{-\frac{n}{2}}$ and $\phi_0' = \tau_0^{2n}$.

3. ANISOTROPIC MODELS

In this section we consider the WDW equation for two anisotropic cosmological models, namely, the Bianchi types I and II. Some time ago the canonical formulation of the Brans–Dicke theory was considered by Toton and by Matzner et al. [7]. The line element for the class of spatially–homogeneous space–times is given by

$$ds^2 = -dt^2 + h_{ab}\omega^a\omega^b, \qquad a, b = 1, 2, 3, \tag{23}$$

where $h_{ab}(t)$ is the metric on the surfaces of homogeneity and ω^a are one–forms. With a topology $R \times G_3$, with G_3 a Lie group of isometries acting transitively on the space–like three–dimensional orbits [8], with the structure constants $C^a{}_{bc} = m^{ad}\epsilon_{dbc} + \delta^a{}_{[b}a_{c]}$, m^{ab} is a symmetric matrix, $a_c \equiv C^a{}_{ac}$ and $\epsilon_{abc} = \epsilon_{[abc]}$; m^{ab} must be transverse to a_b [9]. A basis may be found with $a_b = (a, 0, 0)$ and $m^{ab} = \mathrm{diag}\,[m_{11}, m_{22}, m_{33}]$, where m_{ii} take the values ± 1 or 0. In the Bianchi class A, the Lie algebra is uniquely determined by the rank and signature of m^{ab}. The two cases $(0, 0, 0)$, $(1, 0, 0)$ correspond to the Bianchi types I, II. We take $h_{ab}(t) = e^{2\alpha(t)}\left(e^{2\beta(t)}\right)_{ab}$, where $e^{3\alpha}$ is the spatial volume of the Universe and $\beta_{ab} = \mathrm{diag}\left[\beta_+ + \sqrt{3}\beta_-, \beta_+ - \sqrt{3}\beta_-, -2\beta_+\right]$ gives the anisotropy.

The action for the class A cosmologies simplifies to [6]

$$\begin{aligned} A &= \int dt e^{3\alpha-\Phi}\left[6\dot{\alpha}\dot{\Phi} - 6\dot{\alpha}^2 + 6\dot{\beta}_+^2 + 6\dot{\beta}_-^2 + \omega(\Phi)\dot{\Phi}^2\right. \\ &- \left. 2\lambda(\Phi) + e^{-2\alpha}U(\beta_\pm)\right], \end{aligned} \tag{24}$$

where $U(\beta_\pm) = -e^{-4\alpha}\left(m_{ab}m^{ab} - m^2/2\right)$ is the curvature potential, $m \equiv m^a{}_a$ and indices are raised and lowered with h^{ab} and h_{ab}. Now we take the following changes of variables,

$$dt = e^{\Phi/2}d\tau,\ x = \alpha - \Phi/2,\ y = \int\sqrt{\frac{3 + 2\omega(\Phi)}{12}}d\Phi, \tag{25}$$

and the action is

$$A = \int d\tau\left[6e^{3x}\left\{y'^2 - x'^2 + \beta_+'^2 + \beta_-'^2 - \Lambda(y)\right\} + e^x U(\beta_\pm)\right], \tag{26}$$

where $\Lambda(y) := e^{\Phi}\lambda(\Phi)/3$.

The WDW equation obtained from the above action, for an arbitrary factor ordering, encoded in the parameter B, is

$$\left[\partial_x^2 - B\partial_x - \partial_y^2 - \partial_+^2 - \partial_-^2 - 24e^{4x}U(\beta_\pm) + 144e^{6x}\Lambda(y)\right]\Psi(x, y, \beta_\pm) = 0. \tag{27}$$

3.1. EXACT SOLUTIONS FOR BIANCHI I WITH $\Lambda(Y) = \Lambda_0$

For Bianchi type I cosmological model $U(\beta_\pm) = 0$. We restrict ourselves to the case $\Lambda(y) = \Lambda_o/144 =$ constant,i.e. $\lambda(\phi) \propto \phi$; this potential has been used in classical Bianchi type I vacuum cosmology [10] and for isotropic models with a barotropic fluid[11]; this potential is one of those that could produce inflation[4]. This choice simplifies the WDW equation and allows us to obtain exact solutions,

$$\begin{aligned}\Psi(x,y,\beta_\pm) &= e^{Bx/2}[c_1 J_\nu(\frac{\sqrt{\Lambda_0}}{3}e^{3x}) + c_2 Y_\nu(\frac{\sqrt{\Lambda_0}}{3}e^{3x})][c_3 e^{iky} + c_4 e^{-iky}] \\ & [c_5 e^{ik_+\beta_+} + c_6 e^{-ik_+\beta_+}][c_7 e^{ik_-\beta_-} + c_8 e^{ik_-\beta_-}], \qquad (28)\end{aligned}$$

where $\kappa_0^2 = k_+^2 + k_-^2 + k^2$, $\nu = \sqrt{-B - 4\kappa_0^2}/6$ and the c_i are constants. The following wavefunction can be obtained by superposition or it can be substituted into the WDW equation to check that it is a particular solution

$$\Psi(x,y,\beta_\pm) = e^{mCosh[ny+p\beta_+ + q\beta_- + r]exp(3x)}, \qquad (29)$$

where r is an arbitrary real number; Bm, n, p, q satisfy the relations $B = 3 + \Lambda_0/3$, $n^2 + p^2 + q^2 = 9$, $-(n^2+p^2+q^2)m^2 = \Lambda_0$. For a wormhole $\Lambda_0 < 0$. Taking the negative root for m, we can check that the wavefunction is exponentially damped for large spatial geometry, i.e., when $\alpha \to \infty$ ($x \to \infty$) and also that it does not oscillate when $\alpha \to -\infty$($x \to -\infty$).

3.2. SOLUTIONS FOR BIANCHI I WITH ANOTHER POTENTIAL

We take another potential for which it is possible to obtain exact solutions to WDW equation. In Eq.(26) if we take $d\tau = e^{-x}d\sigma$, the action becomes,

$$A = 6\int d\sigma \left[e^{4x}\left\{y'^2 - x'^2 + \beta_+'^2 + \beta_-'^2\right\} - e^{2x}\Lambda(y)\right]. \qquad (30)$$

We change variables and take a potential,

$$\eta = e^{2x}\text{Cosh}(2y),\ \xi = e^{2x}\text{Sinh}(2y),\ \Lambda(y) = \Lambda_1\text{Cosh}(2y) + \Lambda_2\text{Sinh}(2y). \qquad (31)$$

The corresponding WDW equation is given by

$$\left[\partial_\xi^2 - \partial_\eta^2 + (\eta^2 - \xi^2)(\partial_+^2 + \partial_-^2) - 24\Lambda_1\eta - 24\Lambda_2\xi\right]\Psi(\eta,\xi,\beta_\pm) = 0. \quad (32)$$

Assuming $\Psi(\eta,\xi,\beta_\pm) = E(\eta)X(\xi)P(\beta_+)M(\beta_-)$, the solutions are

$$\begin{aligned}
E &= c_1\ {}_1F_1(a_1,1/2,z_1) + c_2 z_1^{1/2}\ {}_1F_1(a_1+1/2,3/2,z_1),\\
a_1 &= \frac{1}{4}[-k^2 + (\frac{12\Lambda_1}{k_\perp})^2\sqrt{-k_\perp^2}+1],\ z_1 = \sqrt{-k_\perp^2}(\eta + \frac{12\Lambda_1}{k_\perp^2})^2,\\
X &= c_3\ {}_1F_1(a_2,1/2,z_2) + c_4 z_2^{1/2}\ {}_1F_1(a_2+1/2,3/2,z_2),\\
a_2 &= \frac{1}{4}[-k^2 + (\frac{12\Lambda_2}{k_\perp})^2\sqrt{-k_\perp^2}+1],\ z_2 = \sqrt{-k_\perp^2}(\xi + \frac{12\Lambda_2}{k_\perp^2})^2,\\
P &= c_5 e^{ik_+} + c_6 e^{-ik_+},\quad M = c_7 e^{ik_-} + c_8 e^{ik_-}, \qquad (33)
\end{aligned}$$

where ${}_1F_1$ is the confluent hypergeometric function, k_+, k_- and k are arbitrary separation constants and $k_\perp^2 = k_+^2 + k_-^2$.

3.3. EXACT SOLUTIONS FOR BIANCHI II WITH $\Lambda(Y) = 0$

In this case the potential for $m^{ab} = diag(1,0,0)$ is given as follows $U(\beta_\pm) = -e^{4(\beta_+ + \sqrt{3}\beta_-)}/2$, and the WDW equation is given by

$$\left[\partial_x^2 - B\partial_x - \partial_y^2 - \partial_+^2 - \partial_-^2 + 12e^{4(x+\beta_+ + \sqrt{3}\beta_-)}\right]\Psi(x,y,\beta_\pm) = 0, \qquad (34)$$

the wavefunction takes the final form

$$\begin{aligned}
\Psi(x,y,\beta_\pm) &= e^{-\frac{B}{6}(x+\beta_+ + \sqrt{3}\beta_-)}\\
&\left[c_1 K_p(e^{2(x+\beta_+ + \sqrt{3}\beta_-)}) + c_2 I_p(e^{2(x+\beta_+ + \sqrt{3}\beta_-)})\right]\\
&\left[c_5 e^{\frac{1}{4}(\sqrt{3}B+r)(-x+2\beta_+ - \sqrt{3}\beta_-)} + c_6 e^{\frac{1}{4}(\sqrt{3}B-r)(-x+2\beta_+ - \sqrt{3}\beta_-)}\right]\\
&\left[c_3 e^{\frac{1}{12}(B-q)(\sqrt{3}x+\beta_-)} + c_4 e^{\frac{1}{12}(B+q)(\sqrt{3}x+\beta_-)}\right]\\
&[c_7 \sin ky + c_8 \cos ky]\,. \qquad (35)
\end{aligned}$$

where $n^2 = k^2 + l^2 + m^2$, $p = \sqrt{B^2 - 12n^2}/12$, $q = \sqrt{B^2 - 24l^2}$ and $r = \sqrt{3B^2 - 8m^2}$ and the c_i are integration constants.

4. FINAL REMARKS

We have obtained exact solutions for quantum and classical cosmologies in the general scalar–tensor theory of gravitation with arbitrary coupling function $\omega(\Phi)$ and particular potentials $\lambda(\Phi)$. We have constructed quantum wormholes solutions. The Bianchi I,II solutions are for arbitrary large anisotropies. Quantum cosmology using general relativity in homogeneous spacetimes with small anisotropies and a scalar field were considered by Lukash and Schmidt and Amsterdamski [12]. Lidsey

[13] has considered the wavefunctions in a highly anisotropic cosmologies with a massless minimally coupled scalar field. Recently Bachmann and Schmidt [14] have considered the arbitrary anisotropies in Bianchi I quantum cosmology.

References

[1] L.O. Pimentel and C. Mora, *Mod. Phys. Lett.* **A15** (1999) 333; preprints gr-qc/000926/000927/000946/9803025; L.O. Pimentel,"Bianchi I Quantum cosmology in the Bergmann–Wagoner theory", to appear in *Gen. Rel. Grav.* (2001).

[2] P.G. Bergmann, *Int. J. Theor. Phys.* **1** (1968) 25; R. V. Wagoner, *Phys. Rev.* D **1** (1970) 3209; K. Nordtvedt, *Astrophys. J.* **161** (1970) 1059.

[3] C. Brans and R.H. Dicke, *Phys. Rev.* **124** (1961) 925.

[4] L.O. Pimentel and J. Stein–Schabes, *Phys. Lett.* **B216** (1989) 27.

[5] S.W. Hawking and D.N. Page, *Phys. Rev. D* **42** (1990) 2655.

[6] J.E. Lidsey, *Class. Quantum Grav.* **13** (1996) 2449.

[7] E.T. Toton, *J. Math. Phys.* **11** (1970) 1713; R.A. Matzner, M.P. Ryan and E.T. Toton, *Nuovo Cimento* **14B** (1973) 161.

[8] M.P. Ryan and L.C. Shepley, *Homogeneous Relativistic Cosmologies* (Princeton: Princeton University Press, 1975).

[9] G.F.R. Ellis and M.A.H. MacCallum, *Commun. Math. Phys.* **12** (1969) 108.

[10] A. Banerjee, S.B. Duttachoudhury, and N. Banerjee, *Phys. Rev. D* **32** (1985) 3096.

[11] C. Santos and R. Gregory, *Ann. Phys. (N.Y.)* **258** (1997) 111; gr-qc/9611065; L.M. Diaz–Rivera and L.O. Pimentel, *Phys. Rev. D* **32** (1999) 3096; gr-qc/9907016.

[12] V. Lukash, H.–J. Schmidt, *Astron. Nachr.* **309** (1988) 25; H.-J. Schmidt, *J. Math. Phys.* **37** (1996) 1244; gr-qc/9510062; P. Amsterdamski, *Phys. Rev.* D **31** (1985) 3073.

[13] J.E. Lidsey, *Phys. Lett.* **B352** (1995) 207.

[14] M. Bachmann and H.–J. Schmidt, *Phys. Rev. D* **62** (2000) 043515; gr-qc/9912068.

THE BIG BANG IN T^3 GOWDY COSMOLOGICAL MODELS

Hernando Quevedo
Instituto de Ciencias Nucleares
Universidad Nacional Autónoma de México
A. P. 70-543, México, D. F. 04510, México.

Abstract We establish a formal relationship between stationary axisymmetric spacetimes and T^3 Gowdy cosmological models which allows us to derive several preliminary results about the generation of exact cosmological solutions and their possible behavior near the initial singularity. In particular, we argue that it is possible to generate a Gowdy model from its values at the singularity and that this could be used to construct cosmological solutions with any desired spatial behavior at the Big Bang.

Keywords: Quantum cosmology, singularities, big bang, Gowdy spaces.

1. INTRODUCTION

The Hawking–Penrose theorems [1] prove that singularities are a generic characteristic of Einstein's equations. These theorems establish an equivalence between geodesic incompleteness and the blow up of some curvature scalars (the singularity) and allow us to determine the region (or regions) where singularities may exist. Nevertheless, these theorems say nothing about the nature of the singularities. This question is of great interest especially in the context of cosmological models, where the initial singularity (the "Big Bang") characterizes the "beginning" of the evolution of the universe.

The first attempt to understand the nature of the Big Bang was made by Belinsky, Khalatnikov and Lifshitz [2]. They argued that the generic Big Bang is characterized by the mixmaster dynamics of spatially homogeneous Bianchi cosmologies of type VIII and IX. However, there exist counter–examples suggesting that the mixmaster behavior is no longer valid in models with more than three dynamical degrees of freedom [3]. Several alternative behaviors have been suggested [4], but during the last

Exact Solutions and Scalar Fields in Gravity: Recent Developments
Edited by Macias *et al.*, Kluwer Academic/Plenum Publishers, New York, 2001

two decades investigations have concentrated on the so–called asymptotically velocity term dominated (AVTD) behavior according to which, near the singularity, each point in space is characterized by a different spatially homogeneous cosmology [5]. The AVTD behavior of a given cosmological metric is obtained by solving the set of "truncated" Einstein's equations which are the result of neglecting all terms containing spatial derivatives and considering only the terms with time derivatives.

Spatially compact inhomogeneous spacetimes admitting two commuting spatial Killing vector fields are known as Gowdy cosmological models [6]. Recently, a great deal of attention has been paid to these solutions as favorable models for the study of the asymptotic behavior towards the initial cosmological singularity. Since Gowdy spacetimes provide the simplest inhomogeneous cosmologies, it seems natural to use them to analyze the correctness of the AVTD behavior. In particular, it has been proved that all polarized Gowdy models belong to the class of AVTD solutions and it has been conjectured that the general (unpolarized) models are AVTD too [7].

In this work, we focus on T^3 Gowdy cosmological models and present the Ernst representation of the corresponding field equations. We show that this representation can be used to explore different types of solution generating techniques which have been applied very intensively to generate stationary axisymmetric solutions. We use this analogy to apply several known theorems to the case of Gowdy cosmological solutions. In particular, we use these results to show that all polarized T^3 Gowdy models preserve the AVTD behavior at the initial singularity and that "almost" all unpolarized T^3 Gowdy models can be generated from a given polarized seed solution by applying the solution generating techniques. We also analyze the possibility of generating a polarized model if we specify *a priori* any desired value of its Ernst potential at the initial singularity. We also argue that this method could be used to generate a cosmological model starting from its value at the Big Bang.

2. GOWDY T^3 COSMOLOGICAL MODELS

Gowdy cosmological models are characterized by the existence of two commuting spatial Killing vector fields, say, $\eta_I = \partial/\partial\sigma$ and $\eta_{II} = \partial/\partial\delta$ which define a two parameter spacelike isommetry group. Here σ and δ are spatial coordinates delimited by $0 \leq \sigma,\ \delta \leq 2\pi$ as a consequence of the space topology. In the case of a T^3-topology, the line element for unpolarized Gowdy models [6] can be written as

$$ds^2 = e^{-\lambda/2}e^{\tau/2}(-e^{-2\tau}d\tau^2 + d\theta^2) + e^{-\tau}[e^P(d\sigma + Qd\delta)^2 + e^{-P}d\delta^2]\ , \quad (1)$$

where the functions λ, P and Q depend on the coordinates τ and θ only, with $\tau \geq 0$ and $0 \leq \theta \leq 2\pi$. In the special case $Q = 0$, the Killing vector fields η_I and η_{II} become hypersurface orthogonal to each other and the metric (1) describes the polarized T^3 Gowdy models.

The corresponding Einstein's vacuum field equations consist of a set of two second order differential equations for P and Q

$$P_{\tau\tau} - e^{-2\tau} P_{\theta\theta} - e^{2P}(Q_\tau^2 - e^{-2\tau} Q_\theta^2) = 0 , \tag{2}$$

$$Q_{\tau\tau} - e^{-2\tau} Q_{\theta\theta} + 2(P_\tau Q_\tau - e^{-2\tau} P_\theta Q_\theta) = 0 , \tag{3}$$

and two first order differential equations for λ

$$\lambda_\tau = P_\tau^2 + e^{-2\tau} P_\theta^2 + e^{2P}(Q_\tau^2 + e^{-2\tau} Q_\theta^2) , \tag{4}$$

$$\lambda_\theta = 2(P_\theta P_\tau + e^{2P} Q_\theta Q_\tau) . \tag{5}$$

The set of equations for λ can be solved by quadratures once P and Q are known, because the integrability condition $\lambda_{\tau\theta} = \lambda_{\theta\tau}$ turns out to be equivalent to Eqs.(2) and (3).

To apply the solution generating techniques to the T^3 Gowdy models it is useful to introduce the Ernst representation of the field equations. To this end, let us introduce a new coordinate $t = e^{-\tau}$ and a new function $R = R(t, \theta)$ by means of the equations $R_t = te^{2P} Q_\theta$, $R_\theta = te^{2P} Q_t$. Then, the field equations (2) and (3) can be expressed as

$$t^2 \left(P_{tt} + \frac{1}{t} P_t - P_{\theta\theta} \right) + e^{-2P}(R_t^2 - R_\theta^2) = 0 , \tag{6}$$

$$te^P \left(R_{tt} + \frac{1}{t} R_t - R_{\theta\theta} \right) - 2[(te^P)_t R_t - (te^P)_\theta R_\theta] = 0 . \tag{7}$$

Furthermore, this last equation for R turns out to be identically satisfied if the integrability condition $R_{t\theta} = R_{\theta t}$ is fulfilled. We can now introduce the complex Ernst potential ϵ and the complex gradient operator D as

$$\epsilon = te^P + iR , \qquad \text{and} \qquad D = \left(\frac{\partial}{\partial t} , i\frac{\partial}{\partial \theta} \right) , \tag{8}$$

which allow us to write the main field equations in the *Ernst–like representation*

$$Re(\epsilon) \left(D^2 \epsilon + \frac{1}{t} Dt \, D\epsilon \right) - (D\epsilon)^2 = 0 . \tag{9}$$

It is easy to see that the field equations (6) and (7) can be obtained as the real and imaginary part of the Ernst equation (9), respectively.

The particular importance of the Ernst representation (9) is that it is very appropriate to investigate the symmetries of the field equations. In particular, the symmetries of the Ernst equation for stationary axisymmetric spacetimes have been used to develop the modern solution generating techniques [12]. Similar studies can be carried out for any spacetime possessing two commuting Killing vector fields. Consequently, it is possible to apply the known techniques (with some small changes) to generate new solutions for Gowdy cosmological models. This task will treated in a forthcoming work. Here, we will use the analogies which exist in spacetimes with two commuting Killing vector fields in order to establish some general properties of Gowdy cosmological models.

An interesting feature of Gowdy models is its behavior at the initial singularity which in the coordinates used here corresponds to the limiting case $\tau \to \infty$. The asymptotically velocity term dominated (AVTD) behavior has been conjectured as a characteristic of spatially inhomogeneous Gowdy models. This behavior implies that, at the singularity, all spatial derivatives in the field equations can be neglected in favor of the time derivatives. For the case under consideration, it can be shown that the AVTD solution can be written as [8]

$$P = \ln[\alpha(e^{-\beta\tau} + \zeta^2 e^{\beta\tau})] \, , \qquad Q = \frac{\zeta}{\alpha(e^{-2\beta\tau} + \zeta^2)} + \xi \, , \qquad (10)$$

where α, β, ζ and ξ are arbitrary functions of θ. At the singularity, $\tau \to \infty$, the AVTD solution behaves as $P \to \beta\tau$ and $Q \to Q_0 = 1/(\alpha\zeta) + \xi$. It has been shown [9] that that all polarized ($Q = 0$) T^3 Gowdy models have the AVTD behavior, while for unpolarized ($Q \neq 0$) models this has been conjectured. We will see in the following section that these results can be confirmed by using the analogy with stationary axisymmetric spacetimes.

3. ANALOGIES AND GENERAL RESULTS

Consider the line element for stationary axisymmetric spacetimes in the Lewis–Papapetrou form

$$ds^2 = -e^{2\psi}(dt + \omega d\phi)^2 + e^{-2\psi}[e^{2\gamma}(d\rho^2 + dz^2) + \rho^2 d\phi^2] \, , \qquad (11)$$

where ψ, ω, and γ are functions of the nonignorable coordinates ρ and z. The ignorable coordinates t and ϕ are associated with two Killing vector fields $\eta_I = \partial/\partial t$ and $\eta_{II} = \partial/\partial\phi$. The field equations take the form

$$\psi_{\rho\rho} + \frac{1}{\rho}\psi_\rho + \psi_{zz} + \frac{e^{4\psi}}{2\rho^2}(\omega_\rho^2 + \omega_z^2) = 0 , \tag{12}$$

$$\omega_{\rho\rho} - \frac{1}{\rho}\omega_\rho + \omega_{zz} + 4(\omega_\rho\psi_\rho + \omega_z\psi_z) = 0 , \tag{13}$$

$$\gamma_\rho = \rho(\psi_\rho^2 - \psi_z^2) - \frac{e^{4\psi}}{4\rho^2}(\omega_\rho^2 - \omega_z^2) , \tag{14}$$

$$\gamma_z = 2\rho\psi_\rho\psi_z - \frac{1}{2\rho}e^{4\psi}\omega_\rho\omega_z . \tag{15}$$

Consider now the following coordinate transformation $(\rho, t) \to (\tau, \sigma)$ and the complex change of coordinates $(\phi, z) \to (\delta, \theta)$ defined by

$$\rho = e^{-\tau}, \quad t = \sigma, \quad z = i\theta, \quad \phi = i\delta, \tag{16}$$

and introduce the functions P, Q and λ by means of the relationships

$$\psi = \frac{1}{2}(P - \tau), \quad Q = i\omega, \quad \gamma = \frac{1}{2}\left(P - \frac{\lambda}{2} - \frac{\tau}{2}\right) . \tag{17}$$

Introducing Eqs.(16) and (17) into the line element (11), we obtain the Gowdy line element (1), up to an overall minus sign. Notice that this method for obtaining the Gowdy line element from the stationary axisymmetric one involves real as well as complex transformations at the level of coordinates and metric functions. It is, therefore, necessary to demand that the resulting metric functions be real. Indeed, one can verify that the action of the transformations (16) and (17) on the field equations (12)-(15) yields exactly the field equations (2)-(5) for the Gowdy cosmological models. This is an interesting property that allows us to generalize several results known for stationary axisymmetric spacetimes to the case of Gowdy spacetimes.

The counterparts of static axisymmetric solutions ($\omega = 0$) are the polarized ($Q = 0$) Gowdy models. For instance, the Kantowski–Sachs [10] cosmological model is the counterpart of the Schwarzschild spacetime, one of the simplest static solutions. Furthermore, it is well known that the field equations for static axisymmetric spacetimes are linear and there exists a general solution which can be generated (by using properties of harmonic functions) from the Schwarzschild one [11]. According to the analogy described above, this implies the following

Lemma 1: All polarized T^3 Gowdy cosmological models can be generated from the Kantowski–Sachs solution.

The method for generating polarized Gowdy models can be briefly explained in the following way. If we introduce the time coordinate

$t = e^{-t}$, Eq. (2) becomes $\Delta P = 0$ with $\Delta = \partial_{tt} + t^{-1}\partial_t - \partial_{\theta\theta}$. It can easily be verified that the operators $L_1 = t^{-1}\partial_t(t\partial_t)$ and $L_2 = \partial_\theta$ commute with the operator Δ. Then if a solution P_0 is known ($\Delta P_0 = 0$), the action of the operator L_i ($i = 1, 2$) on P_0 generates new solutions $\tilde{P}_i$, i.e., if $\tilde{P}_i = L_i P_0$ then $\Delta \tilde{P}_i = \Delta L_i P_0 = 0$. This procedure can be repeated as many times as desired, generating an infinite number of solutions whose sum with arbitrary constant coefficients represents the general solution. An important property of the action of L_i on a given solution P_0 is that it preserves the behavior of P_0 for $\tau \to \infty$. If we choose the Kantowski–Sachs spacetime as the seed solution P_0, the general solution will be AVTD. Hence, as a consequence of Lemma 1, we obtain

Lemma 2: All polarized T^3 Gowdy cosmological models are AVTD.

We now turn to the general unpolarized ($Q \neq 0$) case. The solution generating techniques have been applied intensively to generate stationary solutions from static ones. In particular, it has been shown that "almost" all stationary axisymmetric solutions can be generated from a given static solution (the Schwarzschild spacetime, for instance) [12]. The term "almost" means that there exist "critical" points where the field equations are not well defined and, therefore, the solution generating techniques can not bee applied. Using the analogy with Gowdy models, we obtain

Lemma 3: "Almost" all unpolarized T^3 Gowdy cosmological solutions can be generated from a given polarized seed solution.

As in the previous case, it can be shown that the solution generating techniques preserve the asymptotic behavior of the seed solution. If we take the Kantowski–Sachs spacetime as seed solution and apply Lemma 3, we obtain

Lemma 4: "Almost" all T^3 unpolarized Gowdy models can be generated from the Kantowski-Sachs solution and are AVTD.

It should be mentioned that all the results presented in Lemma 1 - 4 must be treated as "preliminary". To "prove" them we have used only the analogy between stationary axisymmetric spacetimes and Gowdy cosmological models, based on the transformations (16) and (17). A rigorous proof requires a more detailed investigation and analysis of the symmetries of the equations (2)–(5), especially in their Ernst representation (9).

4. THE BIG BANG

The behavior of the Gowdy T^3 cosmological models at the initial singularity ($\tau \to \infty$) is dictated by the AVTD solution (10). Although this is not the only possible case for the Big Bang, there are physical reasons to believe that this is a suitable scenario for the simplest case of inhomogeneous cosmological models. But an inhomogeneous Big Bang has two different aspects. The first one concerns the temporal behavior as the singularity is approached. The second aspect is related to the spatial inhomogeneities which should be present during the Big Bang. If we accept the AVTD behavior, the first aspect of the problem becomes solved by the AVTD solution (10), which determines the time dependence at the Big Bang. However, the spatial dependence remains undetermined as it is given by the arbitrary functions α, β, ζ and ξ which can be specified only once a solution is known. The question arises whether it is possible to have a solution with any desired spatial behavior at the Big Bang. We will see that the answer to this question is affirmative.

Using the solution generating techniques for stationary axisymmetric spacetimes it has been shown that any solution can be generated from its values on the axis of symmetry ($\rho = 0$) [12]. Specific procedures have been developed that allow us to construct any solution once the value of the corresponding Ernst potential is given at the axis [13]. On the other hand, the analogy with Gowdy models determined by Eqs.(16) and (17) indicates that the limit $\rho \to 0$ is equivalent to $\tau \to \infty$. To see that this equivalence is also valid for specific solutions we write the counterpart of the AVTD solution (10) in the coordinates ρ and z, according to the Eqs.(16) and (17). Then, we obtain

$$\psi = \frac{1}{2}\ln[\alpha(\rho^{1+\beta} + \zeta^2\rho^{1-\beta})], \qquad \omega = \frac{\zeta}{\alpha(\rho^{2\beta}+\zeta^2)} + \xi\ , \qquad (18)$$

where we have chosen the arbitrary functions ζ and ξ such that ω becomes a real function. One can verify that the solution (18) satisfies the corresponding "AVTD equations" (12) and (13) (dropping the derivatives with respect to z) for the stationary case. Consequently, the behavior of stationary axisymmetric solutions at the axis ($\rho \to 0$) is equivalent to the behavior at T^3 Gowdy models at the Big Bang.

Thus, if we consider the asymptotic behavior of the Ernst potential (9) for the AVTD solution (10)

$$\epsilon(\tau \to \infty) \to \epsilon_0 = e^{\tau(\beta-1)}[1 + Q_0 e^{\tau(\beta-1)}]\ , \qquad Q_0 = \frac{1}{\alpha\zeta} + \xi\ , \qquad (19)$$

and specify α, β, ζ and ξ as functions of the spatial coordinate θ, it is possible to derive the corresponding unpolarized ($Q \neq 0$) solution by using the solution generating techniques. In other words, we can construct a Gowdy model with any desired behavior at the Big Bang. This is an interesting possibility which could be used to analyze physically reasonable scenarios for the Big Bang.

Of course, a more detailed study of the solution generating techniques for Gowdy cosmological models is necessary in order to provide a rigorous proof ot these results and to construct realistic models for the Big Bang. Here we have used the analogy between stationary axisymmetric spacetimes and T^3 Gowdy cosmological models to show that this is a task that in principle can be solved.

Acknowledgments

It is a pleasure to dedicate this work to Prof. H. Dehnen and Prof. D. Kramer on their 60-th birthday. I would like to thank O. Obregón and M. Ryan for an ongoing collaboration that contributed to this work. Also, support by the joint German–Mexican project DLR–Conacyt MXI 010/98 OTH — E130–1148, and DGAPA–UNAM, grant 121298, is gratefully acknowledged.

References

[1] S. W. Hawking and G. F. R. Ellis, *The Large Structure of Spacetime* (Cambridge University Press, Cambridge, England, 1973).

[2] V. A. Belinsky, I. M. Khalatnikov and E. M. Lifshitz, *Adv. Phys.* **19** (1970) 525.

[3] M. P. Ryan, Jr., *Ann. Phys. (N.Y.)***65** (1971) 506; H. Ishihara, *Prog. Theor. Phys.* **74** (1985) 490; P. Halpern, *Gen. Rel. Grav.* **19** (1987) 73.

[4] J. D. Barrow and F. Tripler, *Phys. Rep.* **56** (1979) 372.

[5] D. Eardley, E. Liang, and R. Sachs, J. Math. Phys. **13** (1972) 99.

[6] R. H. Gowdy, *Phys. Rev. Lett.* **27** (1971) 826; *ibid* **27** (1971) E1102; *Ann. Phys. (N.Y.)* **83** (1974) 203.

[7] J. Isenberg and V. Moncrief, *Ann. Phys. (N.Y.)* **199** (1990) 84; P. T. Crusciel, J. Isenberg, and V. Moncrief, *Class. Quantum Grav.* **7** (1990) 1671; B. K. Berger, *Ann. Phys. (N.Y.)* **83** (1974) 458; B. Grubisic and V. Moncrief, *Phys. Rev.* **D47** (1993) 2371; O. Obregon and M. P. Ryan, Jr., *Phys. Rev.* **D** (2001) in press.

[8] J. Isenberg and V. Moncrief, Ann. Phys. (N.Y.) **199** (1990) 84.

[9] P. T. Crusciel, J. Isenberg, and V. Moncrief, *Class. Quantum Grav.* **7** (1990) 1671.

[10] A. Krasinski, *Inhomogeneous Cosmological Models* (Cambridge University Press, Cambridge, 1997).

[11] H. Quevedo, *Fortsch. Phys.* **38** (1990) 733; *Phys. Rev. Lett.* **67** (1991) 1050.

[12] W. Dietz and C. Hoenselaers (eds.) *Solutions of Einstein's Equations: Techniques and Results* (Springer-Verlag, Heidelberg, Germany, 1984).

[13] N. R. Sibgatullin, *Oscillations and waves in strong gravitational and electromagnetic fields* (Nauka, Moscow, 1984).

The Universe's composition at the time galaxies formed could be, theoretically, very varied, including baryonic visible and dark matter, non baryonic dark matter, neutrinos, and many cosmological relics stemming from symmetry breaking processes predicted by high energy physics [3]. All these particles, if present, should have played a role in the structure formation. Then, galaxies are expected to possess dark matter components and, in accordance with the rotational curves of stars and gas around the centres of spirals, they are in the form of halos [4] and must contribute to at least 3 to 10 times the mass of the visible matter [3].

Whatever the Universe composition was, protogalaxies were originated due to a spectrum of scale-invariant perturbations [5, 6] that was present within the cosmological background at the beginning of structure formation; the inflationary cosmology is the most convincing scenario that explains its origin [7]. Protogalactic structures began to acquire some momenta, e.g. tidal torques [8], because of local gravitational instabilities to provoke plenty of collisions, mergings, fly-bys, etc, between these original, cosmic structures. As a result of their evolution, galaxies, as we presently know them, must have formed. The dynamics of protogalaxies has been studied intensively, for a review see Ref. [1]. There has been much interest to understand how galaxies acquired their present features, especially how their internal angular momentum (spin) has been gotten, $J/M \sim \mathcal{O}(10^{30})$ cm^2/s. A very important issue about it is how the transfer of angular momentum between protogalaxies took place to give rise to the observed elliptical and spiral galaxies with their known mass and rotational properties. As an initial condition can be thought that protogalaxies were gravitationally isolated. However, there are some indications that the orbital angular momentum in spiral galaxies in pairs is few times larger than their spins, so pairs seem to be not dynamically isolated [9]. Part of this angular momentum could had its origin in the cosmic expansion [10, 11, 12], where it has been computed the torques at the beginning of strong decoupling from the Hubble flow of spherical-symmetric density perturbations. Moreover, observations of spin angular momentum of various thousands of disc galaxies are compatible with the mechanism of generation of spin via tidal torques [13]. Other theoretical and numerical studies of evolution of angular momentum of protogalaxies from tidal torques are in line with observations [14, 15].

In the present work we investigate how the transfer of orbital to spin angular momentum is achieved when two equal, spherical clouds pass by, and in some cases when they collide; this type of interactions are to be expected in the tidal torques scheme. Studies of interacting spherical systems very related to ours have been done using the Newtonian theory of gravity, including a number of topics: mergings [16, 17], mixing processes [18], simulation of sinking satellites [19], and mass and energy lost in tidal interactions [20], among others. However, we made our calculations within the framework of scalar-tensor theories, in the Newtonian limit, to investigate the influence of massive scalar fields on the dynamics.

This paper is organized as follows: In section 2 we present the Newtonian approximation of a typical scalar field theory. In section 3, we present our models of protogalaxies, the initial conditions, and the results. The conclusions are in section 4.

2. SCALAR FIELDS AND THE NEWTONIAN APPROXIMATION

We consider a typical scalar field theory given by the following Lagrangian

$$\mathcal{L} = \frac{\sqrt{-g}}{16\pi}\left[-\phi R + \frac{\omega(\phi)}{\phi}(\partial\phi)^2 - V(\phi)\right] + L_M(g_{\mu\nu}) \tag{1}$$

from which we get the gravity equations,

$$\begin{aligned} R_{\mu\nu} - \frac{1}{2}g_{\mu\nu}R &= \frac{8\pi}{\phi}T_{\mu\nu} + \frac{V}{2\phi}g_{\mu\nu} + \frac{\omega}{\phi^2}\partial_\mu\phi\partial_\nu\phi - \frac{1}{2}\frac{\omega}{\phi^2}(\partial\phi)^2 g_{\mu\nu} \\ &+ \frac{\phi_{;\mu\nu}}{\phi} - \frac{g_{\mu\nu}}{\phi}\Box\phi \end{aligned} \tag{2}$$

and the scalar field equation

$$\Box\phi + \frac{\phi V' - 2V}{3+2\omega} = \frac{1}{3+2\omega}\left[8\pi T - \omega'(\partial\phi)^2\right] \tag{3}$$

We expect to have nowadays small deviations of the scalar fields around the background defined here as $\langle\phi\rangle = 1$. If we define $\bar{\phi} = \phi - 1$ the Newtonian approximation gives [21]

$$R_{00} = \frac{1}{2}\nabla^2 h_{00} = 4\pi\rho - \frac{1}{2}\nabla^2\bar{\phi} \tag{4}$$

$$\nabla^2\bar{\phi} - m^2\bar{\phi} = -8\pi\alpha\rho \tag{5}$$

where we have done

$$\frac{\phi V' - 2V}{3+2\omega} = m^2\bar{\phi} - m^2 k\bar{\phi}^2 + \ldots$$

and $\alpha = 1/(3+2\omega)$.

The solution of these equations is

$$\begin{aligned} \bar{\phi} &= 2\alpha u_\lambda \\ h_{00} &= -2u - 2\alpha u_\lambda \end{aligned} \tag{6}$$

where

$$\begin{aligned} u &= \sum_a \frac{m_a}{|\mathbf{r} - \mathbf{r}_a|} \\ u_\lambda &= \sum_a \frac{m_a}{|\mathbf{r} - \mathbf{r}_a|} \exp\left[-|\mathbf{r} - \mathbf{r}_a|/\lambda\right] \end{aligned} \tag{7}$$

$\lambda = 1/m$, where m is the mass given through the potential. This mass can have a variety of values depending on the particular particle physics model. The potential u is the Newtonian part and u_λ is the dark matter contribution which is of Yukawa type. The total force on a particle of mass m_i is

$$\sum \mathbf{F} = -\frac{1}{2}\nabla h_{00} = m_i \mathbf{a} \quad . \tag{8}$$

3. PROTOGALACTIC CLOUD MODELS AND RESULTS

Original protogalaxies could have very irregular forms, but we use as a first approximation spherical clouds for simplicity, and because this form seems to resemble the global shape of both visible and dark matter of many galaxies, i.e., taking into account their spherical halos [4, 22]. The initial clouds are in polytropic equilibrium with small internal velocities, compared to what they would need to be in dynamical, gravitational equilibrium. This feature avoids a large initial spin, in accordance with the fact that there were no primordial rotational motions in the universe [23, 24, 25, 26]. Then, the clouds are sent to approach each other, and only after their gravitational interaction takes place, spin will be gained.

For the simulations, each protogalaxy is constructed using the Plummer model given by the potential-density pair [27],

$$\Phi_P(r) = -\frac{GM}{\sqrt{r^2 + b^2}} \quad , \quad \rho_P(r) = \left(\frac{3M}{4\pi b^3}\right)\left(1 + \frac{r^2}{b^2}\right)^{-5/2} \tag{9}$$

where G is the gravitational constant, M is the total mass and b is a parameter which determines the dimensions of the cloud. Particle velocities are chosen everywhere isotropic which gives a system initially in steady state. The total energy of the cloud is $E = -(3\pi/64)GM^2b^{-1}$. We are using units in which $G = M = -4E = 1$.

We take two identical 3-D clouds consisting of $N = 2^{11}$ particles. The initial separation between the clouds is 10, and the velocity of the center of masses are $V_1 = (0.1, 0.1, 0)$ and $V_2 = (-0.2, 0.4, 0)$ in our units. The initial velocities are given so that the kinetic energy is a fraction of the potential energy and we consider a range that is consistent with the observed velocities of galaxies in clusters that goes from 50 km/s up to about 1000 km/s. For the present investigation we consider that the protogalaxies moves initially in the plane (x, y), and the angle of both protogalaxies is the same. More general initial conditions will be considered in a future communication.

A three-dimensional hydrodynamic code based on the TREESPH algorithm formulated in Ref. [28] was used for the computations of this paper. The code combines the method of SPH, developed in Refs. [29, 30], with the hierarchical tree algorithm of Ref. [31] for the calculation of the gravitational acceleration forces. The SPH method is a grid-free Lagrangian scheme in which the physical properties of the fluid are determined from the properties of a finite number of particles. In order to reduce the statistical fluctuations resulting from representing a fluid continuum by a finite set of particles, a smoothing procedure is employed in which the mean value of a field quantity is estimated from its local values by kernel interpolation. Thus, the evolution of particle i is determined by solving Euler's equation

$$\frac{d\mathbf{r}_i}{dt} \equiv \mathbf{v}_i \tag{10}$$

$$\frac{d\mathbf{v}_i}{dt} = -\frac{1}{\rho}\nabla p_i - \frac{1}{2}\nabla(h_{00})_i + \mathbf{A}_{visc,i} \ , \tag{11}$$

where p_i and $\mathbf{A}_{visc,i}$ denote, respectively, the gas pressure and the artificial viscous acceleration associated with particle i. This quantities are introduced because we are considering that the protogalaxies are gaseous. The code was modified [32] to include the effect of the scalar field, through the term h_{00} given by Eq. (6).

The simulation of the interaction of two protogalactic models starts with the clouds separated by a distance of 10 and on the x-axis. The selected separation is large enough to ensure that tidal effects are important but small enough that the calculation is possible in a reasonable computing time. Each particle in the initial steady state clouds is given an additional velocity (V_1 or V_2, corresponding to cloud 1 or 2) so that its magnitude is much bigger than the internal velocities they have at the equilibrium described above. In this way, initial clouds are almost spinless, and the given velocities V_1, V_2 imply kinetic energies associated with each cloud. The evolution of the intrinsic angular momentum

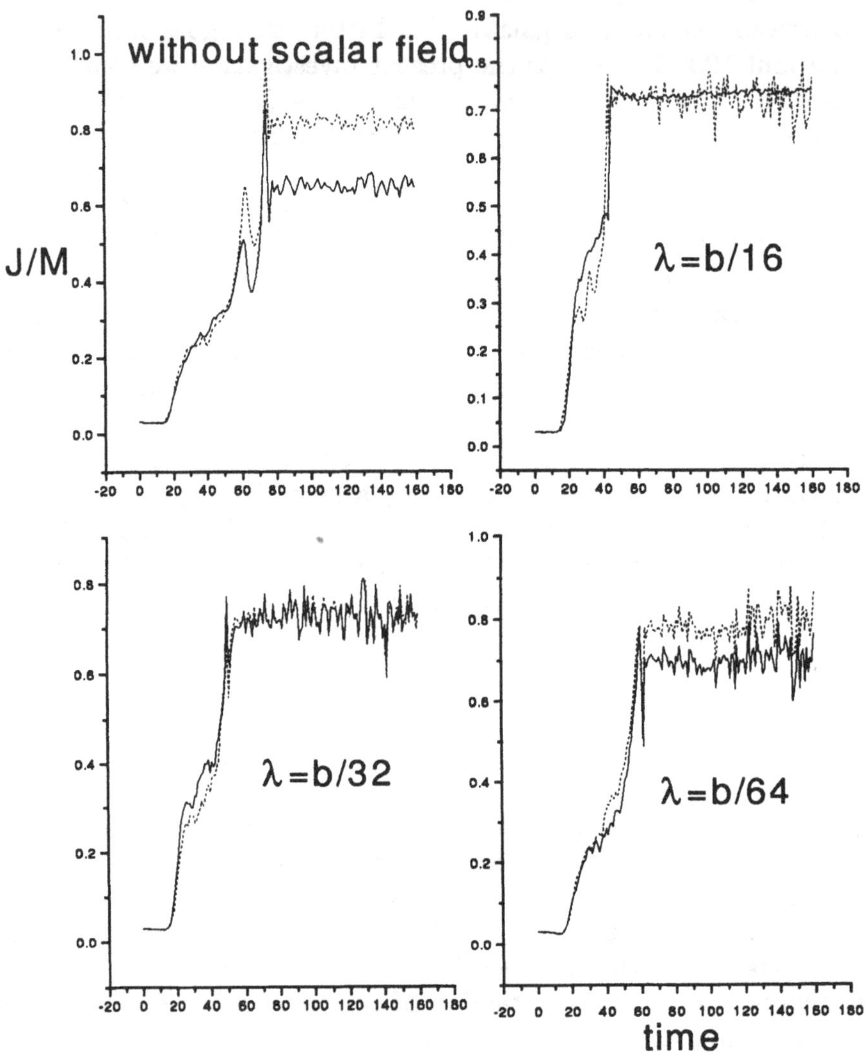

Figure 1. Evolution of the angular momentum without scalar field and with scalar field for different values of λ

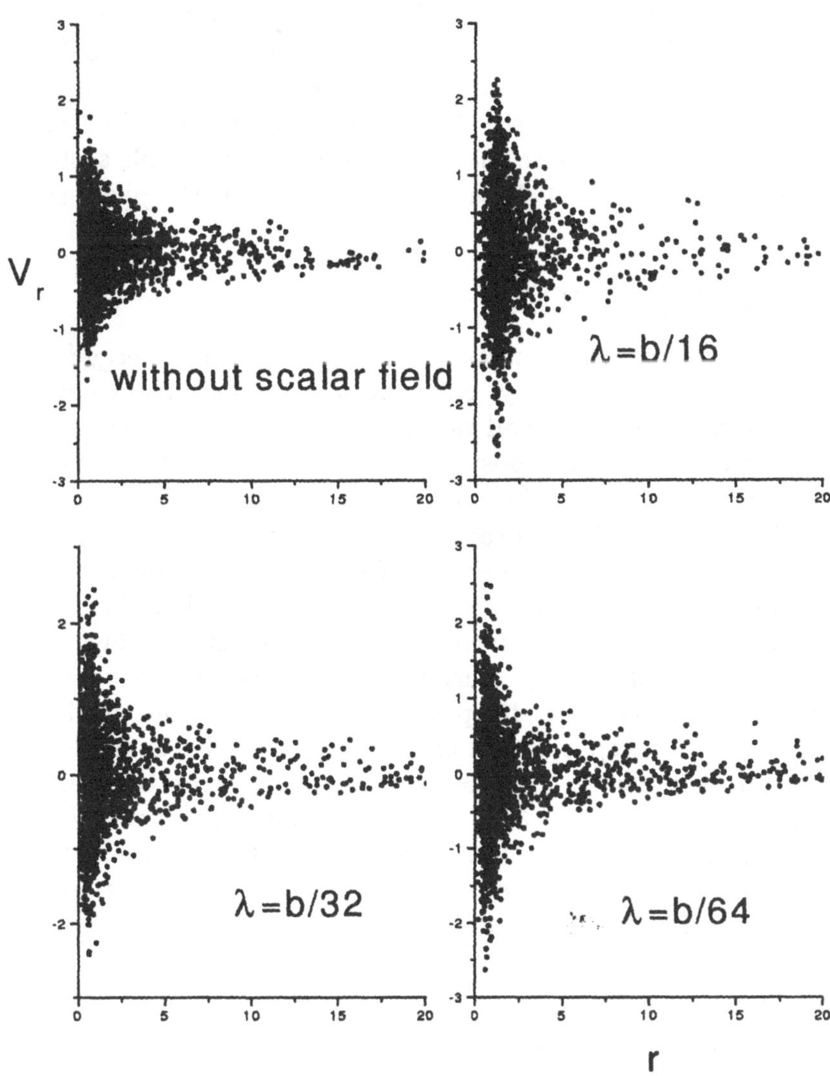

Figure 2. Phase space v_r versus r of the combined system after the transient stage without scalar field and with scalar field for different values of λ

(J/M) with respect to the center of mass of each cloud is shown in Fig. 1. Continuous lines indicate cloud 1 and dashed lines cloud 2. The first plot is without scalar field, the other plots consider values of λ of $b/16$, $b/32$, and $b/64$, respectively. We observe that the intrinsic angular momenta start from their initial values to a constant mean value approximately of 0.75 which in the cgs units is of the order of 10^{30} cm^2/s. This transient stage is slower without scalar field than the ones which consider scalar field. The faster transient occurs when λ is bigger. In Fig. 2 we show plots of phase space v_r versus r of the whole system and for the same cases as Fig. 1. The scalar field extends the phase space in the v_r direction.

4. CONCLUSIONS

We have made other computations varying the kinetic energy from 4 to 1/16 times the potential energy. For large values of the kinetic energy the deflection is small, but for small values there is a considerable deflection, and in some cases we got almost a head-on collision. This is consistent with the known fact that the merging probability in an encounter of two clouds is enhanced significantly when the encounter takes place at relatively low speed (see for instance Ref. [33]). We found that only close encounters and mergings permit the original spinless clouds to gain rotational velocities as is observed in typical galaxies nowadays. Similar studies have being done [34] considering point mass perturbers. In our approach the perturber is itself a protogalaxy and therefore the dynamics is more complicate, especially in close encounters. We have also found that the transient time to spin up the clouds depends on the scalar field. The transient stage is faster than the one without scalar field. When the scalar field is included faster transients occur for bigger values of λ. The phase space v_r versus r of the combined system is also more extended in the v_r direction with scalar field than the one without scalar field.

Acknowledgments

The computations of this paper were performed using the Silicon Graphics Origin2000 computer of the Instituto de Física, Universidad Autónoma de Puebla, México. This work was supported in part by CONACyT grants 33278-E and 33290-E, and by the joint German–Mexican project DLR–Conacyt.

References

[1] J. Barnes and L. Hernquist, *Annu. Rev. Astron. Astrophys.* **30** (1992) 705.

[2] J.E. Barnes 1998, in: *Galaxies: Interactions and Induced Star Formation.* Edited

by R.C. Kennicutt, F. Schweizer, and J.E. Barnes (Springer-Verlag, Berlin, 1998) pp. 275.

[3] E. W. Kolb and M. S. Turner, *The Early Universe*, Addison-Wesley, serie Frontiers in Physics, 1990.

[4] J. P. Ostriker and P.J.E. Peebles, *Ap. J.* **186** (1973) 467.

[5] E.R. Harrison, *Phys. Rev.* **D1** (1970) 2726.

[6] Y. B. Zel'dovich, *MNRAS* **160** (1972) 1.

[7] V. F. Mukhanov, H. A. Feldman and R. H. Brandenberger, *Phys. Reports* **215** (1992) 203.

[8] S.M. Fall and G. Efstathiou, *MNRAS* **193** (1980) 189.

[9] T. Oosterloo, *Astron. Astrophys.* **272** (1993) 389.

[10] R. Caimmi, *Astron. Astrophys.* **223** (1989) 29.

[11] R. Caimmi, *Astron. Astrophys.* **239** (1990) 7.

[12] E. Andriani and R. Caimmi *Astron. Astrophys.* **289** (1994) 1.

[13] H. Sugai and I.M. Iye *MNRAS* **276** (1995) 327.

[14] A.D. Chernin, *Astrom. Astrophys.* **267** (1993) 315.

[15] P. Catelan and T. Theuns, *MNRAS* **282** (1996) 436.

[16] S.D.M. White, *MNRAS* **184** (1978) 185.

[17] S.D.M. White, *MNRAS* **189** (1979) 831.

[18] S.D.M. White, *MNRAS* **191** (1980) 1.

[19] S.D.M. White, *Ap. J.* **274** (1983a) 53.

[20] L.A. Aguilar and S.D.M. White, *Astrophys. J.* **295** (1985) 374.

[21] T. Helbig, *Astrophys. J.* **382** (1991) 223.

[22] S.D.M. White, in: *Internal Kinematics and Dynamics of galaxies*, Editor: E. Athanassoula, Reidel, Dordrecht, The Netherlands, 1983b, pp. 337.

[23] L.M. Ozernoy, A.D. Chernin, *Soviet Astron.* **11** (1968) 907; *ibid* **12** (1968) 901.

[24] Yu.N. Parijskij, *Astrophys. J.* **180** (1973) L47.

[25] E. Boynton and R. B. Partridge, *Astrophys. J.* **181** (1973) 247.

[26] P.J.E. Peebles and J. Silk, *Nature* **346** (1990) 233.

[27] S.J. Aarseth, M. Hénon, and R. Wielen, *Astron. Astrophys.* **37** (1974) 183.

[28] L. Hernquist and N. Katz, *Astrophys. J. S.* **70** (1989) 419.

[29] L. Lucy, *A. J.* **82** (1977) 1013.

[30] R.A. Gingold and J.J. Monaghan, *MNRAS* **181** (1977) 375.

[31] L. Hernquist, *Astrophys. J. S.* **64** (1987) 715.

[32] M.A. Rodríguez-Meza, *PN-SPH code*, ININ report (2000).

[33] J. Makino and P. Hut, *Astrophys. J.* **481** (1997) 83.

[34] P.M.S. Namboodiri and R. K. Kochhar, *MNRAS* **250** (1991) 541.

ADAPTIVE CALCULATION OF A COLLAPSING MOLECULAR CLOUD CORE: THE JEANS CONDITION

Leonardo Di G. Sigalotti
Instituto Venezolano de Investigaciones Científicas (IVIC)
Carretera Panamericana Km. 11, Altos de Pipe, Estado Miranda, Venezuela

Jaime Klapp
Instituto Nacional de Investigaciones Nucleares
Ocoyoacac 52045, Estado de México, México

Abstract In 1997 Truelove et al. introduced the Jeans condition to determine what level of spatial resolution is needed to avoid artificial fragmentation during protostellar collapse calculations. They first found using a Cartesian code based on an adaptive mesh refinement (AMR) technique that a Gaussian cloud model collapsed isothermally to form a singular filament rather than a binary or quadruple protostellar system as predicted by previous calculations. Recently Boss et al. in 2000 using a different hydrodynamics code with high spatial resolution reproduced the filamentary collapse solution of Truelove et al., implying that high resolution coupled with the Jeans condition is necessary to perform reliable calculations of the isothermal protostellar collapse. Here we recalculate the isothermal Gaussian cloud model of Truelove et al. and Boss et al. using a completely different code based on zooming coordinates to achieve the required high spatial resolution. We follow the collapse through 7 orders of magnitude increase in density and reproduce the filamentary solution. With the zooming coordinates, we are allowed to perform an adaptive calculation with a much lower computational cost than the AMR technique and other grid redefinition methods.

Keywords: Molecular collapse, Newtonian codes, Jeans condition.

1. INTRODUCTION

Observations of young stars have revealed a frequency of binary companions that is at least comparable to that found for pre–main–sequence [1] and main–sequence [2] stars. In addition, the detection of multiplicity

Exact Solutions and Scalar Fields in Gravity: Recent Developments
Edited by Macias *et al.*, Kluwer Academic/Plenum Publishers, New York, 2001

among protostellar objects [3, 4, 5] suggests that most binary stars may have formed during the earliest phases of stellar evolution, probably as a consequence of protostellar collapse. In the last decade, fragmentation during the dynamic collapse of molecular cloud cores has emerged as the likely mechanism for explaining the detected binary frequency [6, 7, 8, 9]. However, studying the non–linear process of fragmentation unavoidably requires the use of three-dimensional hydrodynamical collapse calculations, which must rely upon convergence testing and intercode comparisons to validate their results.

The reliability of isothermal collapse calculations has recently been questioned by the results of Ref. [10] where it was performed calculations of a particular Gaussian cloud model employing an adaptive mesh refinement (AMR) code based on Cartesian coordinates. They found that the cloud did not fragment into a binary or quadruple system as in previous calculations [11, 12, 13, 14] but, rather, collapsed into a singular filament consistent with the self-similar solution derived in Ref. [15]. In Ref. [10] it was also found that perturbations arising from the finite-difference discretization of the gravitohydrodynamic equations may induce "artificial" fragmentation in isothermal collapse calculations for which the Cartesian cell size Δx exceeds one-fourth of the local Jeans length $\lambda_J = (\pi c_s^2/\rho G)^{1/2}$. In other words, artificial fragmentation is avoided if the mass within a cell never exceeds 1/64 of the Jeans mass $\rho\lambda_J^3$. While these findings specialize to locally uniform Cartesian grids, in Ref. [16] it was found that for a non-uniform spherical coordinate grid the Jeans mass constraint translates into requiring that any of the four lengths $[\Delta r, r\Delta\theta, r\sin\theta\Delta\phi, (r^2\sin\theta\Delta r\Delta\theta\Delta\phi)^{1/3}]$ do not exceed $\lambda_J/4$. More recent intercomparison calculations [17] have shown that satisfying the Jeans condition at the location of the maximum density combined with sufficiently high spatial resolution is necessary to change the nature of the solution from a binary to a singular filament and achieve agreement with the AMR code results of Ref. [10]. High resolution is therefore necessary to produce reliable calculations of the collapse and fragmentation of molecular cloud cores.

Here we present a further intercode comparison for the same Gaussian cloud model of Refs. [10] and [17] Boss. The calculation was made using a modified version of the spherical-coordinate based code described in [18], which adopts zooming coordinates to achieve the required high spatial resolution. These coordinates were previously used in Ref. [19] to study disk and bar formation during the isothermal collapse phase of a molecular cloud core up to extremely high density contrasts. With the zooming coordinates, the computational cells shrink adaptively in accordance with the cloud collapse and the spatial resolution increases in proportion to the Jeans length as the density enhances. In this way, the four Jeans conditions introduced in Ref. [16] are automatically satisfied. With respect to the AMR framework, the zooming coordinates provide an effective refinement method which works at a much lower computational cost and without any detailed control. We follow the collapse of the Gaussian cloud model through 7 orders of magnitude increase in density without violating the Jeans condition and reproduce the filamentary solution obtained in Refs. [10, 17].

2. INITIAL MODEL AND COMPUTATIONAL METHODS

The initial conditions chosen for our protostellar collapse model corresponds to a centrally condensed, rapidly rotating cloud of $1M_\odot$ and radius $R = 5.0 \times 10^{16}$ cm ≈ 0.016 pc. The central condensation is obtained assuming a Gaussian density profile of the form $\rho(r) = \rho_0 \exp[-(r/b)^2]$, where $\rho_0 = 1.7 \times 10^{-17}$ g cm^{-3} is the initial central density and $b \approx 0.577R$ is a length chosen such that the central density is 20 times the density at the spherical cloud boundary. In addition, a small amplitude ($a = 0.10$), $m = 2$ density perturbation of the form $\rho(r, \phi) = \rho(r)[1 + a\cos(m\phi)]$ is imposed on the background spherically symmetric density distribution. The model has ideal gas thermodynamics at a fixed temperature of 10 K and a chemical composition of $X = 0.769$, $Y = 0.214$, $Z = 0.017$ corresponding to a mean molecular weight of $\mu \approx 2.28$. These parameters lead to an initial ratio of the thermal to the absolute value of the gravitational energy of $\alpha \approx 0.265$. With uniform rotation at the rate $\omega = 1.0 \times 10^{-12}$ s^{-1}, the initial ratio of the rotational to the absolute value of the gravitational energy is $\beta \approx 0.16$. The isothermal approximation is valid for densities roughly in the range of $\sim 10^{-19}$ to $\sim 10^{-13}$ g cm^{-3}. The collapse of the Gaussian cloud is assumed to remain isothermal beyond densities of $\sim 10^{-13}$ g cm^{-3} to check the ability of the code to reproduce the filamentary solution encountered in Refs. [10, 17] working at very high spatial resolution.

Previous low-resolution calculations for this model not obeying the Jeans condition advanced in Ref. [10] produced a quadruple system [11, 12, 13]. Evidently, these low resolution models suffered from artificial fragmentation - unphysical growth of numerical noise caused by insufficient spatial resolution - as first suggested in Ref. [10]. Starting with low resolution ($n_r = 50$ and $n_\phi = 64$), in Ref. [16] it is recalculated the collapse of the isothermal Gaussian cloud by allowing inward motion of the radial grid in such a way as to satisfy the Jeans mass

$$\Delta x = (\Delta x_r \Delta x_\theta \Delta x_\phi)^{1/3} = (r^2 \sin\theta \Delta r \Delta\theta \Delta\phi)^{1/3} < \frac{\lambda_J}{4}, \tag{1}$$

and the three Jeans length conditions

$$\begin{aligned} \Delta x_r &= \Delta r < \frac{\lambda_J}{4}, \\ \Delta x_\theta &= r\Delta\theta < \frac{\lambda_J}{4}, \\ \Delta x_\phi &= r\sin\theta\Delta\phi < \frac{\lambda_J}{4}, \end{aligned} \tag{2}$$

for spherical coordinates, and still obtained fragmentation into a well-defined binary. Only when the spatial resolution was increased to $n_r = 200$ and $n_\phi = 128$, the solution changed behaviour [17]. That is, instead of fragmenting into a binary, the central collapse produced a bar which thereafter condensed into a thin filament in good agreement with the AMR results of Ref. [10]. Thus, it appears that for grids based on spherical coordinates, the four Jeans conditions (1) and (2) are necessary but not sufficient for realistic fragmentation, and that these must be combined with sufficiently high spatial resolution to ensure a converged solution. In Ref. [17] it is argued that one reason for requiring more numerical resolution in spherical coordinate calculations is the occurrence of high-aspect ratio cells near the center, where $\Delta x_r/\Delta x_\theta$ and $\Delta x_r/\Delta x_\phi$ may become large compared with a locally uniform Cartesian grid with unit-aspect ratio.

Here we have recalculated the isothermal Gaussian cloud model using a modified version of the spherical coordinate (r, θ, ϕ) code described in Ref [18] . In order to maintain a numerical resolution higher than that demanded by the Jeans conditions (1) and (2), the gravitohydrodynamic equations are solved in the zooming coordinates, defined by [19]

$$\xi = (\hat{r}, \theta, \phi) = \left[\frac{r}{c_s(t_0 - t)}, \theta, \phi\right], \tag{3}$$

$$\tau = -\ln(t_0 - t), \tag{4}$$

where c_s denotes the isothermal speed of sound and the time difference $t_0 - t$ is in units of the central free-fall time $\tau_{ff} = \sqrt{3\pi/32G\rho_0}$.

With these transformations, the gravitohydrodynamic equations can be written as

$$\frac{\partial d}{\partial \tau} + \nabla_\xi \cdot (d\mathbf{u}) = d, \tag{5}$$

$$\frac{\partial (d\mathbf{u})}{\partial \tau} + \nabla_\xi \cdot (d\mathbf{u} \otimes \mathbf{u}) + \nabla_\xi d + d\nabla_\xi \Psi = 2d\mathbf{u}, \tag{6}$$

$$\nabla_\xi^2 \Psi = d, \tag{7}$$

where ∇_ξ denotes the gradient operator in terms of the zooming coordinates. The mass-density ρ, the fluid velocity $\mathbf{v} = (v_r, v_\theta, v_\phi)$ and the gravitational potential Φ transform according to the following relations

$$d = 4\pi G(t_0 - t)^2 \rho, \tag{8}$$

$$\mathbf{u} = (u_r, u_\theta, u_\phi) = \left[\frac{v_r}{c_s} + \hat{r}, \frac{v_\theta}{c_s}, \frac{v_\phi}{c_s}\right], \tag{9}$$

$$\Psi = c_s^{-2}\Phi, \tag{10}$$

respectively. Note that d, $\mathbf{u}$ and Ψ in the transformed zooming space are non-dimensional quantities.

In the zooming coordinates, the radial length and time are proportional to $t_0 - t$ for fixed $\Delta\hat{r}$ and $\Delta\tau$. This allows the numerical resolution to change adaptively in accordance with the density increase since $\rho \propto (t_0 - t)^{-2}$. The equations of continuity (5) and momentum (6) are almost invariant under the transformation, except for the source terms on the right-hand side, whereas the Poisson equation (7) retains essentially the same form. Therefore, the spherical coordinate code described in Ref. [18] can be used for solving equations (5)-(7) with only minor changes. As in the previous version, the code employs a multi-step solution procedure to advance the fluid variables d and $\mathbf{u}$ through the use of spatially second-order accurate finite-difference methods. Advection of mass is performed using a generalization to spherical coordinates of the van Leer monotonic interpolation scheme. Temporal second-order accuracy is achieved by means of a predictor-corrector treatment of the convective and source hydrodynamical terms. The Poisson equation is solved by a spherical harmonic expansion, including terms up to $l, m = 16$. This procedure allows for a separation of the radial and angular variables such that the three-dimensional equation can be replaced by a set of ordinary differential equations for the amplitudes $\Psi_{lm}(r)$, which are efficiently solved by means of a tridiagonal matrix algorithm. The $\hat{r}$ grid is uniformly spaced and kept fixed during the calculation. Setting $t_0 = 1.40\tau_{ff}$, a radial resolution consisting of $1 + n_{\hat{r}}$=51 points (including the origin $\hat{r} = 0$) is enough to guarantee satisfaction of the Jeans conditions (1) and (2) through 7 orders of magnitude increase in density. The θ grid has $n_\theta = 34$ fairly equidistant points for $0 \leq \theta \leq \pi/2$, with symmetry through the cloud's equatorial plane being used to represent the bottom hemisphere of the cloud. The ϕ grid is also uniformly spaced with $n_\phi = 64$ for $0 \leq \phi < \pi$, corresponding to $n_\phi = 128$ effective points due to the assumption of π-symmetry. Thus, in a Fourier decomposition ($e^{im\phi}$) only even modes can grow. At the external surface of the

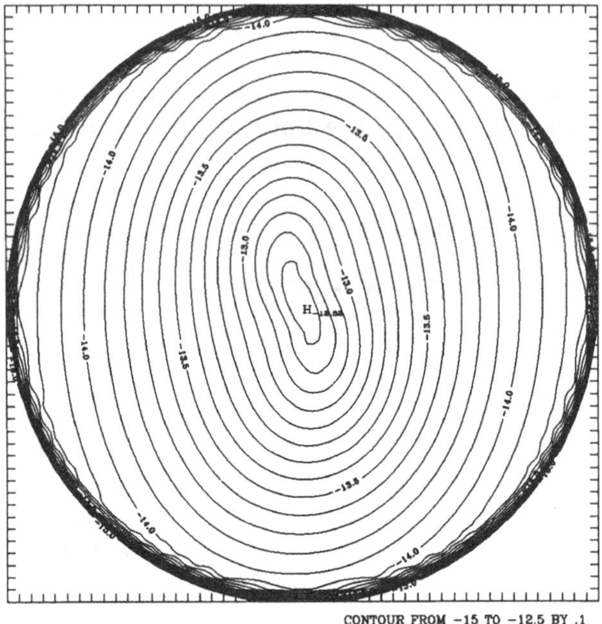

Figure 1. Density contours in the equatorial plane at $t/\tau_{ff} = 1.35$, when $\rho_{max} \approx 19118\rho_0$. The whole computational grid is shown and the density is expressed by its values in the ordinary coordinates. At this time, the grid radius is $\approx 1.76 \times 10^{15}$ cm, corresponding to a factor of 28.4 times smaller than the initial cloud radius. Formation of a central barlike core is evident.

grid we impose an outflow boundary condition such that no disturbances can propagate into the domain of computation from the boundary surface. This boundary condition is naturally implemented because the gas flows out supersonically from the domain of computation. The supersonic outflow is due to the apparent motion with respect to the zooming coordinates.

The present calculation provides a further intercomparison test at high resolution for the isothermal Gaussian cloud using a numerical framework that is completely different to those employed in Refs. [10, 17]. With the present zooming grid resolution, the calculation took about 1.0×10^5 timesteps $\Delta\tau$ to complete the evolution, corresponding to about 30 hr of CPU time on an Origin 2000 machine.

3. RESULTS

The Gaussian cloud model can increase in density by about 4 orders of magnitude before the isothermal approximation begins to fail. In real molecular cloud cores significant non-isothermal heating will take place once $\rho > 10^{-13}$ g cm^{-3}, rendering the isothermal approximation invalid. However, for the purposes of this paper we ignore the effects

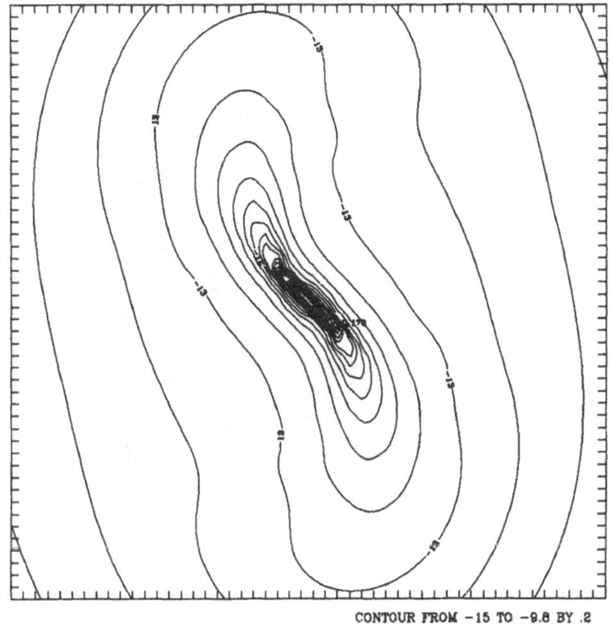

Figure 2. Density contours in the equatorial plane at $t/\tau_{ff} = 1.362$, when $\rho_{max} \approx 1.03 \times 10^7 \rho_0$. A radius of 8.0×10^{14} cm is shown and the density is expressed by its values in the ordinary coordinates. At this time, the grid radius is $\approx 1.37 \times 10^{15}$ cm, corresponding to a factor of 36.6 times smaller than the initial cloud radius. Formation of a dense filamentary structure is evident.

of radiative transfer and assume the collapse to remain isothermal at densities higher than 10^{-13} g cm^{-3} in order to test the ability of the code to reproduce the filamentary collapse solution expected at high spatial resolution. Since the collapse is calculated in the zooming grid, the radial cell sizes in the ordinary coordinates shrink in proportion to the time difference $t_0 - t$. When t_0 is smaller, the central cloud region is zoomed with higher magnification. A value of $t_0 = 1.40\tau_{ff}$ is enough to ensure sufficient high resolution to resolve the Jeans length λ_J through 7 orders of magnitude increase in density.

During the first free-fall time, the cloud collapses slowly to form a rotationally supported disk about the equatorial plane. In this phase, the gas is mainly seen to flow outwards in the zooming coordinates corresponding to a general mild infall in the ordinary coordinates. After about $1.28\tau_{ff}$, the collapse of the central cloud regions becomes progressively more pronounced leading to a more rapid increase of the central density contrast. The zooming radial velocity $u_r = v_r/c_s + \hat{r}$ inside the disk decreases as v_r/c_s increases in magnitude compared to $\hat{r}$. As the central collapse continues to even higher densities, these terms roughly balance making $u_r \approx 0$. When this occurs, the central cloud region

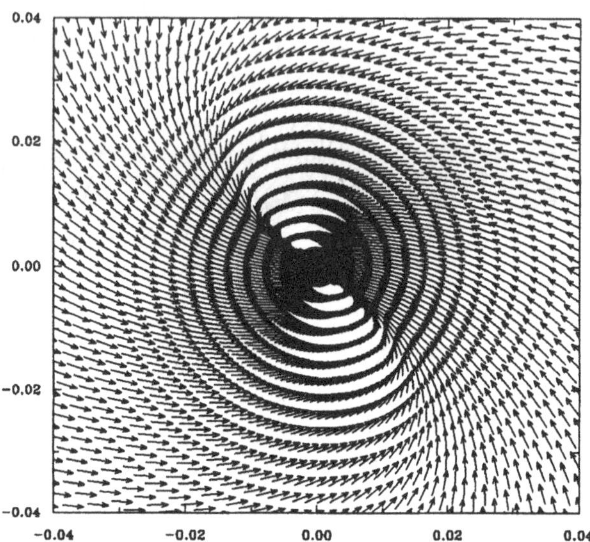

Figure 3. Velocity vectors in the equatorial plane at the same time of Fig. 2. A radius of 4.0×10^{14} cm is shown. Every fourth radial grid point is displayed. The maximum velocity is 0.696 km s^{-1}.

approaches a rotating steady state in the zooming space. In the ordinary coordinates, the central core is seen to rotate non-uniformly and by $1.32\tau_{ff}$, it starts deforming into an elongated bar as a result of the growth of a non-axisymmetric $m = 2$ instability. Figure 1 displays equatorial density contours of the central cloud portion at $1.35\tau_{ff}$, when a dense bar has already formed. In Ref. [19] it is studied the mechanisms responsible for bar formation within a collapsing disk using the zooming coordinates. They found that the growth of the shear component of velocity may induce a non-uniformly rotating disklike core to elongate into a barlike configuration. Subsequent collapse of the central bar is seen to proceed self-similarly. That is, the bar shrinks faster in the direction of the cylindrical radius than along the long axis. As a consequence the collapsing bar becomes progressively narrower and a filamentary structure eventually forms as shown in Figure 2 at $1.362\tau_{ff}$. At this time, the density of the filament is $\sim 1.0 \times 10^7 \rho_0$ and because of the assumption of isothermality it will undergo undefinite collapse upon itself. This solution strongly resembles the self-similar solution for isothermal collapse first found in Ref. [15]. They predicted by linear analysis that when the mass per unit length of an infinitely long cylinder is much greater than the equilibrium value $(M/L)_e = 2c_s^2/G$, fragmentation of the cylinder will not occur as long as the collapse remains isothermal. The details of

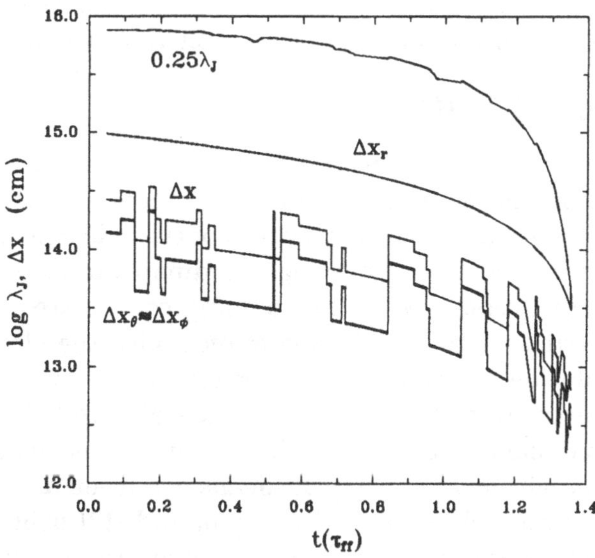

Figure 4. Dependence of the four spherical grid sizes and the Jeans length on evolution time, evaluated at the location of the maximum density in the equatorial plane. The angular resolution in the zooming coordinates is such that $\Delta x_\theta \approx \Delta x_\phi$ at the midplane.

the collapse of the filament upon itself are clarified in Figure 3, which depicts the velocity field in the equatorial plane at the same time of Figure 2.

By $1.362\tau_{ff}$, the first radial point r_1 is at $\sim 3.16 \times 10^{13}$ cm from the origin $r = 0$. This radial resolution is equivalent to having started the calculation with roughly 1700 equidistant radial points in the ordinary space. Figure 4 describes the evolution of the four grid sizes and the Jeans length during the cloud collapse. Evidently, satisfaction of the Jeans conditions combined with sufficiently high spatial resolution allows us to follow the isothermal collapse up to the point where filament formation is expected. The present solution is in very good agreement with the results of the AMR code calculation in Ref. [10] and the refined spherical code calculation [17]. The agreement between these three independent codes, each working at very high spatial resolution, implies that the filamentary solution is the correct one. Although the outcome of the isothermal collapse of the Gaussian cloud is shown to be a singular filament, a physically realistic solution would require the inclusion of non-isothermal heating. In particular, in Ref. [17] it was showed that thermal retardation of the collapse may allow the Gaussian cloud to fragment into a binary protostellar system at the same maximum density where the isothermal collapse yields a thin filament. Thus, fragmentation of the Gaussian cloud model appears to depend sensitively on both the numerical resolution and the detailed thermodynamical treatment.

4. CONCLUSIONS

Here we have presented a further intercode comparison with high spatial resolution for the same Gaussian cloud model of Refs. [10] and [17]. The calculation was made using a variant of the spherical code described in Ref. [18], which employs zooming coordinates to achieve sufficiently high spatial resolution. With the zooming coordinates, the computational cells shrink adaptively in accordance with the cloud collapse so that the resolution increases in proportion to the Jeans length as the density enhances. In this framework, we are able to follow the collapse of the Gaussian cloud model through 7 orders of magnitude increase in central density without violating the Jeans condition and reproduce the filamentary solution obtained in Refs. [10] and [17] using a completely independent numerical method. The present approach can compete with the AMR methodology and other grid refinement techniques to provide unprecedentedly high resolution in collapse calculations.

The isothermal collapse of the Gaussian cloud model is an excellent test case for any gravitational hydrodynamics code and should then be used to check the reliability of the results for the isothermal phase of collapse calculations. This paper represents a further step in this direction and provides extra evidence that the outcome of the isothermal collapse of the Gaussian cloud is the formation of a singular filament rather than the fragmentation into two or more clumps as found in previous low-resolution calculations.

Acknowledgments

The calculation of this paper was performed on the Origin 2000 supercomputer of the Instituto Mexicano del Petroleo (IMP). This work was partially supported by the Consejo Nacional de Investigaciones Científicas y Tecnológicas (CONICIT) of Venezuela and by the Consejo Nacional de Ciencia y Tecnología (CONACyT) of México.

References

[1] R. Mathieu, *ARA&A* **32** (1994) 465.

[2] A. Duquennoy and M. Mayor, *A&A* **248** (1991) 485.

[3] O.P. Lay, J.E. Carlstrom and R.E. Hills, *Astrophys. J.* **452** (1995) L73.

[4] L.W. Looney, L.G. Mundy and W.J. Welch, *Astrophys. J.* **484** (1997) L157.

[5] S. Terebey, D. van Buren, D.L. Padgett, T. Hancock and M. Brundage, *Astrophys. J.* **507** (1998) L71.

[6] A.P. Boss, in: *The Realm of Interacting Binary Stars*, ed. J. Sahade, G. McCluskey, and Y. Kondo, Dordrecht, (Kluwer Academic, 1993a) pp. 355.

[7] P. Bodenheimer, T. Ruzmaikina and R.D. Mathieu, in: *Protostars and Planets III*, ed. E.H. Levy and J.I. Lunine, Tucson: Univ. Arizona Press, 1993, pp. 367.

[8] P. Bodenheimer, A. Burkert, R.I. Klein and A.P. Boss, in: *Protostars and Planets IV*, ed. V.G. Mannings, A.P. Boss and S.S. Russell, Tucson: Univ. Arizona Press, 2000, in press.

[9] L. Di G. Sigalotti and J. Klapp, *Int. J. Mod. Phys.* D (2001), in press.

[10] J.K. Truelove, R.I. Klein, C. F. McKee, J. H. Holliman, L. H. Howell, and J.A. Greenough, *Astrophys. J.* **489** (1997) L179.

[11] A.P. Boss, *Nature* **351** (1991) 298.

[12] A.P. Boss, *Astrophys. J.* **410** (1993b) 157.

[13] J. Klapp, L. Di G. Sigalotti and F. de Felice, *A&A* **273** (1993) 175.

[14] A. Burkert and P. Bodenheimer, *MNRAS* **280** (1996) 1190.

[15] S. Inutsuka and S. M. Miyama, *Astrophys. J.* **388** (1992) 392.

[16] A.P. Boss, *Astrophys. J.* **501** (1998) L77.

[17] A.P. Boss, R. T. Fisher, R.I. Klein and C. F. McKee, *Astrophys. J.* **528** (2000) 325.

[18] L. Di G. Sigalotti, *Astrophys. J. S.* **116** (1998) 75.

[19] T. Matsumoto and T. Hanawa, *Astrophys. J.* **521** (1999) 659.

REVISITING THE CALCULATION OF INFLATIONARY PERTURBATIONS.

César A. Terrero–Escalante
Departamento de Física,
Centro de Investigación y de Estudios Avanzados del IPN,
Apdo. Postal 14–740, 07000, México, D.F., México.

Dominik J. Schwarz
Institut für Theoretische Physik, TU-Wien,
Wiedner Hauptstrass e 8–10, A–1040 Wien, Austria.

Alberto Garcia
Departamento de Física,
Centro de Investigación y de Estudios Avanzados del IPN,
Apdo. Postal 14–740, 07000, México, D.F., México.

Abstract We present a new approximation scheme that allows us to increase the accuracy of analytical predictions of the power spectra of inflationary perturbations for two specific classes of inflationary models. Among these models are chaotic inflation with a monomial potential, power-law inflation and natural inflation (inflation at a maximum). After reviewing the established first order results we calculate the amplitudes and spectral indices for these classes of models at next order in the slow–roll parameters for scalar and tensor perturbations.

1. INTRODUCTION

Inflationary cosmology [1] is facing exciting times due to a new generation of ground and satellite based experiments to be carried out (e.g., the SDSS, MAP and Planck experiments [2]). Upcoming observations will allow us to determine the values of the cosmological parameters with high confidence. To measure the values of these parameters it is necessary to make assumptions on the initial conditions of the density fluctuations that evolved into the observed large scale structure and CMB anisotropies.

In the simplest inflationary scenario a single scalar field ϕ drives the inflationary expansion of the Universe. Cosmological perturbations are generated by the quantum fluctuations of this scalar field and of space-time. Perturbations are adiabatic and gaussian and are characterized by their power spectra. Usually these spectra are described in terms of an amplitude at a pivot scale and the spectral index at this scale. For general models, these quantities are difficult to compute exactly. The state–of–the–art in such calculations are the approximated expressions due to Stewart and Lyth [3], which are obtained up to next–to–leading order in terms of an expansion of the so–called slow–roll parameters. This expansion allows to approximate the solutions to the equations of motion by means of Bessel functions.

To reliably compare analytical predictions with measurements, an error in the theoretical calculations of some percents below the threshold confidence of observations is required. In Ref. [4] it was shown that amplitudes of next–to–leading order power spectra can match the current level of observational precision. However, the error in the spectral index and the resulting net error in the multipole moments of the cosmic microwave background anisotropies might be large due to a long lever arm for wave numbers far away from the pivot scale [5]. A clever choice of the pivot point is essential [5, 6] for todays and future precision measurements. The slow–roll expressions as calculated in [3] are not precise enough for the Planck experiment [5].

In this paper we introduce another formalism that allows us to calculate the amplitudes and indices of the scalar and tensorial perturbations up to higher orders for two specific classes of inflationary models. The first class contains all models that are 'close' to power–law inflation, one example is chaotic inflation with $V \propto \phi^{\alpha}$. Our second class of models is characterized by extremely slow rolling, an example is inflation near a maximum.

2. THE STANDARD FORMULAS

The slow–roll parameters are defined as, [8],

$$\epsilon(\phi) \equiv= \frac{2}{\kappa}\left[\frac{H'}{H}\right]^2, \quad \eta(\phi) \equiv \frac{2}{\kappa}\frac{H''}{H}, \quad \xi(\phi) \equiv \frac{2}{\kappa}\left(\frac{H'H'''}{H^2}\right)^{1/2}, \tag{1}$$

with the equations of motion

$$\frac{\dot{\epsilon}}{H} = 2\epsilon(\epsilon - \eta), \quad \frac{\dot{\eta}}{H} = \epsilon\eta - \xi^2. \tag{2}$$

H is the Hubble rate, dot and prime denote derivatives with respect to cosmic time and ϕ, $\kappa = 8\pi/m_{\rm Pl}^2$, and $m_{\rm Pl}$ is the Planck mass. By definition $\epsilon \geq 0$ and it has to be less than unity for inflation to proceed.

2.1. AMPLITUDES OF INFLATIONARY PERTURBATIONS

We call *standard* those formulas which are considered to be the state–of–the–art in the analytical calculation of perturbations spectra, i.e, those obtained by Stewart and Lyth [3].

The general expression for the spectrum of the curvature perturbations is [3]

$$\mathcal{P}_R{}^{1/2}(k) = \sqrt{\frac{k^3}{2\pi^2}}\left|\frac{u_k}{z}\right| . \tag{3}$$

$u_k(\tau)$ are solutions of thc mode equation [3, 7]

$$\frac{d^2u_k}{d\tau^2} + \left(k^2 - \frac{1}{z}\frac{d^2z}{d\tau^2}\right)u_k = 0 , \tag{4}$$

where τ is the conformal time and z is defined as $z \equiv a\dot{\phi}/H$ with a denoting the scale factor. The potential of the mode equation (4) reads [3, 8]

$$\frac{1}{z}\frac{d^2z}{d\tau^2} = 2a^2H^2\left(1 + \epsilon - \frac{3}{2}\eta + \epsilon^2 - 2\epsilon\eta + \frac{1}{2}\eta^2 + \frac{1}{2}\xi^2\right) . \tag{5}$$

Despite its appearance as an expansion in slow–roll parameters, Eq. (5) is an exact expression.

The crucial point in the Stewart and Lyth calculations is to use the solution for power–law inflation (where the slow–roll parameters are constant and equal each other) as a pivot expression to look for a general solution in terms of a slow–roll expansion. An answer to whether the slow–roll parameters can be regarded as constants that differ from each other is found it looking at the exact equations of motion (2).

The standard procedure is to consider the slow–roll parameters so small that second order terms in any expression can be neglected. Now aH in (5) can be replaced with help of $(aH)^{-1} \simeq -\tau(1-\epsilon)$. Then according with Eq. (2) the slow–roll parameters can be fairly regarded as constants and Eq. (4) becomes a Bessel equation readily solved. From that solution the scalar amplitudes are written as [3]

$$\mathcal{P}_R{}^{1/2}(k) = 2^{\nu-\frac{1}{2}}\frac{\Gamma(\nu)}{\Gamma(\frac{3}{2})}(1-\epsilon)^{\nu-\frac{1}{2}}\frac{1}{m_{Pl}^2}\left.\frac{H^2}{|H'|}\right|_{k=aH} , \tag{6}$$

where ν is given by

$$\nu = \frac{1+\epsilon-\eta}{1-\epsilon} + \frac{1}{2}, \tag{7}$$

and k is the wavenumber corresponding to the scale matching the Hubble radius. Expanding solution (6) with ν given by $\nu = 3/2 + 2\epsilon - \eta$ and truncating the results to first order in ϵ and η the standard general expression for the scalar spectrum is obtained,

$$\mathcal{P}_R{}^{1/2} = \frac{\kappa}{4\pi}\left[1-(2C+1)\epsilon + C\eta\right] \frac{H^2}{|H'|}\bigg|_{k=aH}, \tag{8}$$

where $C = -2 + \ln 2 + \gamma \simeq -0.73$ is a numerical constant, and γ is the Euler constant that arises when expanding the Gamma function. Eq. (8) is called the next-to-leading order expression for the spectrum amplitudes of scalar perturbations, and from it the leading order is recovered by neglecting first order terms for ϵ and η.

The corresponding equation of motion for the tensorial modes is

$$\frac{d^2 v_k}{d\tau^2} + \left(k^2 - \frac{1}{a}\frac{d^2 a}{d\tau^2}\right) v_k = 0, \tag{9}$$

where,

$$\frac{1}{a}\frac{d^2 a}{d\tau^2} = 2a^2 H^2 \left(1 - \frac{1}{2}\epsilon\right). \tag{10}$$

Neglecting any order of ϵ higher than the first one, Eq. (10) can be written as

$$\frac{1}{a}\frac{d^2 a}{d\tau^2} = \frac{1}{\tau^2}\left(\mu^2 - \frac{1}{4}\right), \tag{11}$$

where,

$$\mu = \frac{1}{1-\epsilon} + \frac{1}{2} \simeq \frac{3}{2} + \epsilon. \tag{12}$$

This way, Eq. (9) can be also approximated as a Bessel equation with solution,

$$\mathcal{P}_g^{1/2}(k) = 2^{\nu-\frac{3}{2}} \frac{\Gamma(\nu)}{\Gamma(\frac{3}{2})} (1-\epsilon)^{\nu-\frac{1}{2}} \frac{\sqrt{2}\kappa}{\pi} H \bigg|_{k=aH}. \tag{13}$$

Substituting μ given by Eq. (12) in Eq. (13), expanding on ϵ and truncating to first order, the Stewart–Lyth next-to-leading order result is obtained

$$\mathcal{P}_g^{1/2} = \left[1 - (C+1)\epsilon\right] \frac{\sqrt{2}\kappa}{\pi} H \bigg|_{k=aH}. \tag{14}$$

The leading order equation is recovered by neglecting ϵ.

2.2. THE SPECTRAL INDICES

To derive the expressions for the spectral indices we introduce here a variation of the standard procedure that has the advantage of being more comprehensive while dealing with the order of expressions to be derived using the slow–roll expansion.

First, let us assume the following ansatz for the power spectrum of general inflationary models,

$$\mathcal{P}_R{}^{1/2} = h(\epsilon,\eta,\xi) f(\epsilon,\eta,\xi)\,, \tag{15}$$

where $f(\epsilon,\eta,\xi)$ may be written as a Taylor expansion while $h(\epsilon,\eta,\xi)$ is a general function which, in principle, can be non–continuous in $\epsilon = \eta = \xi = 0$. Note that any function can be decomposed into this form. We must also consider the following function of the slow–roll parameters:

$$\frac{\kappa}{2}\frac{H}{H'}\frac{d\phi}{d\ln k} = -\frac{1}{1-\epsilon} \equiv g(\epsilon,\eta,\xi)\,. \tag{16}$$

In this expression all of the parameters are to be evaluated at values of ϕ corresponding to $k = aH$. The functions f and g can be expanded in Taylor series,

$$\begin{aligned} f(\epsilon,\eta,\xi) &= a_{00} + a_{10}\epsilon + a_{20}\eta + a_{30}\xi \\ &+ a_{11}\epsilon^2 + a_{12}\epsilon\eta + a_{13}\epsilon\xi + a_{22}\eta^2 + a_{23}\eta\xi + a_{33}\xi^2 \cdots, \quad (17) \\ g(\epsilon,\eta,\xi) &= -(1+\epsilon+\epsilon^2+\cdots)\,. \quad (18) \end{aligned}$$

We proceed with the derivation of the equations for the scalar spectral index

$$n_S(k) - 1 \equiv \frac{d\ln\mathcal{P}_R}{d\ln k}\,. \tag{19}$$

From Eq. (15) we obtain

$$\begin{aligned} \frac{d\ln\mathcal{P}_R{}^{1/2}}{d\phi} &= \frac{d\ln h(\epsilon,\eta,\xi)}{d\phi} + \frac{d}{d\phi}\left[P(\epsilon,\eta,\xi) - \frac{P^2(\epsilon,\eta,\xi)}{2}\right. \\ &+ \left.\frac{P^3(\epsilon,\eta,\xi)}{3} - \cdots\right], \quad (20) \end{aligned}$$

where

$$\begin{aligned} P(\epsilon,\eta,\xi) &= \tilde{a}_{10}\epsilon + \tilde{a}_{20}\eta + \tilde{a}_{30}\xi + \tilde{a}_{11}\epsilon^2 + \tilde{a}_{12}\epsilon\eta + \tilde{a}_{13}\epsilon\xi + \tilde{a}_{22}\eta^2 \\ &+ \tilde{a}_{23}\eta\xi + \tilde{a}_{33}\xi^2 + \cdots\,, \quad (21) \end{aligned}$$

and $\tilde{a}_{ij} \equiv a_{ij}/a_{00}$. After differentiation,

$$\frac{d\ln\mathcal{P}_R{}^{1/2}}{d\phi} = \frac{1}{h}\frac{dh}{d\phi} + \left(1 - P(\epsilon,\eta,\xi) + P^2(\epsilon,\eta,\xi) - \cdots\right) P'(\epsilon,\eta,\xi,\epsilon',\eta',\xi')\,, \tag{22}$$

where,

$$\begin{aligned} P' &\equiv \tilde{a}_{10}\epsilon' + \tilde{a}_{20}\eta' + \tilde{a}_{30}\xi' + 2\tilde{a}_{11}\epsilon\epsilon' + \tilde{a}_{12}(\epsilon\eta)' + \tilde{a}_{13}(\epsilon\xi)' \\ &+ 2\tilde{a}_{22}\eta\eta' + \tilde{a}_{23}(\eta\xi)' + 2\tilde{a}_{33}\xi\xi' \end{aligned} \quad (23)$$

plus higher order derivatives. Using Eqs. (16) and (18), and the definitions of the slow–roll parameters we obtain that

$$\frac{n_S - 1}{2} = \left[\sqrt{\frac{2}{\kappa}}\sqrt{\epsilon}\frac{h'}{h} + \left(1 - P + P^2 - \cdots\right)P'\right](1 + \epsilon + \epsilon^2 + \epsilon^3 + \cdots), \quad (24)$$

where P' is written now as,

$$\begin{aligned} P' &= 2\tilde{a}_{10}\epsilon(\epsilon - \eta) + \tilde{a}_{20}(\epsilon\eta - \xi^2) + \tilde{a}_{30}\sqrt{\frac{2}{\kappa}}\epsilon\,\xi' + 4\tilde{a}_{11}\epsilon^2(\epsilon - \eta) + \cdots \\ &+ 2\tilde{a}_{22}\eta(\epsilon\eta - \xi^2)\cdots. \end{aligned} \quad (25)$$

Expression (24) can be used to any order whenever information on the function h and the coefficients of the expansion (17) is available. The standard result of Stewart and Lyth, Eq. (8), has been tested as a reliable approximation for the scalar spectrum of general inflationary models [4]. Then, it is reasonable to assume

$$h(\epsilon, \eta, \xi) = \frac{\kappa}{4\pi}\frac{H^2}{|H'|}\bigg|_{k=aH} = \frac{1}{2\pi}\sqrt{\frac{\kappa}{2}}\frac{H}{\sqrt{\epsilon}}, \quad (26)$$

$$a_{00} = 1, \quad a_{10} = -(2C+1), \quad a_{20} = C. \quad (27)$$

With these assumptions, it is obtained that

$$\sqrt{\frac{2}{\kappa}}\sqrt{\epsilon}\frac{h'}{h} = -2\epsilon + \eta, \quad (28)$$

and to second order (the order for which information on the coefficients for (25) is available from Eq. (8)), expression (24) reduces to

$$n_S - 1 = -4\epsilon + 2\eta - 8(C+1)\epsilon^2 + 2(5C+3)\epsilon\eta - 2C\xi^2, \quad (29)$$

which is the standard result for the scalar spectral index.

The equation of the tensorial index can be derived in a similar way. By definition

$$n_T(k) \equiv \frac{d\ln \mathcal{P}_g}{d\ln k}. \quad (30)$$

The standard equation for tensorial modes is given by Eq. (14). If we now use expression (24) substituting $n_S - 1$ by n_T and

$$h(\epsilon, \eta, \xi) = \frac{\sqrt{2\kappa}}{\pi}H, \quad (31)$$

$$a_{00} = 1, \quad a_{10} = -(C+1), \quad (32)$$

then we obtain

$$\sqrt{\frac{2}{\kappa}}\sqrt{\epsilon}\frac{h'}{h} = -\epsilon\,, \tag{33}$$

and the standard expression,

$$n_T = -2\epsilon - 2(2C+3)\epsilon^2 + 4(C+1)\epsilon\eta\,. \tag{34}$$

3. GENERALIZING THE BESSEL APPROXIMATION

As can be observed from Eq. (2), the standard assumption used to approximate the slow–roll parameters as constants can not be used beyond the linear term in the slow–roll expansion. Hence, from this point of view, the feasibility of using the Bessel equation to calculate the power spectra is limited to this order. Nevertheless, what it is actually needed is the right hand sides of Eqs. (2) being negligible. That can also be achieved if $\upsilon \equiv \epsilon - \eta \ll \epsilon$ (we shall call this condition generalized power-law approximation) or if $\epsilon \ll \upsilon$ (generalized slow–roll approximation). In both cases we require $\dot{\upsilon}H^{-1} = \xi^2 - \epsilon^2 + 3\epsilon\upsilon \ll \min(\epsilon, \upsilon)$.

3.1. GENERALIZED POWER–LAW APPROXIMATION

Power–law inflation is the model which gives rise to the commonly used power–law shape of the primordial spectra, although not always properly implemented [6]. The assumption of a power–law shape of the spectrum has been successful in describing large scale structure from the scales probed by the cosmic microwave background to the scales probed by redshift surveys. It is reasonable to expect that the actual model behind the inflationary perturbations has a strong similarity with power–law inflation. For this class of potentials the precision of the power spectra calculation can be increased while still using the Bessel approximation. If we have a model with $\upsilon \propto \epsilon^n$, then, according with Eq. (2), the first slow–roll parameter can be considered as constant if terms like ϵ^{n+1} and with higher orders are neglected. These considerations implies the right hand side of the second equation in (2) to be also negligible and, this way, η can be regarded as a constant too. The higher the order in the slow–roll parameters, the smaller should be the difference between them, finally leading, for infinite order in the parameters, to the case of power–law inflation. That means that this approach to the problem of calculating the spectra for more general inflationary models is in fact an expansion around the power–law solution. An example is given by chaotic inflation with the potential $V \propto \phi^\alpha$ with $\alpha > 2$. In this case the slow–roll parameters are given by $\epsilon \simeq \alpha/(4N)$ and $\upsilon \simeq 1/(2N)$ [1], i.e., $\epsilon > \upsilon$. N denotes the number of e–folds of inflation.

Let us proceed with the calculations. In fact all that has to be done is to repeat the calculations of the previous sections keeping the desired order in the slow–roll expansions and neglecting all the terms similar to the right hand sides of Eqs. (2). For example, the next–to–next–to–leading order expression for the scalar power spectrum will be obtained from solution (6) but now with ν given by $\nu = 3/2+\epsilon+\upsilon+\epsilon^2$. Expanding and keeping terms up to second order, the final expression is

$$\mathcal{P}_R{}^{1/2}(k) = \frac{\kappa}{4\pi}\left[1-(C+1)\epsilon - C\upsilon + (B-1)\epsilon^2\right]\frac{H^2}{|H'|}\bigg|_{k=aH}, \tag{35}$$

where $B = -2 + C^2/2 + \pi^2/4 \simeq 0.73$. Eq. (8) can be readily recovered from Eq. (35) by neglecting second order terms. For the corresponding expression of the tensorial power spectrum, μ must be $\mu = 3/2 + \epsilon + \epsilon^2$ and,

$$\mathcal{P}_g^{1/2} = \left[1-(C+1)\epsilon + (B-1)\epsilon^2\right]\frac{\sqrt{2}\kappa}{\pi}H\bigg|_{k=aH}. \tag{36}$$

Equation (14) is obtained from Eq. (36) by neglecting second order terms of ϵ.

3.2. GENERALIZED SLOW–ROLL APPROXIMATION

There is no reason to reject the possibility of an inflationary model with $|\upsilon| \gg \epsilon$. As we shall see later, some important models belong to this class. For these cases, the accuracy of the calculations of the amplitudes can also be increased. We shall focus on the next–to–next–to–leading order. With regards of conditions (2), we neglect terms like ϵ^2, $\epsilon\upsilon$ and $\xi^2 - \epsilon^2$ but we keep υ^2 terms. Repeating the calculations under this set of assumptions we obtain for the amplitudes of the scalar spectrum:

$$\mathcal{P}_R{}^{1/2} = \frac{\kappa}{4\pi}\left[1-(C+1)\epsilon - C\upsilon + B\upsilon^2\right]\frac{H^2}{|H'|}\bigg|_{k=aH}. \tag{37}$$

For tensorial perturbations we note that the Bessel function index (12) does not depend on υ hence, no term like υ^2 will arise at any moment of the calculations. This way, the expression for the tensorial amplitudes is given by Eq. (14).

3.3. THE SPECTRAL INDICES

In general, if the approximation neglecting the right hand sides of Eqs. (2) is taken into account, expressions for the spectral indices at any order are easy to be derived noting that in these cases expression (25) always vanishes:

$$\frac{n_i}{2} = \left[\sqrt{\frac{2}{\kappa}}\sqrt{\epsilon}\frac{h_i'}{h_i}\right](1+\epsilon+\epsilon^2+\epsilon^3+\cdots), \tag{38}$$

where, n_i are $n_S - 1$ or n_T and h_i is correspondingly given by Eqs.(26) and (31). This way it is obtained

$$\begin{aligned} n_S(k) - 1 &= (-4\epsilon + 2\eta)(1+\epsilon+\epsilon^2+\epsilon^3+\cdots) \\ &= -2v - 2\epsilon(1+\epsilon+\epsilon^2+\cdots), \end{aligned} \tag{39}$$

$$n_T(k) = -2\epsilon(1+\epsilon+\epsilon^2+\epsilon^3+\cdots). \tag{40}$$

An alternative for more general models. In case it comes out that Eqs. (35), (36) and (37) are also a good approximation for more general inflationary potentials, then in the fashion it was done in Sec. 2.2, more general expressions for the spectral indices can be derived.

Under generalized power–law approximation, for the scalar index we have

$$a_{00} = 1, \quad a_{10} = -(2C+1), \quad a_{20} = C, \quad a_{11} = B - 1, \tag{41}$$

and to third order, expression (24) reduces to

$$\begin{aligned} n_S - 1 &= -4\epsilon + 2\eta - 8(C+1)\epsilon^2 + 2(5C+3)\epsilon\eta - 2C\xi^2 \\ &+ 4(2B - 4C^2 - 6C - 5)\epsilon^3 - 2(4B - 14C^2 - 16C - 9)\epsilon^2\eta \\ &- 2C(5C+2)\epsilon\eta^2 - 4C(C+1)\epsilon\xi^2 + 2C^2\eta\xi^2. \end{aligned} \tag{42}$$

Correspondingly, for the tensorial index,

$$a_{00} = 1, \quad a_{10} = -(C+1), \quad a_{11} = B - 1, \tag{43}$$

and its obtained,

$$\begin{aligned} n_T &= -2\epsilon - 2(2C+3)\epsilon^2 + 4(C+1)\epsilon\eta \\ &+ 2(4B - 2C^2 - 6C - 9)\epsilon^3 - 4(2B - C^2 - 3C - 4)\epsilon^2\eta. \end{aligned} \tag{44}$$

For the scalar amplitudes of the generalized slow–roll approximation,

$$a_{00} = 1, \quad a_{10} = -(2C+1), \quad a_{20} = C, \quad a_{22} = B, \tag{45}$$

and the spectral index reads,

$$\begin{aligned} n_S - 1 &= -4\epsilon + 2\eta - 8(C+1)\epsilon^2 + 2(5C+3)\epsilon\eta - 2C\xi^2 \\ &- 4(4C^2+6C+3)\epsilon^3 + 2(14C^2+16C+5)\epsilon^2\eta \\ &+ 2(2B-5C^2-2C)\epsilon\eta^2 - 4C(C+1)\epsilon\xi^2 - 2(2B-C^2)\eta\xi^2 , \quad (46) \end{aligned}$$

while, as it was already noted, the tensorial index remains the same that the standard one.

4. TESTING THE EXPRESSIONS

To test all the expressions presented in this paper, we will use the well known exact results for amplitudes and indices calculations in the cases of power–law and natural inflation.

Power–law inflation [9] is an inflationary scenario where,

$$a(t) \propto t^p , \quad H(\phi) \propto \exp\left(-\sqrt{\frac{\kappa}{2p}}\,\phi\right), \quad V(\phi) \propto \exp\left(-\sqrt{\frac{2\kappa}{p}}\,\phi\right), \quad (47)$$

with p being a positive constant. It follows from (1) that in this case the slow–roll parameters are constant and equal each other,

$$\epsilon = \eta = \xi = 1/p . \quad (48)$$

Thus, this is the limit case of the generalized power–law approximation. Note that condition $\epsilon < 1$ implies $p > 1$.

For this model, the power spectrum of scalar perturbations is given by [10]

$$\mathcal{P}_R{}^{1/2}(k) = \frac{2^{\nu-\frac{1}{2}}\,\Gamma(\nu)}{m_{Pl}^2\,\Gamma(\frac{3}{2})}\left(1-\frac{1}{p}\right)^{\nu-\frac{1}{2}} \left.\frac{H^2}{|H'|}\right|_{k=aH} , \quad (49)$$

with

$$\nu \equiv \frac{3}{2} + \frac{1}{p-1} .$$

Expanding up to second order for $p \gg 1$ it is obtained that

$$\mathcal{P}_R{}^{1/2}(k) = \frac{\kappa}{4\pi}\left[1-(C+1)\frac{1}{p}+(B-1)\frac{1}{p^2}\right] \left.\frac{H^2}{|H'|}\right|_{k=aH} , \quad (50)$$

in full correspondence with Eq. (35) when relations (48) are taken into account. Testing the tensorial amplitudes can be easily done by checking that Eqs. (35) and (36) indeed satisfy the relation between amplitudes characteristic of power–law inflation, i.e.,

$$\mathcal{P}_g{}^{1/2}(k) = \frac{4}{\sqrt{p}}\mathcal{P}_R{}^{1/2}(k) . \quad (51)$$

For the spectral indices one can see that substitution of relations (48) in Eqs. (38) and (40), as well as in Eqs. (42) and (44), yield

$$\frac{n_S-1}{2}=\frac{n_T}{2}=\frac{1}{1-p}=-\frac{1}{p}\left(1+\frac{1}{p}+\frac{1}{p^2}+\cdots\right). \tag{52}$$

Another inflationary scenario where precise predictions for the spectra amplitudes and indices can be done is natural inflation [11]. In this case, the potential is given by,

$$V=\Lambda^4\left[1\pm\cos(\frac{\phi}{f})\right], \tag{53}$$

where Λ and f are mass scales. For simplicity we choose the plus sign in the remaining calculations. The point here is to analyze inflation near the origin so that the small–angle approximation applies, i.e., $\phi \ll f$. In this case the slow–roll parameters are,

$$\begin{aligned}
\epsilon &= \frac{3}{4}\kappa\left(\sqrt{1+\frac{2}{3}\frac{1}{\kappa f^2}}-1\right)\phi^2 \simeq 0, && (54)\\
\eta &= -\frac{3}{2}\left(\sqrt{1+\frac{2}{3}\frac{1}{\kappa f^2}}-1\right), && (55)\\
\xi^2 &\simeq 0, && (56)\\
\upsilon &\simeq -\eta. && (57)
\end{aligned}$$

As it can be observed, in this approximation natural inflation belongs to the class of models that fulfill the conditions of the generalized slow–roll approximation. With the values of the parameters given above, Eq. (6) can be used with the corresponding $\nu = 3/2+\upsilon$. After expanding and truncating at the proper order

$$\mathcal{P}_R{}^{1/2}=\frac{\kappa}{4\pi}\left[1-C\upsilon+B\upsilon^2\right]\frac{H^2}{|H'|}\bigg|_{k=aH} \tag{58}$$

is obtained, consistent with Eq. (37). With this same set of assumptions, Eq. (40) and (46) are reduced to

$$n_S-1=2\eta, \tag{59}$$

the well known result for natural inflation.

5. CONCLUSIONS

In this paper we introduced a formalism which allows to increase the precision for the theoretical predictions of spectra of inflationary perturbations for two specific classes of inflationary models. The method is based on the ratio between the slow–roll parameters and their respective differences. When the value of the parameters is much greater than the difference between any pair of parameters, a generalization of the standard Bessel approximation takes place. We called this approach generalized power–law approximation. In the opposite case, when the value of the parameters is much less than their difference, a similar approach (generalized slow–roll approximation) can be used. These approximations plus the standard approximation cover a large space of inflationary models.

The obtained expressions were tested against the known exact results for power-law and natural inflation. A more rigorous test allowing to determine the contribution of higher corrections will be carried out in a forthcoming paper.

Acknowledgments

This research was supported by CONACyT Grant: 32138E.

References

[1] A. Linde, *Particle Physics and Inflationary Cosmology* (Harwood, Chur, Switzerland, 1990); A.R. Liddle and D.H. Lyth, *Cosmological Inflation and Large–Scale Structure* (Cambridge University Press, Cambridge, UK, 2000)

[2] http://www.sdss.org/;
http://map.gsfc.nasa.gov/;
http://astro.estec.esa.nl/SA-general/Projects/Planck/

[3] E.D. Stewart and D.H. Lyth, *Phys. Lett.* **B302** (1993) 171.

[4] I.J. Grivell and A.R. Liddle, *Phys. Rev.* **D54** (1996) 12.

[5] J. Martin and D.J. Schwarz, *Phys. Rev.* **D62** (2000) 103520.

[6] J. Martin, A. Riazuelo and D.J. Schwarz, *Astrophys. J.* **543** (2000) L99.

[7] V.F. Mukhanov, Pis'ma. *Zh. Eksp. Teor. Fiz.* **41** (1985) 402.

[8] J.E. Lidsey, A.R. Liddle, E.W. Kolb, E.J. Copeland, T. Barreiro and M. Abney, *Rev. Mod. Phys.* **69** (1997) 373.

[9] F. Lucchin and S. Matarrese, *Phys. Rev.* **D32**, (1985) 1316.

[10] D.H. Lyth and E.D. Stewart, *Phys. Lett.* **B274** (1992) 168.

[11] K. Freese, J.A. Frieman and A.V. Olinto, *Phys. Rev. Lett.* **65** (1990) 3233.

CONFORMAL SYMMETRY AND DEFLATIONARY GAS UNIVERSE

Winfried Zimdahl
Universität Konstanz, PF M678 D-78457 Konstanz, Germany

Alexander B. Balakin
Kazan State University, 420008 Kazan, Russia

Abstract We describe the "deflationary" evolution from an initial de Sitter phase to a subsequent Friedmann–Lemaître–Robertson–Walker (FLRW) period as a specific non–equilibrium configuration of a self-interacting gas. The transition dynamics corresponds to a conformal, timelike symmetry of an "optical" metric, characterized by a refraction index of the cosmic medium which continuously decreases from a very large initial value to unity in the FLRW phase.

Keywords: Conformal symmetry, inflation, cosmology.

1. INTRODUCTION

It is well known that a thermodynamic equilibrium in the expanding universe is only possible under special circumstances. For a simple gas model of the cosmic medium e.g., the relevant equilibrium condition can be satisfied only for massless particles, corresponding to an equation of state $P = \rho/3$, where P is the pressure and ρ is the energy density [1, 2]. This equilibrium condition is equivalent to a symmetry requirement for the cosmological dynamics: The quantity u^a/T, where u^a is the four velocity of the medium and T is its temperature, has to be a conformal Killing vector (CKV) [3, 4]. For deviations from $P = \rho/3$ the conformal symmetry is violated and no equilibrium is possible for the gaseous fluid. If we give up the diluted gas approximation on which this model is based and take into account additional interactions inside the many–particle system, the situation changes. Suitable interaction terms give rise to a generalization of the equilibrium conditions of the gas ("generalized equilibrium") [5, 6, 7, 8, 9]. Among these generalized equilibrium configurations are "true" equilibrium states, i.e. states with vanishing entropy production [8], but there are also specific types of non–equilibrium states. The latter are characterized by an equilibrium distribution function, the moments of which are not conserved, however [9]. The corresponding source terms in the balance equations for the moments generally give rise to the production of particles and entropy. They represent deviations from "standard" equilibrium which are not necessarily small. Configurations with substantial deviations are specific far–from–equilibrium states which are of interest for out–of–equilibrium situations in a cosmological context.

Gas models of the cosmic medium have the advantage that the relation between microscopic dynamics (kinetic theory) and a macroscopic thermo–hydrodynamical description is exactly known. An apparent disadvantage of these models is that they are restricted to "ordinary" matter, i.e. to matter with equations of state in the range $0 < P \leq \rho/3$. However, periods of accelerated cosmological expansion, either inflation in the early universe or a dark energy dominated dynamics at the present epoch, require "exotic" matter, namely a substratum with an effective negative pressure that violates the strong energy condition $\rho + 3P > 0$. Usually, such kind of matter is modeled by the dynamics of scalar fields with suitable potentials. Here we show that the generalized equilibrium concept provides a framework which makes use of the advantages of gas models and at the same time avoids the restriction to equations of state for "ordinary" matter. The point is that the specific non–equilibrium configurations mentioned above may give rise to a negative pressure of the cosmic medium. If the latter is sufficiently large to violate the strong energy condition, it induces an accelerated expansion of the universe. Our main interest here is to understand the "deflationary" [10] transition from an initial de Sitter phase to a subsequent FLRW period as a specific non–equilibrium configuration which is based on the generalized equilibrium concept. Moreover, we show that the non–equilibrium contributions may be mapped onto an refraction index n_r of the cosmic medium which is used to define an "optical" metric $\bar{g}_{ik} = g_{ik} + \left[1 - n_r^{-2}\right] u_i u_k$ [11, 12, 13]. The circumstance that this kind of non–equilibrium retains essential equilibrium aspects is reflected by the fact that u^a/T is a CKV of the optical metric $\bar{g}_{ik}$ during the entire transition period. The out–of–equilibrium process represents a conformal symmetry of the optical metric. As the FLRW stage is approached, the refraction index tends to unity, $n_r \rightarrow 1$, which implies $\bar{g}_{ik} \rightarrow g_{ik}$.

2. FLUID DESCRIPTION

Let the fluid be characterized by a four velocity u^i and a temperature T. The Lie derivative $\pounds_{\frac{u_a}{T}} g_{ik}$ of the metric g_{ik} with respect to the vector $\frac{u^a}{T}$ may be decomposed into contributions parallel and perpendicular to the four-velocity:

$$\pounds_{\frac{u_a}{T}} g_{ik} \equiv \left(\frac{u_i}{T}\right)_{;k} + \left(\frac{u_k}{T}\right)_{;i} = 2Au_iu_k + B_iu_k + B_ku_i + 2\phi h_{ik} + b_{ik} \ , \quad (1)$$

where $h^{ik} = g^{ik} + u^iu^k$, $B_iu^i = 0$, $h_{ik}u^i = b_{ik}u^i = 0$, and $b^i_i = 0$. Units are chosen so that $c = k_B = 1$. With the help of the projections

$$\frac{\dot{T}}{T} = AT = \frac{T}{2}u^iu^k\pounds_{\frac{u_a}{T}} g_{ik} \ , \qquad \Theta = 3T\phi = \frac{T}{2}h^{ik}\pounds_{\frac{u_a}{T}} g_{ik} \ , \quad (2)$$

one obtains the rate of change of the temperature $\frac{\dot{T}}{T}$, where $\dot{T} \equiv T_{,a}u^a$, and the fluid expansion $\Theta \equiv u^a_{;a}$, respectively. The remaining projections are $\dot{u}_m + \frac{\nabla_m T}{T} = -TB_m = Th^i_m u^k \pounds_{\frac{u_a}{T}} g_{ik}$ and $2\sigma_{ab} = Tb_{ab} = T\left[h^i_a h^k_b - \frac{1}{3}h_{ab}h^{ik}\right]\pounds_{\frac{u_a}{T}} g_{ik}$, where $\dot{u}_m$ is the fluid acceleration, $\nabla_a T \equiv h^a_m T_{,a}$ and σ_{ab} is the shear tensor. The temperature change rate $\frac{\dot{T}}{T}$ and the expansion Θ in (2) are given by orthogonal projections of the Lie derivative. However, in a cosmological context one expects a relationship between these two quantities which represents the cooling rate of the cosmic fluid in an expanding universe. To establish the corresponding link we take into account that the fluid is characterized by a particle flow vector N^i and an energy–momentum tensor $T^{ik}_{(eff)}$:

$$N^i = nu^i \ , \quad T^{ik}_{(eff)} = \rho u^iu^k + Ph^{ik} \ , \quad P = p + \Pi \ . \quad (3)$$

Here, n is the particle number density, ρ is the energy density seen by a comoving observer, P is the total pressure with the equilibrium part p and the non–equilibrium contribution Π. The balance equations are

$$N^a_{;a} = \dot{n} + 3Hn = n\Gamma \ , \qquad -u_iT^{ik}_{(eff);k} = \dot{\rho} + 3H(\rho + P) = 0 \ , \quad (4)$$

where Γ is the particle production rate, i.e., we have admitted the possibility of particle number non–conserving processes. Moreover, we have introduced the Hubble rate H by $\Theta \equiv 3H$. Furthermore, we assume equations of state in the general form

$$p = p(n, T) \ , \qquad \rho = \rho(n, T) \ , \quad (5)$$

i.e., particle number density and temperature are the independent thermodynamical variables. Differentiating the latter relation, using the balances (4) and restricting ourselves to a constant entropy per particle, we obtain the cooling rate

$$\frac{\dot{T}}{T} = -3H\Delta\left(\frac{\partial p}{\partial \rho}\right)_n , \qquad \Delta \equiv 1 - \frac{\Gamma}{3H} . \tag{6}$$

Comparing the expressions in (2) and (6) for the temperature evolution we find that they are consistent for

$$\begin{aligned} \pounds_{\frac{u_a}{T}} g_{ik} &= \frac{2\mathrm{H}}{T}\left\{g_{ik} + \left[1 - 3\Delta\left(\frac{\partial p}{\partial \rho}\right)_n\right] u_i u_k\right\} + \frac{2\sigma_{ik}}{T} \\ &\quad - \frac{u_k}{T}\left[\dot{u}_i + \frac{\nabla_i T}{T}\right] - \frac{u_i}{T}\left[\dot{u}_k + \frac{\nabla_k T}{T}\right] . \end{aligned} \tag{7}$$

In the special case $\Delta = 1$, $p = \frac{\rho}{3}$, $\sigma_{ab} = 0$, $\dot{u}_a + \frac{\nabla_a T}{T} = 0$ we recover that $\frac{u^a}{T}$ is a conformal Killing vector (CKV) of the metric g_{ik},

$$\Delta = 1 , \quad p = \frac{\rho}{3} , \quad \sigma_{ab} = 0 , \quad \dot{u}_a + \frac{\nabla_a T}{T} = 0 \quad \Rightarrow \quad \pounds_{\frac{u_a}{T}} g_{ik} = \frac{2\mathrm{H}}{T} g_{ik} . \tag{8}$$

Equation (7) represents a modification of the CKV condition in (8) even in the homogeneous and isotropic case where $\sigma_{ab} = 0$ and $\dot{u}_a + \frac{\nabla_a T}{T} = 0$. Since the CKV property of $\frac{u^a}{T}$ is known to be related to the dynamics of a standard radiation dominated universe, the question arises to what extent the replacement

$$\frac{2H}{T} g_{ik} \to \frac{2H}{T}\left\{g_{ik} + \left[1 - 3\Delta\left(\frac{\partial p}{\partial \rho}\right)_n\right] u_i u_k\right\} \tag{9}$$

in (7) gives rise to a modified cosmological dynamics. In particular, it is of interest to explore whether this modification possibly admits phases of accelerated expansion of the universe. To clarify this question we first recall that the CKV property of $\frac{u^a}{T}$ may be obtained as a "global" equilibrium condition in relativistic gas dynamics. Then we ask whether the more general case corresponding to the modification (9) can be derived in a gas dynamical context as well.

3. GAS DYNAMICS

Relativistic gas dynamics is based on Boltzmann's equation for the one–particle distribution function $f = f(x, p)$,

$$p^i \frac{\partial f}{\partial x^i} - \Gamma^k_{il} p^i p^l \frac{\partial f}{\partial p^k} + mF^i \frac{\partial f}{\partial p^i} = C[f] . \tag{10}$$

Both the collision integral $C[f]$ and the force $F^i = F^i(x,p)$ describe interactions within the many-particle system. While C accounts for elastic binary collisions, we assume that all other microscopic interactions, however involved their detailed structure may be, can be mapped onto an effective one–particle force F^i, which may be used for a subsequent self–consistent treatment. Let us restrict ourselves to the class of forces which admit solutions of Boltzmann's equation that are of the type of Jüttner's distribution function

$$f^0(x,p) = \exp\left[\alpha + \beta_a p^a\right] , \tag{11}$$

where $\alpha = \alpha(x)$ and $\beta_a(x)$ is timelike. For $f \to f^0$ the collision integral vanishes: $C\left[f^0\right] = 0$. Substituting f^0 into Eq. (10) we obtain the condition

$$p^a \alpha_{,a} + \beta_{(a;b)} p^a p^b = -m\beta_i F^i . \tag{12}$$

For the most general expression of the relevant force projection $\beta_i F^i$ [9],

$$\beta_i F^i = \beta_i F^i_a(x)\, p^a + \beta_i F^i_{ab}(x)\, p^a p^b , \tag{13}$$

the condition (12) decomposes into the "generalized" equilibrium conditions

$$\alpha_{,a} = -m\beta_i F^i_a , \qquad \beta_{(k;l)} \equiv \frac{1}{2}\mathcal{L}_{\beta_a} g_{kl} = -m\beta_i F^i_{kl} . \tag{14}$$

With the identification $\beta_a \to \frac{u_a}{T}$ and restricting ourselves to the homogeneous and isotropic case with $\sigma_{ab} = 0$ and $\dot{u}_a + \frac{\nabla_a T}{T} = 0$, we may read off that a force with

$$-m\beta_i F^i_{ab} = \frac{H}{T}\left\{g_{ab} + \left[1 - 3\Delta\left(\frac{\partial p}{\partial \rho}\right)_n\right] u_a u_b\right\} \tag{15}$$

reproduces the previous consistency condition (7) for this case. The latter is recovered as generalized equilibrium condition for particles in a force field of the type (13).

One may interpret this feature in a way which is familiar from gauge theories: Gauge field theories rely on the fact that local symmetry requirements (local gauge invariance) necessarily imply the existence of additional interaction fields (gauge fields). In the present context we impose the "symmetry" requirement (7) (below we clarify in which sense the modification of the conformal symmetry is again a symmetry). Within the presented gas dynamical framework this "symmetry" can only be realized if one introduces additional interactions, here described by an effective force field F^i. Consequently, in a sense, this force field may be regarded as the analogue of gauge fields.

In its quadratic part the force has the structure

$$m\beta_i F^i \propto \frac{H}{T}\left\{m^2 + \left[3\Delta\left(\frac{\partial p}{\partial \rho}\right)_n - 1\right]E^2\right\} , \qquad (16)$$

where $E \equiv -u_a p^a$ is the particle energy. (For a discussion of the linear part see [9].) This force which governs the particle motion depends on macroscopic and microscopic quantities. The macroscopic quantities determine the motion of the individual microscopic particles which themselves are the constituents of the medium. This means, the system is a gas with internal self–interactions.

In a next step we calculate the macroscopic transport equations for the self–interacting gas. The first and second moments of the distribution function are

$$N^i = \int \mathrm{d}P p^i f(x,p) , \qquad T^{ik} = \int \mathrm{d}P p^i p^k f(x,p) . \qquad (17)$$

While N^i may be identified with the corresponding quantity in Eq. (3) the second moment T^{ik} does *not* coincide with the energy–momentum tensor in (3). With $f \to f^0$ we obtain

$$N^a_{;a} = \dot{n} + 3Hn = 3nH(1-\Delta) = n\Gamma , \qquad (18)$$

and

$$u_a T^{ab}_{;b} = -\dot{\rho} - 3H(\rho + p) = -3H(\rho+p)(1-\Delta) . \qquad (19)$$

For $\Delta \neq 1$ there appear "source" terms in the balances. The crucial point now is that these source terms may consistently be mapped according to [9]

$$\Delta = 1 + \frac{\Pi}{\rho + p} \quad \Rightarrow \quad \dot{\rho} + 3H(\rho + p + \Pi) = 0 , \qquad (20)$$

on the effective viscous pressure Π of a conserved energy–momentum tensor of the type of $T^{ik}_{(eff)}$ in (3). In the following the cosmic medium is assumed to be describable by an energy–momentum tensor $T^{ik}_{(eff)}$ where Π is related to Δ by the latter correspondence.

4. COSMOLOGICAL DYNAMICS AND CONFORMAL SYMMETRY

For a homogeneous, isotropic, and spatially flat universe, the relevant dynamical equations are

$$8\pi G\rho = 3H^2 , \quad \dot{H} = -4\pi G(\rho + p + \Pi) \quad \Rightarrow \quad \frac{\Gamma}{3H} = 1 + \frac{2}{3\gamma}\frac{\dot{H}}{H^2} , \qquad (21)$$

where $\gamma = 1 + p/\rho$. A non–vanishing viscous pressure, in our case equivalent to the creation of particles, back reacts on the cosmological dynamics. To solve the equations (21), assumptions about the ratio Γ/H are necessary. Generally, one expects that particle production is a phenomenon which is characteristic for early stages of the cosmic evolution, while lateron a standard FLRW phase is approached. In case H is a decaying function of the cosmic time, this is realized by the simple ansatz $\Gamma/H \propto H$ with the help of which Eqs. (21) for ultrarelativistic matter ($\gamma = 4/3$) may be integrated:

$$\frac{\Gamma}{H} \propto H \quad \Rightarrow \quad H = 2\frac{a_e^2}{a^2 + a_e^2} H_e \quad \Rightarrow \quad \Delta = \frac{a^2}{a^2 + a_e^2} \,. \tag{22}$$

The Hubble rate H changes continuously from $H = 2H_e$, equivalent to $a \propto \exp[Ht]$ at $a \ll a_e$, to $H \propto a^{-2}$, equivalent to $a \propto t^{1/2}$, the standard radiation–dominated universe, for $a \gg a_e$. For $a < a_e$ we have $\ddot{a} > 0$, while $\ddot{a} < 0$ holds for $a > a_e$. The value a_e denotes that transition from accelerated to decelerated expansion, corresponding to $\dot{H}_e = -H_e^2$. For $a \to 0$ we have $\Delta \to 0$, i.e., $\Gamma \to 3H$, while for $a \gg a_e$ we find $\Delta \to 1$, i.e., $\Gamma \to 0$.

Now we ask how such kind of scenario may be realized within the previously discussed gas dynamics. To this purpose we introduce the redefinition $mF^i \to F^i$ of the force since the ultrarelativistic limit corresponds to $m \to 0$. From the second of the generalized equilibrium conditions (14) with (15) and with Δ from (22) one obtains the relevant force term:

$$\pounds_{\frac{u}{T}} g_{ik} = \frac{2H}{T}\left[g_{ik} + \frac{a_e^2}{a^2 + a_e^2} u_i u_k\right] \,, \quad \Rightarrow \quad u_i F^i = -HE^2 \frac{a_e^2}{a^2 + a_e^2} \,. \tag{23}$$

An internal interaction, described by this component of an effective one–particle force F^i which is self–consistently exerted on the microscopic constituents of the cosmic medium, realizes the deflationary dynamics (22).

The explicit knowledge of the self–interacting force allows us to consider the microscopic particle dynamics which corresponds to the deflationary scenario. The equations of motion for the particles are

$$\frac{\mathrm{D}p^i}{\mathrm{d}\gamma} = F^i \,, \tag{24}$$

where γ is a parameter along the particle worldline. Contraction with u_i provides us with a differential equation for the particle energy $E \equiv -u_i p^i$

which in the present case has the solution

$$E \propto T \propto \left[\frac{a_e^2}{a^2 + a_e^2}\right]^{1/2} , \quad \Rightarrow \quad f^0 \propto \exp\left[-\frac{E}{T}\right] = \text{const} , \tag{25}$$

i.e., the equilibrium distribution is indeed maintained which proves the consistency of our approach. The particle energy changes from $E =$ const for $a \ll a_e$ to $E \propto a^{-1}$ for $a \gg a_e$. The set of equations (22)–(25) represents an exactly solvable model of a deflationary transition from an initial de Sitter phase to a subsequent radiation dominated FLRW period, both macroscopically and microscopically. This transition is accompanied by a characteristic change of the "symmetry" condition (14) with (15) from

$$\pounds_{\frac{u_a}{T}} g_{ik} = \frac{2H}{T} h_{ik} \qquad (a \ll a_e) , \tag{26}$$

characterizing a "projector-conformal timelike Killing vector" [7], to

$$\pounds_{\frac{u_a}{T}} g_{ik} = \frac{2H}{T} g_{ik} \qquad (a \gg a_e) . \tag{27}$$

In order to clarify in which sense the modification (9) of the conformal symmetry is a symmetry again, we introduce the "optical metric" (cf. [11, 12, 13])

$$\bar{g}_{ik} = g_{ik} + \left[1 - \frac{1}{n_r^2}\right] u_i u_k , \quad n_r^2 = 1 + \frac{a_e^2}{a^2} = \Delta^{-1} , \tag{28}$$

where n_r plays the role of a refraction index of the medium which in the present case changes monotonically from $n_r \to \infty$ for $a \to 0$ to $n_r \to 1$ for $a \gg a_e$. For $a = a_e$ we have $n_r(a_e) = \sqrt{2}$. Optical metrics are known to be helpful in simplifying the equations of light propagation in isotropic refractive media. With respect to optical metrics light propagates as in vacuum. Here we demonstrate that such type of metrics, in particular their symmetry properties, are also useful in relativistic gas dynamics. Namely, it is easy to realize that the first relation (23) is equivalent to

$$\pounds_{\frac{u_a}{T}} \bar{g}_{ik} = \frac{2H}{T} \bar{g}_{ik} . \tag{29}$$

The quantity $\frac{u^a}{T}$ is a CKV of the optical metric $\bar{g}_{ik}$. This clarifies in which sense the modification of the conformal symmetry is again a symmetry. The transition from a de Sitter phase to a FLRW period may be regarded as a specific non–equilibrium configuration which microscopically is characterized by an equilibrium distribution function and macroscopically by the conformal symmetry of an optical metric with a time dependent refraction index of the cosmic medium. In the following sections we demonstrate how the presented scenario may be related to other approaches. Firstly, we point out that it is equivalent to a phenomenological vacuum decay model and secondly, that it may be translated into an equivalent dynamics of a scalar field with a specific potential term.

5. PHENOMENOLOGICAL VACUUM DECAY

Combining the Friedmann equation in (21) with the Hubble parameter in (22), we obtain the energy density

$$\rho = \frac{3H_e^2}{2\pi} m_P^2 \left[\frac{a_e^2}{a^2 + a_e^2} \right]^2 , \tag{30}$$

where we have replaced G by the Planck mass m_P according to $G = 1/m_P^2$. Initially, i.e., for $a \ll a_e$, the energy density is constant, while for $a \gg a_e$ the familiar behaviour $\rho \propto a^{-4}$ is recovered. The temperature behaves as $T \propto \rho^{1/4}$. According to (25), the same dependence is obtained for the particle energy E. The evolution of the universe starts in a quasistationary state with finite initial values of temperature and energy density at $a \ll a_e$ and approaches the standard radiation dominated universe for $a \gg a_e$. This scenario may alternatively be interpreted in the context of a decaying vacuum (cf. [14, 15]). The energy density (30) may be split into $\rho = \rho_{(v)} + \rho_{(r)}$ where

$$\begin{aligned} \rho_{(v)} &= \frac{3H_e^2}{2\pi} m_P^2 \left[\frac{a_e^2}{a^2 + a_e^2} \right]^3 , \\ \rho_{(r)} &= \frac{3H_e^2}{2\pi} m_P^2 \left(\frac{a}{a_e} \right)^2 \left[\frac{a_e^2}{a^2 + a_e^2} \right]^3 . \end{aligned} \tag{31}$$

The part $\rho_{(v)}$ is finite for $a \to 0$ and decays as a^{-6} for $a \gg a_e$, while the part $\rho_{(r)}$ describes relativistic matter with $\rho_{(r)} \to 0$ for $a \to 0$ and $\rho_{(r)} \propto a^{-4}$ for $a \gg a_e$. The balance equation for $\rho_{(r)}$ is

$$\dot{\rho}_{(r)} + 4H\rho_{(r)} = -\dot{\rho}_{(v)} , \tag{32}$$

while $\rho_{(v)}$ changes according to

$$\dot{\rho}_{(v)} + 3H \left(\rho_{(v)} + P_{(v)} \right) = 0 , \tag{33}$$

where the effective pressure $P_{(v)}$ obeys an equation of state

$$P_{(v)} = \frac{a^2 - a_e^2}{a^2 + a_e^2} \rho_{(v)} \ . \tag{34}$$

For $a \ll a_e$ this pressure approaches $P_{(v)} = -\rho_{(v)}$. Effectively, this component behaves as a vacuum contribution. For $a \gg a_e$ it represents stiff matter with $P_{(v)} = \rho_{(v)}$. Acording to (32) the radiation component may be regarded as emerging from the decay of the initial vacuum. One may introduce a radiation temperature $T_{(r)}$ characterized by a dependence $T_{(r)} \propto \rho_{(r)}^{1/4}$. This temperature starts at $T_{(r)} = 0$ for $a = 0$, then increases to a maximum value $T_{(r)}^{max} = (32/27)^{1/4} T_{(e,r)}$ and finally decreases as a^{-1} for large values of a.

6. EQUIVALENT SCALAR FIELD DYNAMICS

Inflationary cosmology is usually discussed in terms of scalar fields with suitable potentials. The common picture consists of an initial "slow roll" phase with a dynamically dominating (approximately constant) potential term, which generates a de Sitter like exponential expansion, connected with an "adiabatic supercooling". During the subsequent "reheating" which in itself is a highly complicated non–equilibrium period, the scalar field is assumed to decay into "conventional" matter. The entire entropy in the presently observable universe is produced during the reheating phase according to these scenarios. As to the mechanism of entropy production the "standard" (i.e. based on scalar field dynamics) inflationary picture essentially differs from the fluid particle production scenario presented here, since in our approach the entropy is produced already during the period of accelerated expansion. On the other hand, as far as the behavior of the scale factor is concerned, there exists a close correspondence between the role of a negative fluid pressure (due to the production of particles) and a suitable scalar field potential. What counts here is the magnitude of the effective negative pressure, independently of its origin (scalar field potential or particle production). What one would like to have is an interconnection between both these different lines of describing the early universe. This would allow us to switch from the scalar field to the fluid picture and vice versa. To this purpose we start with the familiar identifications

$$\rho = \frac{1}{2}\dot{\phi}^2 + V(\phi) \ , \quad P \equiv p + \Pi = \frac{1}{2}\dot{\phi}^2 - V(\phi) \ , \tag{35}$$

or,

$$\rho + P = \dot{\phi}^2 \ , \quad \rho - P = 2V(\phi) \ . \tag{36}$$

With the expressions (30), (20), and (22), we obtain

$$\dot{\phi}^2 = 4\frac{H_e^2}{2\pi}m_p^2\frac{a^2}{a_e^2}\left[\frac{a_e^2}{a^2+a_e^2}\right]^3 . \tag{37}$$

For $\dot{\phi}$ we write $\dot{\phi} = \phi'\dot{a} = \phi' a H$, where $\phi' \equiv \mathrm{d}\phi/\mathrm{d}a$. It follows that

$$\phi' = \pm\frac{m_P}{2\pi}\frac{1}{\sqrt{a^2+a_e^2}} . \tag{38}$$

Integration of this equation yields

$$\phi - \phi_0 = \frac{m_P}{\sqrt{2\pi}}\mathrm{Arsh}\frac{a}{a_e} \quad \Rightarrow \quad \frac{a}{a_e} = \sinh\left[\sqrt{2\pi}\frac{\phi-\phi_0}{m_P}\right] , \tag{39}$$

where we have restricted ourselves to $\phi \geq \phi_0$. The potential is given by

$$V = \frac{H_e^2}{2\pi}m_P^2\left[\frac{a^2}{a_e^2}+3\right]\left[\frac{a_e^2}{a^2+a_e^2}\right]^3 , \tag{40}$$

or,

$$V = \frac{H_e^2}{2\pi}m_P^2\frac{\cosh^2\left[\sqrt{2\pi}\frac{\phi-\phi_0}{m_p}\right]+2}{\cosh^6\left[\sqrt{2\pi}\frac{\phi-\phi_0}{m_p}\right]} = \frac{H_e^2}{2\pi}m_P^2\frac{3-2\tanh^2\left[\sqrt{2\pi}\frac{\phi-\phi_0}{m_p}\right]}{\cosh^4\left[\sqrt{2\pi}\frac{\phi-\phi_0}{m_p}\right]} . \tag{41}$$

This expression was found by Maartens et al. [16] as exactly that potential which produces a cosmological dynamics of the type (22). In terms of the scalar field the cosmic evolution starts with $V_0 = \left(3H_e^2/2\pi\right)m_P^2$ and $\dot{\phi}_0 = 0$. At $a = a_e$ the value of the potential is reduced to $V_e = \dot{\phi}_e^2 = \left(H_e^2/4\pi\right)m_P^2$. A scalar field description with this potential implies the same cosmological dynamics as a self–interacting gas model in which the particles self–consistently move under the influence of an effective one–particle force characterized by the second equation in (23). The above relations allow us to change from the fluid to the scalar field picture and vice versa at any time of the cosmological evolution. In principle, it is possible to "calculate" the potential if the self–interacting fluid dynamics is known.

7. CONCLUSIONS

We have presented an exactly solvable gas dynamical model of a deflationary transition from an initial de Sitter phase to a subsequent radiation dominated FLRW period. The entire transition dynamics represents a specific non–equilibrium configuration of a self–interacting gas.

Although connected with entropy production, this configuration is characterized by an equilibrium distribution function of the gas particles. Macroscopically, this evolution is connected with a smooth transition from a "symmetry" condition $\mathcal{L}_{\frac{u_a}{T}} g_{ik} = \frac{2H}{T} h_{ik}$ (where $\frac{u^a}{T}$ is a "projector–conformal timelike Killing vector" [7]) during the de Sitter stage to the conformal symmetry $\mathcal{L}_{\frac{u_a}{T}} g_{ik} = \frac{2H}{T} g_{ik}$ as the FLRW dynamics is approached. Mapping the non-equilibrium contributions onto an effective refraction index n_r of the cosmic matter, the deflationary transition appears as the manifestation of a timelike conformal symmetry of an optical metric $\bar{g}_{ik} = g_{ik} + \left[1 - n_r^{-2}\right] u_i u_k$ in which the refraction index n_r changes smoothly from a very large value in the de Sitter period to unity in the FRLW phase. The deflationary gas dynamics may alternatively be interpreted as a production process of relativistic particles out of a decaying vacuum. Furthermore, there exists an equivalent scalar field description with an exponential type potential which generates the same deflationary scenario as the self–interacting gas model.

Acknowledgments

This work was supported by the Deutsche Forschungsgemeinschaft.

References

[1] J.M. Stewart, *Non–equilibrium Relativistic Kinetic Theory* (Springer, New York, 1971).

[2] J. Ehlers, in: *General Relativity and Cosmology*, ed. by B.K. Sachs (Academic Press, New York, 1971).

[3] G.E. Tauber and J.W. Weinberg, *Phys. Rev.* **122** (1961) 1342.

[4] N.A. Chernikov, *Sov. Phys. Dokl.* **7** (1962) 428.

[5] W. Zimdahl, *Phys. Rev.* **D57** (1998) 2245.

[6] W. Zimdahl and A.B. Balakin, *Class. Quantum Grav.* **15** (1998) 3259.

[7] W. Zimdahl and A.B. Balakin, *Phys. Rev.* **D58** (1998) 063503.

[8] W. Zimdahl and A.B. Balakin, *Gen. Rel. Grav.* **31** (1999) 1395.

[9] W. Zimdahl and A.B. Balakin, *Phys. Rev. D* (2001) to appear. (Preprint astro-ph/0010630)

[10] J.D. Barrow, *Phys. Lett.* **B180** (1986) 335.

[11] W. Gordon, *Ann. Phys.* (NY) **72** (1923) 421.

[12] J. Ehlers, *Z. Naturforschg.* **22a** (1967) 1328.

[13] V. Perlick, *Ray Optics, Fermat's Principle, and Applications to General Relativity* (Springer, Berlin, 2000).

[14] J.A.S. Lima and J.M.F. Maia, *Phys. Rev.* **D49** (1994) 5597.

[15] E. Gunzig, R. Maartens, and A.V. Nesteruk, *Class. Quantum Grav.* **15** (1998) 923; A.V. Nesteruk, *Gen. Rel. Grav.* **31** (1999) 983.

[16] R. Maartens, D.R. Taylor, and N. Roussos, *Phys. Rev.* **D52** (1995) 3358.

IV

EXPERIMENTS AND OTHER TOPICS

STATICITY THEOREM FOR NON–ROTATING BLACK HOLES WITH NON–MINIMALLY COUPLED SELF–INTERACTING SCALAR FIELDS

Eloy Ayón–Beato
Departamento de Física
Centro de Investigación y de Estudios Avanzados del IPN
Apdo. Postal 14–740, C.P. 07000, México, D.F., México.

Abstract Self–interacting scalar field configurations which are non–minimally coupled ($\zeta \neq 0$) to the gravity of a strictly stationary black hole with non–rotating horizon are studied. It is concluded that for analytical configurations the corresponding domain of outer communications is static.

1. INTRODUCTION

All the available "no–hair" theorems for non–rotating stationary black holes are based in a staticity hypothesis (see [1, 2, 3, 4, 5] for recent revisions on the subject), i.e., that the asymptotically timelike Killing field $\boldsymbol{k}$, coinciding at the horizon $\mathcal{H}^+$ with its null generator, must be hypersurface orthogonal in all the domain of outer communications $\langle\langle \mathscr{J} \rangle\rangle$. In this way, in order to establish the uniqueness of the final state of the gravitational collapse of all these systems it is needed to make use of *staticity theorems.*

Generalizing a previous result by Lichnerowicz for space–times without horizons [6], the staticity theorem corresponding to vacuum black holes was proved by Hawking [7, 8], assuming the existence of strict stationarity, i.e., that the Killing field $\boldsymbol{k}$ is not only timelike ($V \equiv -(\boldsymbol{k}|\boldsymbol{k}) \geq 0$) at infinity, but in all the domain of outer communications $\langle\langle \mathscr{J} \rangle\rangle$.

The extension of this proof to the case of electrovac black holes was proposed by Carter [9], and holds only assuming a condition more restrictive than that one of strict stationarity. This condition occurs to be unphysical, since it is violated even for the black holes of the

Exact Solutions and Scalar Fields in Gravity: Recent Developments
Edited by Macias *et al.*, Kluwer Academic/Plenum Publishers, New York, 2001

Reissner–Nordström family when they have electric charges in the interval $4M/5 < Q < M$.

Using a Hamiltonian approach, both staticity theorems, for the vacuum and electrovac black holes, have been proved by Sudarsky and Wald [10, 11, 12] without the previous restrictive hypotheses and using only a maximal slice, the existence of which was proved later by Chruściel and Wald [13]. More recently, this last result has been extended, using the same technics, to the axi–dilaton gravity coupled to electromagnetism which is derived from string theory to low energies [14].

For minimally coupled scalar models, staticity theorems has been also proved by Heusler under the strict stationarity hypothesis [15, 16]. In this paper we will extend the results by Heusler to the case of non–minimal coupling, assuming again strict stationarity and also analyticity of the scalar fields, establishing in this way that a non–rotating strictly stationary black hole, corresponding to a self–interacting non–minimally coupled scalar field, is static.

For minimally coupled scalar fields there is no need to impose the existence of analyticity for static configurations, since the elliptical nature of the corresponding Einstein–Scalar system guarantees that all the fields are analytical in appropriate coordinates (see the remarks of [17] about the non–vacuum case). The situation is rather different for non–minimally coupled systems, in this case the relevant equations are not necessarily elliptic, and consequently the existence of analyticity must be imposed as a supplementary assumption (see the related discussion in [18]). However, the analyticity hypothesis cannot be considered a too strong one in the study of stationary black holes, since the classification of them rests implicitly in this condition through the Hawking strong rigidity theorem [7, 8]. This theorem establishes that the event horizon of a stationary black hole is a Killing horizon, i.e., there exist a Killing field coinciding at the event horizon with the null generators of it (for non–rotating horizons it is the same that the stationary Killing field $\boldsymbol{k}$, but for rotating ones it is different). In order to prove this celebrated theorem, Hawking recurred to the existence of analyticity on all the fields constituting the stationary black hole configuration.

Recent attempts to replace this condition by the more natural one of smoothness have given only positive results for the region interior to the event horizon [19], hence they are of little use in the subject of classification of stationary black–hole exteriors.

The result we will establish is of great utility to eliminate the staticity supposition from the "no–hair" theorems which has been proved for non–minimally coupled scalar fields in presence of non–rotating stationary black holes, see e.g. [20] for the conformal case, and [21, 22, 23, 2] for other results recently derived for more general coupling, where it is not only assumed staticity but also spherical symmetry.

2. THE STATICITY THEOREM FOR NON–MINIMALLY COUPLED SCALAR FIELDS

Let us consider the action for a self–interacting scalar field non–minimally coupled to gravity

$$\mathcal{S} = \frac{1}{2}\int dv \left(\frac{1}{\kappa}R - (\nabla_\mu\Phi\nabla^\mu\Phi + U(\Phi)) - \zeta\, R\,\Phi^2\right), \tag{1}$$

where ζ is a real parameter (the values $\zeta = 0$ and $\zeta = 1/6$ correspond to minimal and conformal coupling, respectively).

The variations of this action with respect to the metric and the scalar field, respectively, give rise to the Einstein equations

$$\left(1 - \kappa\zeta\,\Phi^2\right) R_\nu{}^\mu = \kappa\left(\nabla_\nu\Phi\nabla^\mu\Phi + \frac{1}{2}\delta_\nu{}^\mu\left(U(\Phi) - \zeta\,\Box\Phi^2\right) - \zeta\,\nabla_\nu\nabla^\mu\Phi^2\right), \tag{2}$$

and the nonlinear Klein–Gordon equations

$$\Box\Phi - \frac{1}{2}\frac{dU(\Phi)}{d\Phi} - \zeta\,R\,\Phi = 0. \tag{3}$$

From this system of equations it must follows the existence of staticity in the case of strictly stationary black holes with a non–rotating horizon. As it was quoted at the beginning, staticity means that the stationary Killing field $\boldsymbol{k}$ is hypersurface orthogonal, which is equivalent, by the Frobenius theorem, to the vanishing of the twist 1–form

$$\omega_\alpha \equiv \frac{1}{2}\eta_{\alpha\beta\mu\nu}k^\beta\nabla^\mu k^\nu, \tag{4}$$

where $\boldsymbol{\eta}$ is the volume 4–form.

In order to exhibit the existence of staticity we will find the explicit dependence of $\boldsymbol{\omega}$ in terms of Φ, by solving the following differential equations which must be satisfied by the twist 1–form [9]

$$\eta^{\mu\nu\alpha\beta}\nabla_\alpha\omega_\beta = 2k^{[\mu}\mathcal{R}^{\nu]}, \tag{5}$$

where

$$\mathcal{R}^\mu \equiv k^\nu R_\nu{}^\mu,$$

is the Ricci vector, that can be evaluated from Einstein equations (2). Using the stationarity of the scalar field

$$\$_k\ (\Phi) \equiv k^\nu \nabla_\nu \Phi = 0,$$

and the identity

$$k^\nu \nabla_\nu \nabla^\mu \Phi^2 = -\nabla^\mu k^\nu \nabla_\nu \Phi^2,$$

the Ricci vector can be written as

$$\left(1 - \kappa\zeta\,\Phi^2\right)\mathcal{R}^\mu = \kappa\left(\frac{1}{2}k^\mu\left(U(\Phi) - \zeta\Box\Phi^2\right) + \zeta\,\nabla^\mu k^\nu \nabla_\nu \Phi^2\right). \qquad (6)$$

Replacing this expression in (5), using the Killing vector definition

$$\nabla^\mu k^\nu = \nabla^{[\mu} k^{\nu]},$$

it is obtained the following identity

$$\left(1 - \kappa\zeta\,\Phi^2\right)\eta^{\mu\nu\alpha\beta}\nabla_\alpha \omega_\beta = 2\,\kappa\zeta\,k^{[\mu}\nabla^\nu k^{\alpha]}\nabla_\alpha \Phi^2, \qquad (7)$$

which, taking into account definition (4), can be rewritten as

$$\left(1 - \kappa\zeta\,\Phi^2\right)\eta^{\mu\nu\alpha\beta}\nabla_\alpha \omega_\beta = \frac{2}{3}\kappa\zeta\,\eta^{\mu\nu\alpha\beta}\nabla_\alpha \Phi^2\,\omega_\beta, \qquad (8)$$

or equivalently, in the language of differential forms as follows

$$d\omega = d\left(\ln\left[\left(1 - \kappa\zeta\,\Phi^2\right)^{-2/3}\right]\right)\wedge\omega, \qquad (9)$$

where obviously the above expression is valid only in the regions where $\Phi^2 \neq 1/\kappa\zeta$. The expression (9) can be written also as

$$d\left(\left(1 - \kappa\zeta\,\Phi^2\right)^{2/3}\omega\right) = 0, \qquad (10)$$

where it is understood once more that the equality is valid only when $\Phi^2 \neq 1/\kappa\zeta$.

We will analyze now the value of the left hand side of (10) in the regions where $\Phi^2 = 1/\kappa\zeta$, situation which is only possible for positive values of the non–minimal coupling ($\zeta > 0$). We claim that in the analytic case these regions are composed from a countable union of lower dimensional surfaces, hence, by the continuity of the left hand side of (10), the involved 1–form is also closed in regions where $\Phi^2 = 1/\kappa\zeta$.

Lets examine the argument in detail. As it was mentioned at the introductory Section, we suppose that the scalar field Φ and the metric

g are analytical in appropriated coordinates. First it must be noticed that $\Phi^2 \not\equiv 1/\kappa\,\zeta$ in the whole of $\langle\langle\mathcal{J}\rangle\rangle$, i.e., that the square of the scalar field does not take the value $1/\kappa\,\zeta$ in every point of the domain of outer communications. This is based in that the converse is in contradiction with the fact that the asymptotic value of the effective gravitational constant

$$G_{\text{eff}} \equiv \frac{G}{(1-\kappa\,\zeta\,\Phi^2)},$$

must be positive and finite due to the know attractive character of gravity at the asymptotic regions [22]. Since $\Phi^2 \not\equiv 1/\kappa\,\zeta$, it can be shown (see Ref. [26]) from the analyticity of Φ, that if the inverse images of the real values $\{\pm 1/\sqrt{\kappa\,\zeta}\}$ under the function Φ,

$$L_\pm \equiv \Phi^{-1}\left(\left\{\pm 1/\sqrt{\kappa\,\zeta}\right\}\right),$$

are nonempty, they are composed of a countable union of many 1–dimensional, 2–dimensional and 3–dimensional analytical submanifolds of $\langle\langle\mathcal{J}\rangle\rangle$. At first sight, 0–dimensional (point–like) submanifolds are also admissible, but in our case they are excluded by the stationarity. For a proof of the quoted results in $\mathbb{R}^3$ see e.g. [26], the extension to $\mathbb{R}^4$ does not present any problem.

A direct implication of these results is that in principle the equality (10) is valid just in $\langle\langle\mathcal{J}\rangle\rangle\setminus(L_+\cup L_-)$, but by the continuity of the left hand side of (10) in the whole of $\langle\langle\mathcal{J}\rangle\rangle$, and in particular through the lower dimensional surfaces that constitute $L_\pm$, the left hand side of (10) vanishes also in $L_\pm$.

Provided that expression (10) is valid in all the domain of outer communications $\langle\langle\mathcal{J}\rangle\rangle$, it follows from the simple connectedness of this region [25], and the well–known Poincaré lemma, the existence of a global potential U in the whole of $\langle\langle\mathcal{J}\rangle\rangle$ such that

$$\left(1-\kappa\,\zeta\,\Phi^2\right)^{2/3}\omega = dU. \tag{11}$$

The previous potential U is constant in each connected component of the event horizon $\mathcal{H}^+$. This follows from the fact that on the one hand $\omega = 0$ in $\mathcal{H}^+$, since this region is a Killing horizon whose normal vector coincides with $\boldsymbol{k}$, and on the other hand, as it was previously mentioned (see [7, 8]), in order to that $\mathcal{H}^+$ be a Killing horizon the scalar field must be analytical, especially at the horizon, hence Eq. (11) implies that U is constant in each connected component of the horizon.

The same result is achieved as well for the asymptotic regions, because any stationary black hole with a bifurcate Killing horizon admits a maximal hypersurface asymptotically orthogonal to the stationary Killing field $\boldsymbol{k}$ [13]. Furthermore, it has been shown [24] that a stationary black hole can be globally extended to other enlarged one possessing a bifurcate Killing horizon.

In what follows we will show that besides the fact that the potential U is constant at the horizon and the asymptotic regions, it will be also constant in the whole of $\langle\langle \mathcal{J} \rangle\rangle$. For the case of minimal coupling ($\zeta = 0$) this results implies directly the staticity (11) as it has been previously proved by Heusler [15]. We will extend his proof to the case of non–minimally coupled to gravity scalar fields.

For every function f and 1–form $\boldsymbol{\Omega}$, the following identity is satisfied (see the appendix in [16] for the details of the remaining calculations)

$$\boldsymbol{d}^{\dagger}\left(f\boldsymbol{\Omega}\right) = f\boldsymbol{d}^{\dagger}\boldsymbol{\Omega} - \left(\boldsymbol{d}f|\boldsymbol{\Omega}\right),$$

here $\boldsymbol{d}^{\dagger} = *\boldsymbol{d}*$ stands for the co–differential operator. Applying this expression to the function U and the 1–form $\boldsymbol{\Omega}/V^2$, together with (11), the following can be obtained

$$\boldsymbol{d}^{\dagger}\left(U\frac{\boldsymbol{\omega}}{V^2}\right) = -\frac{\left(1-\kappa\zeta\,\Phi^2\right)^{2/3}\left(\boldsymbol{\omega}|\boldsymbol{\omega}\right)}{V^2}, \qquad (12)$$

where the forthcoming identity has been used (the proof of it can be seen in Ref. [16])

$$\boldsymbol{d}^{\dagger}\left(\frac{\boldsymbol{\omega}}{V^2}\right) = 0.$$

Following the same procedure used by Heusler in [15, 16] we will integrate (12) over a spacelike hypersurface Σ, with volume form $\boldsymbol{i_k\eta}$. Taking in consideration that for any stationary 1–form $\boldsymbol{\Omega}$ ($\$_{\boldsymbol{k}}\boldsymbol{\Omega} = 0$)

$$\boldsymbol{d}^{\dagger}\boldsymbol{\Omega}\,\boldsymbol{i_k\eta} = -\boldsymbol{d}*\left(\boldsymbol{k}\wedge\boldsymbol{\Omega}\right),$$

holds [16], the following relation is obtained applying the Stokes theorem to (12),

$$\int_{\partial\Sigma} U*\left(\boldsymbol{k}\wedge\frac{\boldsymbol{\omega}}{V^2}\right) = \int_{\Sigma}\frac{\left(1-\kappa\zeta\,\Phi^2\right)^{2/3}\left(\boldsymbol{\omega}|\boldsymbol{\omega}\right)}{V^2}\,\boldsymbol{i_k\eta}. \qquad (13)$$

Using now the identity [16]

$$2*\left(\boldsymbol{k}\wedge\frac{\boldsymbol{\omega}}{V^2}\right) = \boldsymbol{d}\left(\frac{\boldsymbol{k}}{V}\right),$$

the expression (13) can be brought to the form

$$\frac{1}{2}\int_{\partial\Sigma} U\boldsymbol{d}\left(\frac{\boldsymbol{k}}{V}\right) = \int_{\Sigma}\frac{\left(1-\kappa\zeta\,\Phi^2\right)^{2/3}\left(\boldsymbol{\omega}|\boldsymbol{\omega}\right)}{V^2}\,\boldsymbol{i_k\eta}. \qquad (14)$$

The boundary $\partial\Sigma$ is constituted at its interior by the event horizon $\mathcal{H}^+ \cap \Sigma$, and at the infinity by the asymptotic regions. Since the potential U is constant over each one of the connected components of these boundaries, it can be pulled out from each one of the corresponding boundary integrals in the left hand side of (14).

The asymptotic regions and the connected components of the horizon are all topological 2–spheres [8, 25], by this reason the left hand side of (14) vanishes; from Stokes theorem, the integral of an exact form over a manifold without boundary is zero. Thereby, it is satisfied that

$$\int_\Sigma \frac{\left(1 - \kappa \zeta \, \Phi^2\right)^{2/3} (\omega|\omega)}{V^2} \, i_k \eta = 0. \tag{15}$$

The integrand in (15) is non–negative, due to the fact that ω is a spacelike 1–form since it is orthogonal by definition to the timelike field k (4). Hence, it follows that (15) is satisfied if and only if the integrand vanishes in Σ, and by stationarity it also vanishes in all the domain of outer communications $\langle\langle \mathcal{J} \rangle\rangle$, consequently

$$\left(1 - \kappa \zeta \, \Phi^2\right)^{2/3} (\omega|\omega) = 0. \tag{16}$$

From the previous conclusion (16), it follows that $\omega = 0$ in $\langle\langle \mathcal{J} \rangle\rangle \setminus (L_+ \cup L_-)$, but by the continuity of ω in all of $\langle\langle \mathcal{J} \rangle\rangle$, and in particular through the lower dimensional surfaces that constitute the regions $L_\pm$, ω vanishes also in $L_\pm$ and accordingly in all the domain of outer communications $\langle\langle \mathcal{J} \rangle\rangle$. Hence, the staticity theorem is proved.

3. CONCLUSIONS

Finally, it is concluded that for a non–rotating strictly stationary black hole with a self–interacting scalar field non–minimally coupled to gravity, the corresponding domain of outer communications is static if analytic field configurations are considered. As in the minimal case [15] this result remains valid when no horizon is present.

Acknowledgments

The author thanks Alberto García by its incentive in the study of this topics and by its support in the course of the investigation. This research was partially supported by the CONACyT Grant 32138E. The author also thanks all the encouragement and guide provided by his recently late father: Erasmo Ayón Alayo, *Ibae Ibae Ibayen Torun.*

References

[1] M. Heusler, *Black Hole Uniqueness Theorems* (Cambridge Univ. Press, Cambridge 1996).

[2] J.D. Bekenstein, *Proceedings of Second Sakharov Conference in Physics, Moscow*, eds. I.M. Dremin, A.M. Semikhatov, (World Scientific, Singapore 1997) pp. 761.

[3] M. Heusler, *Living Rev. Rel.*, **1**, (1998) 6, http://www.livingreviews.org/Articles/Volume1/1998-6heusler.

[4] J.D. Bekenstein, *Proceedings of the 9th Brazilian School of Cosmology and Gravitation*, Rio de Janeiro, Brazil, gr-qc/9808028 (1998).

[5] B. Carter, *Proceedings of 8th Marcel Grossmann Meeting, Jerusalem*, ed. T. Piran (World Scientific, Singapore 1999).

[6] A. Lichnerowicz, *Théories Relativistes de la Gravitation et de L'Électromagnétisme* (Masson & Cie, Paris 1955).

[7] S.W. Hawking, *Commun. Math. Phys.*, **25** (1972) 152.

[8] S.W. Hawking and G. F. Ellis, *The Large Scale Structure of Space–Time* (Cambridge Univ. Press 1973).

[9] B. Carter, in: *Gravitation in Astrophysics (Cargèse Summer School 1986)*, eds. B. Carter and J.B. Hartle (Plenum, New York 1987).

[10] D. Sudarsky and R.M. Wald, *Phys. Rev.*, **D46** (1992) 1453.

[11] D. Sudarsky and R.M. Wald, *Phys. Rev.*, **D47** (1993) R3209.

[12] R.M. Wald, in: *Proceedings: Directions In General Relativity*, eds. B.L. Hu, M.P. Ryan, Jr., C.V. Vishveshwara, T.A. Jacobson (New York, Cambridge Univ. Press, 1993).

[13] P.T. Chruściel and R.M. Wald, *Commun. Math. Phys.*, **163** (1994) 561.

[14] M. Rogatko, *Phys. Rev.*, **D58** (1998) 044011.

[15] M. Heusler, *Class. Quant. Grav.*, **10** (1993) 791.

[16] M. Heusler and N. Straumann, *Class. Quant. Grav.*, **10** (1993) 1299.

[17] H. Müller zum Hagen, *Proc. Camb. Phil. Soc.*, **67** (1970) 415.

[18] T. Zannias, *J. Math. Phys.*, **39** (1998) 6651.

[19] H. Friedrich, I. Racz, and R.M. Wald, *Commun. Math. Phys.*, **204** (1999) 691.

[20] T. Zannias, *J. Math. Phys.*, **36** (1995) 6970.

[21] A. Saa, *J. Math. Phys.*, **37** (1996) 2346.

[22] Mayo, A.E., Bekenstein, J.D., *Phys. Rev.*, **D54** (1996) 5059.

[23] A. Saa, *Phys. Rev.*, **D53** (1996) 7377.

[24] I. Racz and R.M. Wald, *Class. Quant. Grav.*, **13** (1996) 539.

[25] P.T. Chruściel and R.M. Wald, *Class. Quant. Grav.*, **11** (1994) L147.

[26] H. Müller zum Hagen and D.C. Robinson, and H.J. Seifert, *Gen. Rel. Grav.*, **4** (1973) 53.

QUANTUM NONDEMOLITION MEASUREMENTS AND NON–NEWTONIAN GRAVITY

A. Camacho *

Departamento de Física

Instituto Nacional de Investigaciones Nucleares.

Apartado Postal 18–1027, México, D. F., México.

Abstract In the present work the detection, by means of a nondemolition measurement, of a Yukawa term, coexisting simultaneously with gravity, has been considered. In other words, a nondemolition variable for the case of a particle immersed in a gravitational field containing a Yukawa term is obtained. Afterwards the continuous monitoring of this nondemolition parameter is analyzed, the corresponding propagator is evaluated, and the probabilities associated with the possible measurement outputs are found. The relevance of these kind proposals in connection with some unified theories of elementary particles has also been underlined.

Keywords: Quantum measurements, non–Newtonian gravity, Yukawa term.

1. INTRODUCTION

The equivalence principle (EP) is one of the fundamental cornerstones in modern physics, and comprises the underlying symmetry of general relativity (GR) [1]. At this point we must be more precise and state that EP has three different formulations, namely the weak, the medium strong, and finally, the very strong equivalence principle. In order to avoid misunderstandings, here we follow [1], namely weak equivalence principle (WEP) means *the motion of any freely falling test particle is independent of its composition and structure*, medium strong form (MSEP) means *for every pointlike event of spacetime, there exists a sufficiently small neighborhood such that in every local, freely falling frame in that neighborhood, all the nongravitational laws of physics obey the laws of special relativity*. Replacing *all the nongravitational laws of physics* with

*E–mail: acamacho@nuclear.inin.mx

all the laws of physics we have the very strong form of the equivalence principle (VSEP).

The proposals that confront the predictions of GR with measurement outputs include already a large amount of experiments, for instance, the gravitational time dilation measurement [2], the gravitational deflection of electromagnetic waves [3], the time delay of electromagnetic waves in the field of the sun [4], or the geodetic effect [5]. The discovery of the first binary pulsar PSR1913+16 [6] allowed not only to probe the propagation properties of the gravitational field [7], but it also offered the possibility of testing the case of strong field gravity [8]. Of course, all these impressive experiments are an indirect confirmation of the different EP.

Another important experimental direction comprises the attempts to test, directly, WEP. Though these efforts are already more than a century old [9], the interest in this area has not disappeared. Recently [10], WEP has been tested using a rotating 3 ton ^{238}U attractor around a compact balance containing Cu and Pb test bodies. The differential acceleration of these test bodies toward the attractor was measured, and compared with the corresponding gravitational acceleration. Clearly, this proposal is designed to test WEP at classical level, i.e., gravity acts upon a classical system. At quantum realm the gravitational acceleration has been measured using light pulse interferometers [11], and also by atom interferometry based on a fountain of laser–cooled atoms [12]. Of course, the classical experiment by Colella, Overhauser, and Werner (COW) [13], is also an experiment that explores the effects at quantum level of gravity, and shows that at this level the effects of gravity are not purely geometric [14].

The interest behind these experiments stems from the fact that various theoretical attempts to construct a unified theory of elementary particles predict the existence of new forces, and they are usually not described by an inverse–square law, and of course, they violate one of the formulations of EP. By studying these violations one could determine what interaction was producing these effects [15].

Among the models that in the direction of noninverse–square forces currently exist we have Fujii's proposal [16], in which a "fifth force", coexisting simultaneously with gravity, comprises a Yukawa term, $V(r) = -G_{\infty}\frac{mM}{r}\left(1+\alpha e^{-\frac{r}{\lambda}}\right)$, here G_{∞} describes the interaction between m and M in the limit case $r \rightarrow \infty$, i.e., $G = G_{\infty}(1+\alpha)$, where G is the Newtonian gravitational constant. This kind of deviation terms arise from the exchange of a single new quantum of mass m_5, where the Compton wavelength of the exchanged field is $\lambda = \frac{\hbar}{m_5 c}$ [15], this field is usually denoted dilaton.

The experiments, already carried out, that intend to detect a Yukawa term have already imposed some limits on the parameters α and λ. For instance, if $10^{-4}m \leq \lambda \leq 10^{-3}m$, then $\alpha \sim 10^{22}$ [17], if $\lambda = 200\mu M$, then $\alpha \leq 8 \times 10^{7}$ [18] (for a more complete report see [15]).

To date, after more than a decade of experiments [19], there is no compelling evidence for any kind of deviations from the predictions of Newtonian gravity. But Gibbons and Whiting (GW) phenomenological analysis of gravity data [20] has proved that the very precise agreement between the predictions of Newtonian gravity and observation for planetary motion does not preclude the existence of large non–Newtonian effects over smaller distance scales, i.e., precise experiments over one scale do not necessarily constrain gravity over another scale. GW results conclude that the current experimental constraints over possible deviations did not severly test Newtonian gravity over the 10–1000m distance scale, usually denoted as the "geophysical window".

The idea in this work is two–fold: firstly, the effects of a Yukawa term upon a quantum system (the one is continuously monitored) will be calculated; secondly, new theoretical predictions for one of the models in the context of quantum measurement theory will be found. In order to achieve these two goals we will obtain a nondemolition variable for the case of a particle subject to a gravitational field which contains a Yukawa term such that λ has the same order of magnitude of the radius of the earth. At this point it is noteworthy to mention that the current experiments set constraints for λ for ranges between 10km and 1000km [10], but the case in which $\lambda \sim$ Earth's radius remains rather unexplored. Afterwards, we will consider, along the ideas of the so–called restricted path integral formalism (RPIF) [21], the continuous monitoring of this nondemolition parameter, and calculate, not only, the corresponding propagators, but also the probabilities associated with the different measurement outputs.

2. YUKAWA TERM

Suppose that we have a spherical body with mass M and radius R. Let us now consider the case of a Yukawa form of gravitational interaction [16], hence the gravitational potential of this body reads

$$V(r) = -G_{\infty}\frac{M}{r}\left(1 + \alpha e^{-\frac{r}{\lambda}}\right). \tag{1}$$

Let us now write $r = R + z$, where R is the body's radius, and z the height over its surface. If $R/\lambda \sim 1$ (which means that the range of this Yukawa term has the same order of magnitude as the radius of our spherical body), and if $z << R$, then we may approximate the Lagrangian of a particle of mass m as follows

$$L = \frac{\vec{p}^{\,2}}{2m} + G_{\infty}\frac{mM}{R}\left([1+\alpha] - [\frac{1+\alpha}{R} + \frac{\alpha}{2\lambda}]z + [\frac{1+\alpha}{2R^2} + \frac{\alpha}{2R\lambda} + \frac{\alpha}{2\lambda^2}]z^2\right). \tag{2}$$

3. QUANTUM MEASUREMENTS

Nowadays one of the fundamental problems in modern physics comprises the so–called quantum measurement problem [22]. Though there are several attempts to solve this old conundrum (some of them are equivalent [23]), here we will resort to RPIF [21], because it allows us to calculate, in an easier manner, propagators and probabilities. RPIF explains a continuous quantum measurement with the introduction of a restriction on the integration domain of the corresponding path integral. This last condition can also be reformulated in terms of a weight functional that has to be considered in the path integral. Clearly, this weight functional contains all the information about the interaction between measuring device and measured system. This model has been employed in the analysis of the response of a gravitational antenna not only of Weber type [21], but also when the measuring process involves a laser–interferometer [24]. We may also find it in the quest for an explanation of the emergence of some classical properties, as time, in quantum cosmology [25].

Suppose now that our particle with mass m goes from point N to point W. Hence its propagator reads

$$U(W,\tau'';N,\tau') = \left(\frac{m}{2\pi i\hbar T}\right)\exp\left\{\frac{im}{2\hbar T}\left[(x_W - x_N)^2 + (y_W - y_N)^2\right]\right\} \times \int_{z_N}^{z_W} d[p]d[z(t)]\exp\left\{\frac{i}{\hbar}\int_{\tau'}^{\tau''}\left[\frac{p^2}{2m} + (1+\alpha)\frac{G_{\infty}mM}{R} + Fz + \frac{m}{2}\Omega^2 z^2\right]dt\right\}. \tag{3}$$

Here we have introduced the following definitions

$$F = -G_{\infty}\frac{mM}{R}\left[\frac{1+\alpha}{R} + \frac{\alpha}{2\lambda}\right], \tag{4}$$

$$\omega^2 = -2\frac{G_\infty M}{R}\left[\frac{1+\alpha}{2R^2} + \frac{\alpha}{2\lambda R} + \frac{\alpha}{2\lambda^2}\right]. \tag{5}$$

$T = \tau'' - \tau'$, and $\sqrt{(x_W - x_N)^2 + (y_W - y_N)^2}$ denotes the projection on the body's surface of the distance between points W and N. We also have that $G = G_\infty[1+\alpha]$, and G is the Newtonian gravitational constant [15]. In our case, $[\frac{1+\alpha}{2R^2} + \frac{\alpha}{2\lambda R} + \frac{\alpha}{2\lambda^2}] > 0$, hence, $\omega = i\Omega$, where $\Omega \in \Re$.

Suppose now that the variable $A(t)$ is continuously monitored. Then we must consider, along the ideas of RPIF, a particular expression for our weight functional, i.e., for $w_{[a(t)]}[A(t)]$. As was mentioned before, the weight functional $w_{[a(t)]}[A(t)]$ contains the information concerning the measuring process.

At this point we face a problem, namely, the choice of our weight functional. In order to solve this difficulty, let us mention that the results coming from a Heaveside weight functional [26] and those coming from a gaussian one [27] coincide up to the order of magnitude. These last remarks allow us to consider a gaussian weight functional as an approximation of the correct expression. But a sounder justification of this choice stems from the fact that there are measuring processes in which the weight functional possesses a gaussian form [28]. In consequence we could think about a measuring device whose weight functional is very close to a gaussian behaviour.

Therefore we may now choose as our weight functional the following expression

$$\omega_{[a(t)]}[A(t)] = \exp\left\{-\frac{2}{T\Delta a^2}\int_{\tau'}^{\tau''}[A(t) - a(t)]^2\,dt\right\}, \tag{6}$$

here Δa represents the error in our measurement, i.e., it is the resolution of the measuring apparatus.

4. QUANTUM NONDEMOLITION MEASUREMENTS

The basic idea around the concept of quantum nondemolition (QND) measurements is to carry out a sequence of measurements of an observable in such a way that the measuring process does not diminish the predictability of the results of subsequent measurements of the same observable [29]. This concept stems from the work in the context of gravitational wave antennae. Indeed, the search for gravitational radiation demands measurements of very small displacements of macroscopic bodies [30]. Braginsky *et al.* [31] showed that there is a quantum limit, the so called "standard quantum limit", which is a consequence of Heisenberg uncertainty principle, the one limits the sensitivity of the corresponding measurement (the original work [31] involves the sensitivity of a gravitational antenna). This work allowed also the introduction of the idea of a QND measurement, in which a variable is measured in such a way that the unavoidable disturbance of the conjugate observable does not disturb the evolution of the chosen variable [32].

Let us now suppose that in our case $A(t) = \rho p + \sigma z$, where ρ and σ are functions of time. In this particular case, the condition that determines when $A(t)$ is a QND variable may be written as a differential equation [21]

$$\frac{df}{dt} = \frac{f^2}{m} - m\Omega^2, \tag{7}$$

where $f(t) = \sigma/\rho$. It is readily seen that a solution to (7) is

$$f(t) = -m\Omega \tanh(\Omega t). \tag{8}$$

Choosing $\rho(t) = 1$, we find that in our case a possible QND variable is

$$A(t) = p - m\Omega z \tanh(\Omega t). \tag{9}$$

5. QND AND NON–NEWTONIAN GRAVITY: PROPAGATORS AND PROBABILITIES

With our weight functional choice (expression (6)) the new propagator involves two gaussian integrals, and can be easily calculated [33]

$$U_{[a(t)]}(W, \tau''; N, \tau') = \left(\frac{m}{2\pi i\hbar T}\right) \exp\left\{\frac{im}{2\hbar T}\left[(x_W - x_N)^2 + (y_W - y_N)^2\right]\right\}$$
$$\exp\left\{\frac{i}{\hbar}(1+\alpha)\frac{G_\infty mM}{R}T\right\} \exp\left\{\frac{-T\Delta a^2 + i2m\hbar}{4m^2\hbar^2 + T^2\Delta a^4}\int_{\tau'}^{\tau''} a^2(t)dt\right\}$$
$$\times \ \exp\{\frac{-i\hbar}{2m\Omega^2}\int_{\tau'}^{\tau''}\left[\frac{F}{\hbar} + \frac{4m^2\hbar\Omega a}{4m^2\hbar^2 + T^2\Delta a^4}\tanh(\Omega t) + i\frac{2ma\Omega T\Delta a^2}{4m^2\hbar^2 + T^2\Delta a^4}\right]^2$$
$$\times \ \left[\frac{4m^2\hbar^2[1+\tanh^2(\Omega t)] + T^2\Delta a^4 - i2m\hbar T\Delta a^2\tanh^2(\Omega t)}{4m^2\hbar^2[1+\tanh^2(\Omega t)]^2 + T^2\Delta a^4}\right] dt\}. \tag{10}$$

The probability, $P_{[a(t)]}$, of obtaining as measurement output $a(t)$ is given by expression $P_{[a(t)]} = |U_{[a(t)]}|^2$ [21]. Hence, in this case

$$P_{[a(t)]} = \exp\left\{\frac{-2T\Delta a^2}{4m^2\hbar^2 + T^2\Delta a^4}\int_{\tau'}^{\tau''} a^2(t)dt\right\}$$
$$\times \exp\left\{\frac{\hbar}{m\Omega^2}\int_{\tau'}^{\tau''}\left[2I_1I_2I_3 + I_4(I_2^2 - I_1^2)\right]dt\right\}. \tag{11}$$

Here the following definitions have been introduced

$$I_1 = \frac{F}{\hbar} + \frac{4m^2\hbar\Omega a}{4m^2\hbar^2 + T^2\Delta a^4}\tanh(\Omega t), \tag{12}$$

$$I_2 = \frac{2ma\Omega T\Delta a^2}{4m^2\hbar^2 + T^2\Delta a^4}\tanh(\Omega t), \tag{13}$$

$$I_3 = \frac{4m^2\hbar^2[1+\tanh^2(\Omega t)] + T^2\Delta a^4}{4m^2\hbar^2[1+\tanh^2(\Omega t)]^2 + T^2\Delta a^4}, \tag{14}$$

$$I_4 = \frac{2m\hbar T\Delta a^2\tanh^2(\Omega t)}{4m^2\hbar^2[1+\tanh^2(\Omega t)]^2 + T^2\Delta a^4}. \tag{15}$$

6. CONCLUSIONS

In this work we have considered a Yukawa term coexisting with the usual Newtonian gravitational potential (expression (1)). Assuming that the range of this new interaction has the same order of magnitude than the Earth's radius a QND variable was obtained for a particle with mass m (expression (9)), and its corresponding propagator was evaluated. Afterwards, it was assumed that this QND variable was continuously monitored, and, along the ideas of RPIF, the propagator and probability associated with the possible measurement outputs were also calculated, expressions (14) and (15), respectively.

Another interesting point around expressions (10) and (11) comprises the role that the mass parameter plays in them. It is readily seen that if we consider two particles with different mass, say m and $\tilde{m}$, then they render different propagators and probabilities. This last fact means that "gravity" is, in this situation, not purely geometric. This is no surprise, the presence of a new interaction, coexisting with the usual Newtonian gravitational potential, could mean the breakdown of the geometrization of the joint interaction (Newtonian contribution plus Yukawa term). The detection of this interaction would mean the violation of WEP, but not necessarily of VSEP [15]. Nevertheless, the possibilities of using quantum measurement theory to analyze the possible limits of VSEP at quantum level do not finish here, indeed, the possible incompatibility between the different formulations of EP and measurement theory can also be studied along the ideas of this theory [42]. Of course, more work is needed around the validity at quantum level of the different formulations of EP. Indeed, as has already been proved [43], in the presence of a gravitational field, the generalization to the quantum level of even the simplest kinematical concepts, for instance the time of flight, has severe conceptual difficulties.

This proposal would also render new theoretical predictions that could be confronted (in the future) against the experiment, and therefore we would obtain a larger framework that could allow us to test the validity of RPIF [44]. As was mentioned before, this comprises our second goal in the present work.

Acknowledgments

The author would like to thank A.A. Cuevas–Sosa for his help. It is also a pleasure to thank R. Onofrio for bringing reference [29] to my attention. This work was partially supported by Conacyt (Mexico) grant no. I35612-E.

References

[1] I. Ciufolini and J.A. Wheeler, "*Gravitation and Inertia*", (Princeton University Press, Princeton, New Jersey, 1995).

[2] R.F. C. Vessot, L.W. Levine, *Gen. Rel. and Grav.* **10** (1979) 181; R.F.C. Vessot, *et al.*, *Phys. Rev. Lett.* **45** (1980) 2081.

[3] R.N. Truehaft and S.T. Lowe, *Phys. Rev. Lett.* **62** (1989) 369.

[4] T.P. Krisher, J.D. Anderson, and A.H. Taylor, *Astophys. J.* **373** (1991) 665.

[5] J.O. Dickey *et al.*, *Science* **265** (1994) 482.

[6] R.A. Hulse and J.H. Taylor, *Astrophys. J. Lett.* **195** (1975) L51.

[7] J.H. Taylor, *Class. Quant. Grav.* **10** (1993) S167.

[8] T. Damour and J.H. Taylor, *Phys. Rev.* **D45** (1992) 1840.

[9] R. v. Eötvös, in: *Roland Eötvös Gesammelte Arbeiten*, P. Selényi ed., (Akademiai Kiado, Budapest, 1953).

[10] G.L. Smith, C.D. Hoyle, J.H. Gundlach, E.G. Adelberger, B.R. Heckel, and H. E. Swanson, *Phys. Rev.* **D61** (1999) 022001.

[11] M. Kasevich and S. Chu, *Appl. Phys.* **B54** (1992) 321.

[12] A. Peters, K.Y. Chung, and S. Chu, *Nature* **400** (1999) 849.

[13] R. Colella, A.W. Overhauser, and S.A. Werner, *Phys. Rev. Lett.* **34** (1975) 1472; K.C. Littrell, B.E. Allman, and S.A. Werner, *Phys. Rev.* **A56** (1997) 1767.

[14] J. J. Sakurai, "*Modern Quantum Mechanics*", (Addison–Wesley Publishing Company, Reading, Mass.,1995).

[15] E. Fishbach and C.L. Talmadge, "*The Search for Non–Newtonian Gravity*", (Springer–Verlag, New York, 1999).

[16] F. Fujii, *Nature* **234** (1971) 5.

[17] G. Carugno, Z. Fontana, R. Onofio, and C. Rizzo, *Phys. Rev.* **D55** (1997) 6591.

[18] R. Onofrio, *Mod. Phys. Lett.* **A15** (1998) 1401.

[19] D.R. Long, *Phys. Rev.* **D9** (1974) 850; F.D. Stacey, *Rev. Mod. Phys.* **59** (1987) 157.

[20] G.W. Gibbons and B.F. Whiting, *Nature* **291** (1981) 636.

[21] M.B. Mensky, "*Continuous Quantum Measurements and Path Integrals*", (IOP, Bristol and Philadelphia, 1993).

[22] R. Omnes, "*The interpretation of quantum mechanics*", (Princeton University Press, Princeton, 1994).

[23] C. Presilla, R. Onofrio, and U. Tambini, *Ann. Phys.* (NY) **248** (1996) 95.

[24] A. Camacho, *Int. J. Mod. Phys.* **A14** (1999) 1997.

[25] M.B. Mensky, *Class. Quan. Grav.* **7** (1990) 2317; A. Camacho, in: *Proceedings of the International Seminar: Current Topics in Mathematical Cosmology*, M. Rainer and H.-J. Schimdt, eds., (World Scientific Publishing Co., Singapore, 1998); A. Camacho and A. Camacho–Galván, *Nuov. Cim.* **B114** (1999) 923.

[26] M.B. Mensky, *Phys. Rev.* **D20** (1979) 384.

[27] M.B. Mensky, *Sov. Phys. JETP.* **50** (1979) 667.

[28] M.B. Mensky, *Physics–Uspekhi* **41** (1998) 923.

[29] M.F. Bocko and R. Onofrio, *Rev. Mod. Phys.* **68** (1996) 755.

[30] K.S. Thorne, *Rev. Mod. Phys.* **52** (1980) 299.

[31] V.B. Braginsky, Yu.I. Vorontsov, and V.D. Krivchenkov, *Sov. Phys. JETP.* **41** (1975) 28.

[32] V.B. Braginsky, Yu.I. Vorontsov, and F.Ya. Khalili, *Pis'ma Zh. Eksp. Teor. Fiz.* **73** (1978) 296.

[33] W. Dittrich and M. Reuter, "*Classical and Quantum Dynamics)*", (Springer–Verlag, Berlin, 1996).

[34] R. Thompson, in: *Latin–American School of Physics XXXI ELAF*, S. Hacyan, R. Jáuregui, and R. López–Peña, eds., (American Institut of Physics, Woodbury, New York, 1999).

[35] W. Paul, *Rev. Mod. Phys.* **62** (1990) 531; H. Dehmelt, *Rev. Mod. Phys.* **62** (1990) 525.

[36] M.F. Bocko and W.W. Johnson, *Phys. Rev. Lett.* **48** (1982) 1371.

[37] V.B. Braginsky and F.Ya. Khalili, *Rev. Mod. Phys.* **68** (1996) 1.

[38] J.F. Roch, G. Roger, P. Grangier, J.-M. Courty, and S. Reynaud, *Appl. Phys.* **B55** (1992) 291.

[39] I. Marzoli and P. Tombesi, *Europhys. Lett.* **24** (1993) 515.

[40] A. Camacho, *Phys. Lett.* **A256** (1999) 339; A. Camacho, *Phys. Lett.* **A262** (1999) 110.

[41] A. Camacho, *Int. J. Mod. Phys.* **D**, in press.

[42] A. Camacho, *Mod. Phys. Lett.* **A14** (1999) 275; A. Camacho, *Mod. Phys.* Lett. **A14** (1999) 2545; A. Camacho, *Mod. Phys. Lett.* **A15** (2000) 1461.

[43] L. Viola and R. Onofrio, *Phys. Rev.* **D55** (1997) 455; R. Onofrio and L. Viola, *Mod. Phys. Lett.* **A12** (1997) 1411.

[44] J. Audretsch, M. B. Mensky, and V. Namiot, *Phys. Letts.* **A237** (1997) 1; A. Camacho and A. Camacho–Galván, *Phys. Letts.* **A247** (1998) 373; A. Camacho, *Phys. Lett.* **A277** (2000) 7.

ON ELECTROMAGNETIC THIRRING PROBLEMS

Markus King
Institute of Theoretical Physics
University of Tübingen, D–72076 Tübingen, Germany

Herbert Pfister
Institute of Theoretical Physics
University of Tübingen, D–72076 Tübingen, Germany

Keywords: Thirring problem, exact solutions.

1. INTRODUCTION

The standard Thirring problem describes the (non–local) influence of rotating masses on the inertial properties of space–time, especially the so–called dragging of inertial frames inside a rotating mass shell relative to the asymptotic frames. It is then a natural question to ask whether and how properties other than inertial ones are also influenced by rotating masses, and the first (non–inertial and non–gravitational) properties which come here to one's mind, are surely electromagnetic phenomena.

Indeed there exists a whole series of papers [1, 2, 3, 4, 5] on models where a charge q sits inside a slowly rotating mass shell (mass M, radius R), and where the rotationally induced magnetic field is calculated and discussed, typically for small ratios M/R and/or q/R. An especially remarkable but confusing fact is that Cohen's results [2] are in complete agreement with Machian expectations whereas Ehlers and Rindler [4] interpret their results to be "in fact Mach–negative or, at best, Mach–neutral".

We examine a class of models consisting of two spherical shells of radii a and $R \geq a$, the first one carrying a nearly arbitrary charge q but no rest mass, the second one being nearly arbitrary massive but electrically neutral. The only restriction on the parameters is that the systems should be free of singularities and horizons. To these shells we apply small but otherwise arbitrary "stirring" angular velocities ω^I and ω^{II},

Exact Solutions and Scalar Fields in Gravity: Recent Developments
Edited by Macias *et al.*, Kluwer Academic/Plenum Publishers, New York, 2001

respectively corresponding angular momenta J^I and J^{II}. We provide explicit solutions of the relevant coupled Einstein–Maxwell equations exactly in M/R and q/R, and in first order of the angular velocities. By performing the additional approximations pertaining to refs. [1, 5], we generally confirm their mathematical results but we find, in contrast to [4], that all results are in complete agreement with Machian expectations. Furthermore we calculate some new interaction effects between strong gravitational and electromagnetic fields, and we consider in detail the collapse limit of the two-shell system.

2. THE STATIC TWO–SHELL MODEL

Mathematically, our models of two concentric spherically symmetric shells are given by three pieces of the Reissner–Nordström metric

$$ds^2 = -F(\rho)d\tau^2 + F(\rho)^{-1}d\rho^2 + \rho^2 d\Omega^2\,, \tag{1}$$

with $F(\rho) = 1 - \frac{2\mathcal{M}}{\rho} + \frac{q^2}{\rho^2}$, and $d\Omega^2 = d\vartheta^2 + \sin^2\vartheta d\varphi^2$: one for the region outside the exterior shell, one for the region between both shells, and one for the interior region, for which both parameters $\mathcal{M}$ and q have to be zero in order to guarantee regularity at the origin $\rho = 0$. However, a matching of these Reissner–Nordström metrics with different mass- and charge-parameters obviously would not be continuous at the shell positions. A global continuous metric is however desirable for the physical interpretation of the dragging effects, and was also used in the papers [2] and [4]. It can be reached by a transformation of the metric (1) to the isotropic form

$$ds^2 = -e^{2U(r)}dt^2 + e^{2V(r)}\left(dr^2 + r^2 d\Omega^2\right)\,. \tag{2}$$

Identification of (2) and (1) results in

$$r(\rho) = \frac{1}{2D}\left(\sqrt{\rho^2 - 2\mathcal{M}\rho + q^2} + \rho - \mathcal{M}\right)\,, \tag{3}$$

with an arbitrary constant D. In the following we often change between the variable ρ (in which the field equations and their solutions are simplest) and the variable r (necessary for the conditions at the shells and for the physical interpretation of the results). In the exterior region $r \geq R$ we choose $D_1 = 1$, identify t with τ, and set $\mathcal{M} = M$. If we then use the dimensionless variables

$$\alpha = \frac{M}{2R}\,; \quad \beta = \frac{a}{R} \leq 1\,; \quad \gamma = \frac{q}{2R}\,; \quad x = \frac{r}{R}\,, \tag{4}$$

the potentials in the exterior region $x \geq 1$ read:

$$V_1(x) = \log\left[\frac{(x+\alpha)^2-\gamma^2}{x^2}\right] \; ; \; U_1(x) = \log\left[\frac{x^2-\alpha^2+\gamma^2}{(x+\alpha)^2-\gamma^2}\right] . \tag{5}$$

In the region $a \leq r \leq R$ we set $\mathcal{M} = \widehat{M}$, and $\widehat{\alpha} = \frac{\widehat{M}}{2R}$. The charge parameter should have the same value q as in the exterior region because the shell at $r = R$ is uncharged. In order to enable continuous potentials $U(r)$ and $V(r)$ at $r = R$, the constant D_2 from (3 is generally different from 1, and a nontrivial time transformation $t = C_2\tau$ is necessary in this region. Then the potentials in the region $\beta \leq x \leq 1$ read:

$$V_2(x) = \log\left[\frac{(D_2x+\widehat{\alpha})^2-\gamma^2}{D_2x^2}\right] \; ; \; U_2(x) = \log\left[\frac{D_2^2x^2-\widehat{\alpha}^2+\gamma^2}{C_2\left((D_2x+\widehat{\alpha})^2-\gamma^2\right)}\right] . \tag{6}$$

In the interior region $r \leq a$, we have, due to $\mathcal{M} = 0$ and $q = 0$: $r = \rho/D_3$, $t = C_3\tau$, and the potentials

$$V_3(x) = \log D_3 = \text{const.}; \quad U_3(x) = -\log C_3 = \text{const.} \tag{7}$$

We now specify the parameter $\widehat{\alpha}$: In our models, the function of the inner shell is mainly to provide a charge, and not so much to provide additional mass. It seems therefore reasonable to simplify our models (in accordance with [2, 3] by setting the rest mass density of the inner shell to zero. If we write Einstein's field equations in the form $G^\mu_\nu = 8\pi\,(T^\mu_\nu + S^\mu_\nu)$, with S^μ_ν being the electromagnetic energy–momentum tensor, then T^μ_ν in our two–shell models consists of two parts: $\tau^\mu_\nu(a)\delta(r-a)$, and $\tau^\mu_\nu(R)\delta(r-R)$, and the τ^μ_ν are expressed by the r–derivatives V_i' and U_i' in the form

$$8\pi\tau^0_0(a) = 2e^{-2V_3(a)}\left(V_2'(a)-V_3'(a)\right) \; ; \tag{8}$$

$$8\pi\,\tau^2_2(a) = 8\pi\tau^3_3(a) = e^{-2V_3(a)}\left(V_2'(a)-V_3'(a)+U_2'(a)-U_3'(a)\right) . \tag{9}$$

Equivalent relations are valid at the position $r = R$. With the useful abbreviations $\delta = (1+\alpha)^2-\gamma^2$ and $\Delta_\pm = \frac{1}{2}(\sqrt{\beta^2\delta^2+4\gamma^2(1-\beta)^2} \pm \beta\delta)$ the condition $\tau^0_0(a) = 0$ leads to $\widehat{\alpha} = \Delta_-/(1-\beta)^2$. Herewith, the continuity conditions for $V(r)$ and $U(r)$ at $r = a$ and $r = R$ fix the constants C_i, D_i as:

$$D_2 = \frac{(1-\beta)^2\Delta_+-\Delta_-}{\beta(1-\beta)^2}; \; D_3 = \frac{\Delta_+}{\beta}; \; C_2 = C_3 = \frac{(1-\beta)\Delta_+-(1+\beta)\Delta_-}{\beta(1-\beta)\left(1-\alpha^2+\gamma^2\right)} . \tag{10}$$

In order that the metric (2) with the potentials (5), (7) describes static two–shell models, and does e.g. not have horizons, the parameters α, β, γ have to fulfil some inequalities: In order that $V_1(x)$ and $U_1(x)$ are real for all $x \geq 1$, we have to have:

$$|\gamma| - 1 \leq \alpha \leq \sqrt{1+\gamma^2}\,. \tag{11}$$

The constants D_i and C_i from formulas (10) obviously are all real. However, in order that the charged shell is interior to the mass shell, that the origin $r = 0$ is interior to the charged shell, and that time runs in the same direction in all regions, all these constants have to be (at least) non–negative. Whereas D_3 is automatically non–negative, the conditions $D_2 \geq 0$, and $C_2 = C_3 \geq 0$ sharpen the inequalities (11) to

$$\sqrt{\gamma(\gamma+2-\beta)} - 1 \leq \alpha \leq \sqrt{1+\gamma^2}\,. \tag{12}$$

According to formulas (8)–(9) we get for the non–zero components of the matter tensor–densities τ^μ_ν :

$$\tau^2_2(a) = \tau^3_3(a) = -, \frac{\beta\widehat{\alpha}}{4\pi R\,\Delta^3_+}\,; \tag{13}$$

$$\tau^0_0(R) = \frac{(1+\alpha)(1-\beta) + \Delta_+ - \delta}{2\pi R\delta^3(1-\beta)}\,; \tag{14}$$

$$\tau^2_2(R) = \tau^3_3(R) = \frac{1}{4\pi R\delta^2(1-\alpha^2+\gamma^2)}\left(1 - \frac{D_2}{C_2}\right)\,. \tag{15}$$

Later we will apply a small rotation to the mass shell at $r = R$, and we will discuss the resulting physical effects under Machian aspects, i.e. the mass shell will be seen as a substitute for the overall masses in the universe. In order that this makes sense, the energy tensor $\tau^\mu_\nu(R)$ of the mass shell has at least to fulfil the weak energy conditions

$$\tau^0_0(R) \leq 0\,, \quad \text{and} \quad \tau^3_3(R) - \tau^0_0(R) \geq 0\,. \tag{16}$$

(The charged shell at $r = a$ violates most energy conditions!) The detailed analysis of these inequalities is algebraically quite involved. The condition $\tau^0_0(R) \leq 0$ leads to a further sharpening of the lower limit of α in inequality (12):

$$\alpha \geq \frac{1}{2}\left(\sqrt{\beta^2+4\gamma^2} - \beta\right)\,, \quad \text{or} \quad \beta \geq \frac{2\alpha+1-\delta}{\alpha}\,. \tag{17}$$

In the over–extreme Reissner–Nordström case $\gamma^2 > \alpha^2$ the second condition in (16) leads to a further sharpening of the inequality $\beta \geq \frac{2\alpha+1-\delta}{\alpha}$ in part of the admissible (α, δ)–region. (For details see [6].)

The electromagnetic field tensor $F_{\mu\nu}$ belonging to the Reissner–Nordström metric (2) has only a radial component $-F_{tr} = E_r = \frac{q}{r^2}\, e^{U-V} H(r-a)$, where $H(r-a)$ is the Heaviside function (the charge q is concentrated solely on the inner shell at $r = a$!), and the potentials (5) or (6) have to be inserted in the respective regions. The charge density of the interior shell results from the inhomogeneous Maxwell equation $\frac{1}{\sqrt{-g}}\,(\sqrt{-g}\; F^{t\mu})_{,\mu} = 4\pi j^t = 4\pi\sigma$ as

$$\sigma(x) = \frac{q\beta C_3}{4\pi R^3 \Delta_+^3}\, \delta(x-\beta)\,. \tag{18}$$

3. FIRST ORDER ROTATION OF THE SHELLS

Rotations of the shells from Sec. 2 in first order of angular velocity ω (denoting the velocity ω^I of the inner charged shell and/or the velocity ω^{II} of the exterior mass shell) are described by an extension of the metric (2):

$$ds^2 = -e^{2U(r)}dt^2 + e^{2V(r)}\left[dr^2 + r^2 d\vartheta^2 + r^2 \sin^2\vartheta (d\varphi - \omega A(r)dt)^2\right], \tag{19}$$

with neglect of the ω^2–terms. By the very definition of an angular velocity, a stationary rotating system is invariant under the common substitutions $t \to -t$, and $\omega \to -\omega$, and since the metric functions U, V, and A are independent of t, they have to be even functions of ω. Therefore, in first order of ω the functions U and V can be taken from Sec. 2, and the ω–independent function A is the only new metric function. (Centrifugal deformations of the spherical shells start only at order ω^2!) The relevant Einstein equation takes its simplest form in the variable ρ:

$$G_3^0 = -\frac{\omega \sin^2\vartheta}{2\rho^2}\frac{d}{d\rho}\left(\rho^4 \frac{d}{d\rho}A(\rho)\right) = 8\pi\left(T_3^0 + S_3^0\right), \tag{20}$$

with $S_3^0 = \frac{1}{4\pi} F_\lambda^0\, F_3^\lambda$. Since there are no electric currents in the r– and ϑ–direction in our models, the magnetic field component $B_\varphi = F_{r\vartheta}$ is zero, and there remains only one homogeneous Maxwell equation (with $\cdot$ denoting the ϑ–derivative)

$$\frac{d}{d\rho}B_\rho + \dot{B}_\vartheta = 0\,, \tag{21}$$

and one inhomogeneous Maxwell equation

$$4\pi j^{\varphi} = \frac{1}{\rho^2 \sin^2\vartheta}\frac{d}{d\rho}\left(F(\rho)B_{\vartheta}\right) - \frac{1}{\rho^4 \sin\vartheta}\left(\frac{B_\rho}{\sin\vartheta}\right)^{\cdot} + \frac{\omega}{\rho^2}\frac{d}{d\rho}\left(\rho^2 A E_\rho\right). \tag{22}$$

Eq. (20) together with $-F_{\tau\rho} = E_\rho = q/\rho^2$ enforces the ansatz $B_\vartheta = \omega q f(r)\sin^2\vartheta$ with a dimensionless function $f(r)$. Eq. (21) then results in the form $B_r = \omega q R g(r)\sin\vartheta\cos\vartheta$ with $f(r) = -\frac{R}{2}g'(r)$. These forms constitute the dipole character of the (rotationally induced) magnetic field. Eqs. (20) and (22) constitute two coupled ordinary differential equations for the functions $A(\rho)$ and $g(\rho)$. In the interior $r \le a$ these equations simplify drastically, and give the results

$$A_3(\rho) = \mu_3\,; \quad g_3(\rho) = \frac{\eta_3}{D_3^2}\frac{\rho^2}{R^2}, \tag{23}$$

with dimensionless constants μ_3 and η_3, i.e. the interior stays flat, and the magnetic field has (in Cartesian coordinates) only a (constant) z–component $B_z = \frac{\omega q}{R}\eta_3$.

In the intermediate and exterior regions, due to $B_\vartheta \sim f(r) \sim g'(r)$, one integration of Eq. (20) is trivial:

$$\frac{d}{d\rho}A(\rho) = \frac{1}{\rho^4}\left(2q^2 R g(\rho) - 4\mathcal{M}R^2\lambda\right), \tag{24}$$

with a dimensionless integration constant λ. Insertion of (23) into (22), together with $j^\varphi = 0$ outside the charged shell, results in

$$\frac{d}{d\rho}\left[F(\rho)\frac{d}{d\rho}g(\rho)\right] - \frac{2}{\rho^2}\left(1 + \frac{2q^2}{\rho^2}\right)g(\rho) = -\frac{8\mathcal{M}R\lambda}{\rho^4}. \tag{25}$$

The general solution of Eq. (25) has the form $g(\rho) = \lambda\,\hat{g}(\rho) + \eta\,\bar{g}(\rho) + \zeta\,\bar{\bar{g}}(\rho)$, with dimensionless integration constants η and ζ, and where $\hat{g}(\rho)$ is a special inhomogeneous solution, and $\bar{g}(\rho), \bar{\bar{g}}(\rho)$ are independent homogeneous solutions of (24). Luckily, there exist simple polynomial solutions

$$\hat{g}(\rho) = \frac{4R}{3\rho}\,; \; \bar{g}(\rho) = \frac{1}{R^2}\left(\rho^2 - 3q^2 + \frac{2q^4}{\mathcal{M}\rho}\right) \tag{26}$$

and $\bar{\bar{g}}(\rho)$ can then be found from $\bar{g}(\rho)$ by d'Alembert's reduction procedure:

$$\bar{\bar{g}}(\rho) = \frac{3\mathcal{M}^2 R}{4(\mathcal{M}^2 - q^2)^2}\left[\frac{2q^2}{3\rho}\left(1 + \frac{2q^2}{\mathcal{M}^2}\right) - \rho - \mathcal{M} + R^2\bar{g}(\rho)S(\rho;\mathcal{M},q)\right],$$

with

$$S(\rho;\mathcal{M},q)=\begin{cases}\frac{1}{\sqrt{q^2-\mathcal{M}^2}}\operatorname{arccot}\left(\frac{\rho-\mathcal{M}}{\sqrt{q^2-\mathcal{M}^2}}\right) & \text{, for } q^2>\mathcal{M}^2\\ \frac{1}{2\sqrt{\mathcal{M}^2-q^2}}\log\left(\frac{\rho-\mathcal{M}+\sqrt{\mathcal{M}^2-q^2}}{\rho-\mathcal{M}-\sqrt{\mathcal{M}^2-q^2}}\right) & \text{, for } q^2<\mathcal{M}^2.\end{cases} \tag{27}$$

(These solutions have already been given in [7].) $\bar{\bar{g}}(\rho)$ behaves asymptotically as $R\rho^{-1}$ (independent of $\mathcal{M}$ and q!), and since $\bar{g}(\rho)$ diverges there, the constant η_1 in the exterior region has to be zero. Insertion of $g(\rho)$ into Eq. (24) results in

$$A(\rho)=\frac{2q^2}{R^2}\left[\lambda\hat{A}(\rho)+\eta\bar{A}(\rho)+\zeta\bar{\bar{A}}(\rho)\right]+\frac{4\mathcal{M}R^2\lambda}{3\rho^3}+\mu\,, \tag{28}$$

with a dimensionless integration constant μ_2 ($\mu_1=0$, if we demand $A(\rho)$ to vanish asymptotically, i.e. if any rotation in our model is defined relative to static observers at infinity), and with

$$\hat{A}(\rho)=-\frac{R^4}{3\rho^4}\,;\;\bar{A}(\rho)=R\left(-\frac{1}{\rho}+\frac{q^2}{\rho^3}-\frac{q^4}{2\mathcal{M}\rho^4}\right)\,; \tag{29}$$

$$\bar{\bar{A}}(\rho)=\frac{3\mathcal{M}R^4}{8(\mathcal{M}^2-q^2)^2}\left[-\frac{1}{\rho}+\frac{\mathcal{M}}{\rho^2}+\frac{q^2+2\mathcal{M}^2}{3\rho^3}-\frac{q^2(\mathcal{M}^2+2q^2)}{3\mathcal{M}\rho^4}+\right.$$
$$\left.+\left(1+\frac{2\mathcal{M}}{R}\bar{A}(\rho)\right)S(\rho;\mathcal{M},q)\right]. \tag{30}$$

In the extreme Reissner–Nordström case $q^2=\mathcal{M}^2$ the functions $\bar{\bar{g}}(\rho)$ and $\bar{\bar{A}}(\rho)$ simplify drastically to polynomial forms.

In the process of integration of the functions $g(\rho)$ and $A(\rho)$ we had to introduce a total of 8 integration constants: μ_2, μ_3, λ_1, λ_2, η_2, η_3, ζ_1 and ζ_2. These have now to be fixed by the continuity–, respectively discontinuity–conditions of the functions g and A and their derivatives at the shell positions. As stated in Sec. 2, the metric function $A(r)$ (expressed in the isotropic variable r!) is continuous at both shell positions $r=a$ and $r=R$. In analogy to classical electrodynamics also the radial magnetic field component $g(r)$ is continuous there. (No magnetic charges at the shells!) This comprises 4 linear, homogeneous equations for the integration constants. The discontinuities of the radial derivatives $g'(r)$ are, by classical electrodynamics, given by the currents j^φ (in the local inertial systems!) at the shell positions, and since the mass shell at $r=R$ is uncharged, $g'(r=R)$ is continuous.

Due to the Einstein equation (20), the discontinuities of $A'(r)$ at the shell positions are fixed by the angular momenta $J^I(a)$ and $J^{II}(R)$ of these shells, and these are the invariant and independent sources of all rotation effects in our models. (At least in first order of rotation we see no mechanism how angular momentum could be transferred from one shell to the other. In contrast, the angular velocities of the two shells are not independent because the rotation of one shell leads by dragging to a non–zero angular velocity also of the other shell.)

The eigenvalue equations $T^{\mu}_{\nu} u^{\nu} = -\varrho_0 u^{\mu}$, with $u^{\mu} = u^0 (1,0,0,\omega)$ (independent of ϑ, due to rigid rotation of the shells!) allow to express the components τ_3^0 (at $r = a$ and $r = R$) by the expressions $\tau_3^3 - \tau_0^0$, appearing in the energy conditions (16), what has to be compared with τ_3^0 resulting from an integration of Eq. (20). Furthermore, we define (in accordance with [4]) the material of the charged shell to be insulating, i.e. the four velocity of the charge carriers has the same form $u^{\mu}(a) = u^0(a)(1,0,0,\omega(a))$ as for the matter elements. This defines together with the charge density σ from (18) the current $j^{\varphi}(a)$. For simplicity we formulate the resulting 4 conditions (partly inhomogeneous, linear equations for the integration constants) on $g'(r)$ and $A'(r)$ at $r = a$ and $r = R$ separately for the cases I ($J^I(a) \neq 0$, $J^{II}(R) = 0$) and II ($J^I(a) = 0$, $J^{II}(R) \neq 0$). Due to the linearity of our first order perturbation calculation, the general case $J^I(a) \neq 0$, $J^{II}(R) \neq 0$ results from linear superposition of the cases I and II, e.g. $\omega A = \omega^I A^I + \omega^{II} A^{II}$, $\mathbf{B} = \mathbf{B}^I + \mathbf{B}^{II}$. It should also be said that mathematically the quantities $J^I(a)$ and $J^{II}(R)$ are not a convenient starting point because in the metric (19) and in the expressions $u^{\mu} = u^0 (1,0,0,\omega)$ the angular velocities and not the angular momenta appear. We therefore start with "stirring" angular velocities ω^I and ω^{II} which only at the end are uniquely connected with the angular momenta by

$$J = \int_0^{\infty} dr \int_0^{\pi} d\vartheta \int_0^{2\pi} d\varphi \sqrt{-g}\left[T_3^0 + S_3^0\right] = \frac{2}{3} M R^2 \omega \cdot \lambda_1 \tag{31}$$

for both cases I and II. (Obviously the constant λ_1 describes the deviation from the corresponding Newtonian relation!) The solution of the system of the 8 linear continuity–/discontinuity–conditions for the 8 integration constants is in principle simple (the matrix has many zeroes!) but the results are algebraically quite involved, mainly due to the complicated form of $\bar{\bar{g}}(\rho)$ and $\bar{\bar{A}}(\rho)$. (The detailed results are given in [6].)

4. RESULTS

In Sec. 3 it was shown how the functions $g(r)$ and $A(r)$ can be calculated exactly in M/R and q/R. Since however the detailed formulas are algebraically quite complicated, it is appropriate first to consider expansions of the results in powers of $\gamma = q/2R$, and partly also of $\alpha = M/2R$, this the more since the papers [1, 2, 3, 4, 5] present only first order results in γ and/or α. The dragging function A is even in q, and therefore receives contributions only from orders γ^0, γ^2, ... The magnetic field is an odd function in q, but since a factor q was taken out in the definition of the function g, also g has only terms of order γ^0, γ^2, ...

4.1. RESULTS IN FIRST ORDER OF THE CHARGE Q

Here only $g(\rho)$ in zeroth order of q is relevant. From Eqs. (26) and (27) we get: $\bar{g}_1(\rho) = \bar{g}_2(\rho) = \frac{\rho^2}{R^2}$; $\bar{\bar{g}}_1(\rho) = -\frac{3R}{4M}[1 + \frac{\rho}{M} + \frac{\rho^2}{2M^2}\log(1 - \frac{2M}{\rho})]$; $\bar{\bar{g}}_2(\rho) = \frac{R}{\rho}$. In the case II the terms $\lambda_2^{II}\hat{g}_2(\rho)$ and $\zeta_2^{II}\bar{\bar{g}}_2(\rho)$ exactly cancel, and therefore the field $g_3^{II}(\rho) = \frac{\eta_3^{II}}{D_3^2}\frac{\rho^2}{R^2}$ from Eq. (23) extends unchanged until $r = R$, i.e. the magnetic field $\mathbf{B}^{II}$ is constant (in z–direction) in the whole region $r \leq R$, which is a direct consequence of the constant dragging term in this part of space–time. In the exterior $r > R$ the field $\mathbf{B}^{II}$ has the usual dipole form with a falloff like r^{-3}. The whole field $\mathbf{B}^{II}$ is independent of β, i.e. of the position of the charged shell. In the region $r \leq R$ the magnetic field should not be expressed in coordinate time t but in proper time $\tau = t/C_3$. Then the decisive factor $\tilde{\eta}_3^{II} = \eta_3^{II}/C_3$ for the strength of this constant field is

$$\tilde{\eta}_3^{II} = \frac{3(1-\alpha^2)(2-\alpha)}{2\alpha(3-\alpha)} \times \frac{(1+\alpha)^2(1+\alpha^2) - \frac{8}{3}\alpha^2 - (1+\alpha)^4(1-\alpha)^2\frac{1}{2\alpha}\log\frac{1+\alpha}{1-\alpha}}{-1+2\alpha+\alpha^2+(1+\alpha)^3(1-\alpha)\frac{1}{2\alpha}\log\frac{1+\alpha}{1-\alpha}}. \quad (32)$$

The positivity of $\tilde{\eta}_3^{II}$ in the region $0 < \alpha < 1$ is expected on Machian grounds: An inertial (!) observer at $r < R$ sees the charged shell static and the mass shell rotating with $-A_3^{II}\omega^{II}$, what is the Mach–equivalent situation to a charge rotating with $+A_3^{II}\omega^{II}$ in a static "cosmos". The behaviour $\tilde{\eta}_3^{II} \approx \frac{4}{3}\alpha$ for small α, i.e. $B_z^{II}(r \leq R) = (2Mq/3R^2)\,\omega^{II}$ in the weak–field approximation, is in mathematical agreement with the results in [1] and [4]. In the collapse limit $\alpha \to 1$ we have $\tilde{\eta}_3^{II} = 0$ because then inside the mass shell total dragging is realized ($A^{II}(r \leq R) \equiv 1$), i.e. no relative rotation between the two shells, and therefore no magnetic field. That $\mathbf{B}^{II}$ is non–zero for $0 < \alpha < 1$, although in the case II we have nowhere electric currents (in the local inertial system) shows exemplary the non–local character of Machian effects.

In the case I, respectively if both sources $J^I(a)$ and $J^{II}(R)$ are active, we discuss (in connection with published literature) two interesting examples: Cohen [2] considers only the collapse limit $\alpha = 1$. In the interior and in the intermediate region it is again advisable to express the magnetic field through the angular velocity $\bar{\omega}^I = C_3\omega^I$. Then we get in agreement with [2]:

$$B_z(r \leq a) = \frac{8q}{3a}\left(1 - \left(\frac{a}{R}\right)^3\right)\bar{\omega}^I . \tag{33}$$

In the region $a \leq r \leq R$ we get the "corresponding" exterior dipole field with a "cosmic correction factor" $\left(1 - \left(\frac{r}{R}\right)^3\right)$. Therefore our models show in principle a "cosmic" influence on local systems, which however is very small for $a, r \ll R$. $\bar{\omega}^I$ has here the meaning of the relative angular velocity between both shells, and therefore Eq. (33) is a completely Machian result. In the words of Cohen [2]: "One cannot distinguish (even with electromagnetic fields reaching beyond the mass shell) whether the charged shell is rotating or the mass shell is rotating in the opposite direction." Eq. (33) contains no contribution from $J^{II}(R)$. This is due to the fact that for an interior observer (measuring in his proper time) in the collapse limit any term with finite $\bar{\omega}^{II}$ belongs to $\omega^{II} = 0$ and $J^{II}(R) = 0$. In contrast, for an exterior observer $J^{II}(R)$ and ω^{II} can have finite values, and the corresponding magnetic field is, as expected due to the no–hair theorem, the Kerr–Newman field (in first order of q and ω), not given by Cohen.

Ehlers and Rindler [4] consider (only in the weak field limit) a situation where the charged shell is (artificially) kept fixed relative to the asymptotic inertial frame. This can in our notation be realized by the relation $\omega^I = -(4M/3R)\omega^{II}$. Then we get

$$B_z^I(r \leq a) = \frac{2q}{3a}\omega^I , \tag{34}$$

and the standard exterior dipole field of classical electrodynamics for $r > a$. This is in mathematical agreement with the field $\mathbf{B}^I$ in [4]. But in contrast to Ehlers and Rindler we like to argue that this field fulfils all Machian expectations: A locally non–rotating, i.e. an inertial observer at $r \leq a$ is dragged by the mass shell with $(4M/3R)\omega^{II} = -\omega^I$, i.e. he sees the charged shell rotating with $+\omega^I$, and therefore expects exactly the field (34.

4.2. RESULTS IN SECOND ORDER OF THE CHARGE Q

Here only the dragging function $A(r)$ is relevant. In the case II we get for the constant dragging factor inside the charged shell:

$$A_3^{II} = \frac{4\alpha(2-\alpha)}{(1+\alpha)(3-\alpha)}\left[1-\gamma^2(1-\alpha)\cdot h(\alpha,\beta)\right], \tag{35}$$

with a positive function $h(\alpha,\beta)$. The γ–independent term is of course the well–known dragging factor of an uncharged mass shell (given in different notation e.g. in [8]). The correction term vanishes in the collapse limit $\alpha \to 1$, leading to $A_3^{II} = 1$ (total dragging), and otherwise is negative: If for fixed total mass M we increase the charge q from zero to a finite (small) value, this means that we "substitute" some mass density of the exterior mass shell by electrostatic energy, thereby reducing the overall dragging. In the region between both shells we get

$$A_2^{II}(r) = \frac{\mu_3^{II}}{C_3} + \frac{8\gamma^2(1-\alpha)\eta_3^{II}}{(1+\alpha)^7}\left[\frac{2R}{3a} - \left(1-\frac{a^2}{3r^2}\right)\frac{R}{r}\right]. \tag{36}$$

Due to the continuity of $A^{II}(r)$ and $A'^{II}(r)$ at $r = a$, $A_2^{II}(r)$ starts at $r = a$ horizontally from the value A_3^{II} (Eq. (35), then increases according to the function $-(1-\frac{a^2}{3r^2})\frac{R}{r}$, but always (until $r = R$) stays below the zero order term $A_3^{II}(q=0)$. Now an increasing dragging function $A_2^{II}(r)$ is at first sight hardly comprehensible. The explanation seems to come from the (positive and non–rotating!) electrostatic energy density $S_0^0(r)$: The degree of dragging is determined by the ratio between the rotating mass–energy and the non–rotating mass–energy. If then (for fixed M!) a small part of the (rotating) exterior shell is "substituted" by electrostatic energy, the dragging constant inside the charged shell is reduced: $A_3^{II}(q \neq 0) < A_3^{II}(q=0)$. For $r = r_0 > a$, the part $S_0^0(r < r_0)$ has only reduced effect because quite generally the dragging due to masses falls off like r^{-3} in their exterior. Therefore $A_2^{II}(r)$ increases in the region $a \leq r \leq R$. The value $A_2^{II}(r = R, q \neq 0)$ is however still smaller than $A_2^{II}(r = R, q = 0)$ because there is still electrostatic energy S_0^0 outside the mass shell. It is important to state that the r–dependent correction term $\sim \gamma^2$ in Eq. (36) survives also in the weak field limit (first order of α). This implies, contrary to the claim of Ehlers and Rindler [4] that the deviation of the full metric from the flat metric is not the sum of the corresponding contributions from the pure Thirring problem ($q = 0$, first order in α) and the Reissner–Nordström problem.

Concerning q^2–corrections in the case I we only like to remark that here quite unusual relations between $J^I(a)$ and ω^I result, and the rotating interior shell can produce an anti–dragging phenomenon, an example of which was also given in [7]. But all these effects are only caused by the somewhat unphysical properties of the massless, charged shell, which e.g. violates most energy conditions.

4.3. RESULTS EXACT IN MASS M AND CHARGE Q

As already remarked, the explicit formulas for the magnetic fields $\mathbf{B}$ and for the dragging functions A in the case of general values M and q are too complicated for extracting much obvious physical interpretation. Therefore we restrict ourselves here to a discussion of only two special, physically interesting cases.

We observe that in the case II all 8 integration constants are proportional to the expression $\Delta\tau = \tau_3^3(R) - \tau_0^0(R)$, appearing in the energy conditions (16). Therefore, the magnetic field $\mathbf{B}^{II}$ and the dragging function A^{II} are zero for $\Delta\tau = 0$, and change sign if $\Delta\tau$ changes sign, e.g. change from dragging to anti–dragging. This emphasizes the importance of the discussion of the energy conditions for the mass shell in Sec. 2. A similar behavior shows up in the analysis of the Thirring problem with cosmological boundary conditions [9].

The collapse of our two–shell system, i.e. the appearance of a horizon, sets in for $\gamma^2 = \alpha^2 - 1$. It turns out that then in both cases I and II we have $A(r \leq R) \equiv 1$, and $\mathbf{B}(r \leq R) \equiv 0$. We see that the important results by Brill and Cohen [8] that inside a rotating collapsing mass shell one has total dragging of the inertial systems, transfers also to our highly charged two–shell system, and we have a complete realization of Machian expectations: only the rotation relative to the "universe" counts. Moreover, all "physical" magnetic fields vanish inside the mass shell. In the exterior region $r > R$ the essential non–zero integration constant in the collapse limit is $\lambda_1^{II} = 6(1+\alpha)^2$, and one gets exactly the Kerr–Newman field in lowest order of J^{II}.

References

[1] K.D. Hofmann, *Z. Phys.* **166** (1962) 567.

[2] J. Cohen, *Phys. Rev.* **148** (1966) 1264.

[3] J. Ehlers and W. Rindler, *Phys. Lett.* **A32** (1970) 257.

[4] J. Ehlers and W. Rindler, *Phys. Rev.* **D4** (1971) 3543.

[5] J. Cohen, J. Tiomno, and R. Wald, *Phys. Rev.* **D7** (1973) 998.

[6] M. King and H. Pfister, *Phys. Rev. D* (to be published).

[7] C. Briggs, J. Cohen, G. DeWoolfson, and L. Kegeles, *Phys. Rev.* **D23** (1981) 1235.

[8] D. Brill and J. Cohen, *Phys. Rev.* **143** (1966) 1011.

[9] C. Klein, *Class. Quantum Grav.* **10** (1993) 1619.

ON THE EXPERIMENTAL FOUNDATION OF MAXWELL'S EQUATIONS

C. Lämmerzahl*
Institute for Experimental Physics, Heinrich–Heine–University Düsseldorf, 40225 Düsseldorf, Germany

Abstract A short review of the characteristic features of the Maxwell equations in vacuum are recalled. Then some of the laboratory experiments and astrophysical observations aimed to test the physics of the electromagnetic field are described. These tests and observations include dispersion, birefringence, the Coulomb $1/r$ potential, tests of Special Relativity, charge conservation, and the superposition principle.

Keywords: Maxwell equations, dispersion, birefringence, charge conservation, fine structure constant, test of Special Relativity

1. INTRODUCTION

The importance of the Maxwell equations describing all electromagnetic phenomena cannot be overestimated: all observations use at least at some stage photons, the electromagnetic interaction between particles play a dominant role in everydays experience, and it determines the properties of atoms and thus the properties of rods and clocks which are essential in establishing Special and General Relativity, for example. In addition, only a very restricted form of Maxwell's equations is compatible with the equivalence principle, so that the specific form of Maxwell's equations are also in this way intimately connected with the structure of space–time and thus, again, with General Relativity.

The special relativistic Maxwell equations in vacuum are given by ($F_{0i} = E_i$, $F_{ij} = B_k$ (cyclic) where E_i and B_i are the electric and magnetic field, respectively)

*E–mail: claus.laemmerzahl@uni-duesseldorf.de

source eqns	$4\pi j^a = \partial_b F^{ab}$	$\nabla \times \boldsymbol{B} - \frac{1}{c}\frac{\partial}{\partial t}\boldsymbol{E} = \frac{4\pi}{c}\boldsymbol{j}$	$\nabla \boldsymbol{E} = 4\pi\rho$
homogeneous eqns	$\partial_{[a}F_{bc]} = 0$	$\nabla \times \boldsymbol{E} = -\frac{1}{c}\frac{\partial}{\partial t}\boldsymbol{B}$	$\nabla \boldsymbol{B} = 0$

The source equations as well as the homogeneous equations split into dynamical equations and constraint equations. The most important features of this set of equations are

- the superposition principle (the sum of two solutions of Maxwell's equations is again a solution)
- charge conservation ($\partial_a j^a = 0$)
- no dispersion (all frequencies propagate with the same speed)
- no birefringence (both polarization states propagates with the same speed)
- no damping (a plane electromagnetic wave always has the same amplitude)
- the Coulomb potential q/r as solution for a point charge.

Though all these features are very well confirmed by experiments, there are some approaches which modify these equations. There is, for example, Ritz's theory [1] where the velocity of light depends on the velocity of the source. The theory of Born and Infeld [2] intends to eliminate the infinities occurring for the electromagnetic field of a point charge. As each field, also the Maxwell field has to be quantized which gives in an approximation procedure QED–modifications of the classical Maxwell equations, [3, 4]. There are additional QED–modifications in curved space–time [5]. Also the approach of Podolsky [6] gets rid of the infinities of the electromagnetic field for point charges. And most recently, quantum gravity induced modifications of the classical Maxwell equations, that is, modifications due to the fluctuations of space–time, were proposed [7, 8]. Because of so many well–motivated suggestions to modify the classical Maxwell equations, it is very important to improve the accuracy of experiments and to invent new experiments aimed to test the validity of the Maxwell equations and its quantized version.

2. QUANTUM GRAVITY INSPIRED MODIFICATION OF MAXWELL'S EQUATIONS

There are two main approaches to a quantization of gravity: canonical quantization and string theory. The latter approach also aimes at a unification of all interactions. In both schemes modifications of the effective equations governing the dynamics of the electromagnetic field is proposed [8, 9]. Despite that fact that all physical laws should be experimentally verified with increasing accuracy, the new quantum gravity induced modifications of the Maxwell equations give additional motivation for testing Maxwell's equations. Moreover, since Maxwell's equations are so intimately connected with the special and general relativistic structure of structure of space–time, all tests of Special and General Relativity are in most cases at the same time also tests of the structure of Maxwell's equations.

2.1. LOOP GRAVITY INDUCED CORRECTIONS

In loop gravity, averaging over some quasiclassical quantum state, a so–called "weave"–state, which includes the state of the geometry as well as of the electromagnetic field, gives the effective Maxwell equations [8]

$$\partial_t \boldsymbol{E} = -\boldsymbol{\nabla} \times \boldsymbol{B} + 2\chi l_P \boldsymbol{\nabla}^2 \boldsymbol{B} \tag{1}$$

$$\partial_t \boldsymbol{B} = \boldsymbol{\nabla} \times \boldsymbol{E} - 2\chi l_P \boldsymbol{\nabla}^2 \boldsymbol{E}\,. \tag{2}$$

The corresponding wave equation is

$$(\partial_t^2 - \Delta)\boldsymbol{E} - 4\chi l_p \boldsymbol{\nabla}^2 \boldsymbol{\nabla} \times \boldsymbol{E} = 0\,. \tag{3}$$

The two helicity states $\boldsymbol{E}_\pm = (\widehat{\boldsymbol{E}}_1 \pm i\widehat{\boldsymbol{E}}_2)e^{i(\omega_\pm t + \boldsymbol{k}\cdot\boldsymbol{x})}$ lead to different dispersion relations

$$\omega_\pm = \sqrt{k^2 \mp 4\chi l_P k^3}\,. \tag{4}$$

There are two points which need to be discussed: First, with (2) also a homogeneous Maxwell equation is modified and, second, the appearance of higher order derivatives means that there are photons which propagate with infinite velocity.

1 The deviation from the homogeneous Maxwell equations has the important consequence that the unique description of charged particle interference is no longer true. If quantum mechanical equations are minimally coupled to the electromagnetic potential, that is by the replacement $\partial \rightarrow \partial - \frac{ie}{\hbar c}A$, then the phase shift in charged particle interferometry is $\delta\phi = \frac{ie}{\hbar c}\oint_C A$ for a closed path C. If one applies Stokes' law in the case of a trivial space–time topology in that region, then this is equivalent to $\delta\phi = \frac{ie}{\hbar c}\int_{\partial C} F$ with $F = dA$ where integration is over some 2–dimensional surface bounded by the closed path C. The fact that the result should not depend on

the chosen surface is secured by the homogeneous Maxwell equations $dF = 0$.

Therefore, if in the case of (1,2) we still want to have a unique description of charged particle interference, then the minimal coupling procedure is no longer true.

2 The appearance of higher order spatial derivatives implies that in the corresponding frame of reference there are photons propagating with infinite velocity. As a consequence, Lorentz–invariance is violated.

3 It is clear that the appearance of higher order spatial derivatives also implies the appearance of higher order time derivatives in other frames of reference.

2.2. STRING THEORY INDUCED MODIFICATIONS

In string theory the gravitational field is given by D–branes which interact with propagating photons via an effective recoil velocity $\bar{\boldsymbol{u}}$ [9]. This recoil effect appears as a modified space–time metric, which influences the Maxwell equations. These equations are then given by

$$\nabla \cdot \boldsymbol{E} + \bar{\boldsymbol{u}} \cdot \partial_t \boldsymbol{E} = 0 \tag{5}$$

$$\nabla \cdot \boldsymbol{B} = 0 \tag{6}$$

$$\nabla \times \boldsymbol{B} - (1 - \bar{u}^2)\partial_t \boldsymbol{E} + \bar{\boldsymbol{u}} \times \partial_t \boldsymbol{B} + (\bar{\boldsymbol{u}} \cdot \nabla)\boldsymbol{E} = 0 \tag{7}$$

$$\nabla \times \boldsymbol{E} = -\partial_t \boldsymbol{B}\,. \tag{8}$$

In this approach the homogeneous equations remain unmodified. The resulting wave equations are

$$0 = (\partial_t^2 - \Delta)\boldsymbol{E} - 2(\bar{\boldsymbol{u}} \cdot \nabla)\partial_t \boldsymbol{E} \tag{9}$$

$$0 = (\partial_t^2 - \Delta)\boldsymbol{B} - 2(\bar{\boldsymbol{u}} \cdot \nabla)\partial_t \boldsymbol{B} \tag{10}$$

which lead to the dispersion relation

$$\omega = \pm k + (\bar{\boldsymbol{u}} \cdot \boldsymbol{k}) + \mathcal{O}(\bar{u}^2)\,. \tag{11}$$

The corresponding group velocity for light is

$$\boldsymbol{c} = \pm \frac{\boldsymbol{k}}{k} + \bar{\boldsymbol{u}}\,. \tag{12}$$

From string theoretical considerations, the recoil velocity $\bar{\boldsymbol{u}}$ can be shown to depend linearly on the energy ω of the photon, $\bar{\boldsymbol{u}} \sim \omega$. Therefore, in this case obtain an energy dependent group velocity of the photons.

2.3. THE GENERAL STRUCTURE OF MODIFIED DISPERSION RELATIONS

In both cases, that is in Eqs.(4) and (11), the general structure of dispersion relation for the propagation in one direction reads [10]

$$k^2c^2 = E^2\left(1+\xi\left(\frac{E}{E_{QG}}\right)+\mathcal{O}\left(\frac{E}{E_{QG}}\right)^2\right). \tag{13}$$

where E is the energy of the photon. The parameter ξ depends on the underlying theory and can be derived to be $\sim 3/2$ for string theory [11] and ~ 4 for loop gravity [8]. The quantum gravity energy scale E_{QG} is, of course, of the order of the Planck energy E_P. From the above dispersion relation we derive the velocity of light

$$c_{QG} = c\left(1-\xi\frac{E}{E_{QG}}\right). \tag{14}$$

Therefore, the difference of the velocity of light for high energy photons and low energy photons, $\Delta c = c(E) - c(E \to 0)$ is given by

$$\frac{\Delta c}{c} = \xi\frac{E}{E_{QG}}. \tag{15}$$

Exactly this quantity will be measured in, e.g., astrophysical observations, see below.

3. TESTS OF MAXWELL'S EQUATIONS

In order to make statements about the experimental status of Maxwell's equations, one can just look experimentally for violations of predictions from Maxwell's equations. A better strategy is first to build up a theoretical model capable to describe as many deviations form the standard Maxwell theory as possible and afterwards to analyze the experimental results and astrophysical observations with this model. Only in the latter case it is possible to make quantitative statements about the order of agreement of the Maxwell equations with experiments and observations.

One theoretical frame which provides a complete description of all tests of Maxwell's equations and which allows for a violation of all the distinctive features characteristic for the validity of the standard Maxwell equation, is given by the generalized Maxwell equations

$$0 = \partial_{[a}F_{bc]} \qquad 4\pi j^a = \underbrace{\chi^{abcd}\partial_b F_{cd}}_{\text{structure of null cones}} + \underbrace{\chi^{acd}F_{cd}}_{\text{mass-term}}. \tag{16}$$

For the usual Maxwell equations we have $\chi^{abcd} = \eta^{a[c}\eta^{d]b}$ and $\chi^{abc} = 0$ where η^{ab} is the Minkowski metric. The homogeneous equations are not modified because this will lead to the serious difficulties described above. The first term in the inhomogeneous equations describes the high–energy behaviour of the electromagnetic field and thus the null or light cones. This term leads to birefringence and an anisotropic propagation of light. The second term plays the role of a mass of the photon and leads to a modification of the Coulomb $1/r$ potential and also gives polarization dependent effects. Since any deviation from the usual Maxwell equations is small, we introduce a $\delta\chi abcd = \chi^{abcd} - \eta^{a[c]}\eta^{d]b}$ and write

$$0 = \partial_{[a}F_{bc]} \qquad 4\pi j^a = \partial_b F^{ab} + \delta\chi^{abcd}\partial_b F_{cd} + \chi^{acd}F_{cd}\,. \tag{17}$$

There are two prominent special cases of the above general model: The first one is the coupling of the Maxwell field to a pseudovector ϕ_a,

$$4\pi j^a = \partial_b F^{ab} + \phi_b \overset{*}{F}{}^{ab} \tag{18}$$

where $\overset{*}{F}{}^{ab} = \frac{1}{2}\varepsilon^{abcd}F_{cd}$ is the dual of the electromagnetic field strength (ε^{abcd} is the totally antisymmetric Levi–Civita tensor). If the pseudovector has a potential, $\phi_a = \partial_a\phi$, then this potential is called axion. This axion is a candidate for the dark matter in the universe and is part of the low energy limit of string theory.

The other model describes a vector valued mass of the photon

$$4\pi j^a = \partial_b F^{ab} + \theta_b F^{ab} \tag{19}$$

and is discussed in [17]. Here θ_a is a vector which time–part θ_0 gives rise to charge non–conservation and which space–part $\theta_{\hat{a}}$ leads to an anisotropic speed of light and to an anisotropic Yukawa–modification of the Coulomb–potential.

Not included in the above model are the modifications of Maxwell equations derived within canonical quantum gravity (2) because this approach modifies the homogeneous equation. Also not included is the usual model of a scalar photon mass which is described by a Proca equation. While the approach with a vector valued mass is manifest gauge invariant, the approach using the Proca equation breaks gauge invariance. Within our gauge invariant formalism it is not possible to have a scalar photon mass.

4. LABORATORY TESTS

4.1. DISPERSION

This class of experiments amounts to a comparison of the velocities of photons of different energies. Such experiments have been carried through by Brown and coworkers [18]. No difference has been found, so that from the accuracy of the apparatus they got the estimate $(c(7\text{ GeV}) - c(\text{eV}))/c(\text{eV}) \leq 8 \times 10^{-6}$. Much better estimates result from astrophysical observations.

4.2. CHARGE CONSERVATION

Direct tests. A laboratory test of charge conservation in the sense of looking for a decay of the electron into neutral particles has been carried through. For this process the limit is $5.3 \times 10^{-21}\ \text{y}^{-1}$ [19]. There is also a lot of experimental work on the neutrality of atoms which of course proof only that the ratio of the charges of electrons and protons remains conserved.

Time dependence of the fine structure constant. In the case of charge non–conservation, the charge of all particles depends on time so that especially for the charge of the electron $de/dt \neq 0$ which then implies for the fine structure constant $d\alpha/\alpha = 2\alpha(1/e)(de/dt)$. If we assume a specific time–dependence of the form $de/dt = \zeta e$, then $d\alpha/\alpha = 2\zeta$. Thus experiments on the time–dependence on the fine–structure constant give estimates on ζ or on parts of χ^{abc} which are not totally antisymmetric.

A measurement of a hypothetical time dependence of the fine structure constant can be obtained by comparing different time or length standards which depend in a different way on α. In the laboratory experiments treated in [20] variations of the fine structure constant can be detected by comparing rates between clocks based on hyperfine transitions in alkali atoms with different atomic numbers. The comparison of the H–maser with a Hg+–clock resulted in $d\alpha/\alpha \leq 3.7 \times 10^{-14}$, so that $\zeta \leq 1.9 \times 10^{-14}$. In a proposal [21] a monolithic resonator is employed where the fine structure constant is related to the dispersion in the resonator. They claim to be able to improve the estimate to $\dot{\alpha}/\alpha \leq 4 \times 10^{-15}\ \text{y}^{-1}$.

4.3. FIELD OF POINT CHARGE

In the laboratory the electrostatic field of a point charge can be measured with very high accuracy while the magnetic field of a magnetic moment can be determined not as good.

Cavendish experiments. The first experiment aimed to test the inverse square law $E \sim e^2/r^2$ or the Coulomb potential $V = e^2/r$ was carried through by Cavendish [22]. The principle of this and many following experiments is to test the following consequence of Maxwell's equation: inside a metallic sphere, which has been given a certain electric potential V_0, there is no electric field. That means that another metallic sphere inside the sphere of potential V_0 should have the same potential. This is only possible if the potential for a charged point particle has the form $\sim 1/r$, that is, if the potential is a harmonic function. If the equation, namely the Poisson equation $\Delta V = 4\pi\rho$, which determines the potential from the charge distribution and the boundary conditions, has a different form, then the solution is different from $1/r$ and does not show up this property.

The most recent experiment of this kind has been carried through by Williams, Faller, and Hill [23]. Until now, the theoretical interpretation of this kind is experiments is restricted to three models only, namely (i) to the model of a scalar photon mass m_{photon}, see [23, 24] and references therein, which gives a potential of the form

$$V = \frac{e^2}{r}\left(1 + \beta e^{m_{\text{photon}} r}\right) , \tag{20}$$

(ii) to a specific ansatz for a deviation from the Coulomb law, $1/r^{1+\varepsilon}$, and (iii) to the context of a vector valued photon mass θ_a [25]. No deviation from Coulomb's law has been found and from the accuracy of the apparatus one gets the estimates $m_{\text{photon}} \leq 1.6 \times 10^{-47}$ g or $m_{\text{photon}} \leq 5 \times 10^{-8}\ \text{m}^{-1}$, $\varepsilon \leq 6 \times 10^{-16}$, $\beta \leq 10^{-10}$, and $\theta_{\hat{a}} \leq 10^{-47}$ g.

Behaviour of atomic clocks. Any modification of the solution for the potential of a point charge should modify the properties of atoms, especially the energy eigenvalues. As a consequence, the Rydberg constant will be modified. Not much has been calculated in that context. The modification for a pure Yukawa modification of the Coulomb potential (20) leads to modified transition frequencies. If one compares a transition for small quantum numbers, n, m, and large quantum numbers N, M, then the Rydberg constant may be different, since the strength of the electric field near the atomic core is different from the strength far from it. This difference is calculated to be [26]

$$\frac{\Delta R}{R} = 2\beta\left(F_{N',N} - F_{n',n}\right) \qquad \text{with} \qquad F_{n'n} = \frac{\frac{1}{n'^2}e^{a_0 m n'^2} - \frac{1}{n^2}e^{a_0 m n^2}}{\frac{1}{n'^2} - \frac{1}{n^2}} \tag{21}$$

where a_0 is Bohr's radius. From corresponding spectroscopic experiments one derives $m \leq 10^{-10}\ \text{m}^{-1}$ for $\beta \leq 10^{-10}$.

4.4. TESTS OF SPECIAL RELATIVITY

The main experiments for Special Relativity test the constancy of the speed of light: Michelson–Morley–experiments [13, 14] test the directional constancy, the isotropy, of c, and Kennedy–Thorndike–experiments [15, 16] and experiments like that of Alväger and coworkers [27] the independence of c from the velocity of the laboratory and of the source. According to Robertson [28] from the Michelson–Morley and Kennedy–Thorndike experiments together with experiments yielding the time dilation factor $1/\sqrt{1-v^2}$ one can derive Special Relativity. The most modern version of these experiments give $|\Delta c/c| \leq 3 \times 10^{-15}$ for the isotropy of light [14] and $|\Delta c/c| \leq 2 \times 10^{-13}$ for the independence of the velocity of light from the velocity of the source [16].

Since the properties of the used interferometer arms or cavities are influenced by the electromagnetic field, too, one has to be careful about the interpretation of these experiments: It could be shown [17] that within the model (19) under certain conditions no effect in these experiments can be detected though there is an anisotropic speed of light. However, these conditions, though being not exotic, are not fulfilled by the above mentioned experiments so that the usual interpretation is still valid.

4.5. SUPERPOSITION PRINCIPLE

For a non–linear theory like the Heisenberg–Euler or Born–Infeld theory the superposition principle is no longer valid. That the effective underlying theory for the electromagnetic field should be non–linear has been demonstrated through inelastic photon–photon scattering [29]. But also other effects should occur: In the frame of the Heisenberg–Euler theory the effect of birefringence of a propagating weak electromagnetic wave in strong electromagnetic fields has first been calculated by Adler [30]. This effect now seems to be in the range of detectability [31]. Also in the case of the Born–Infeld theory the dispersion of propagating waves has been derived and shown to be measurable [32, 33].

5. ASTROPHYSICAL OBSERVATIONS

Although in astrophysical observations the observer cannot manipulate the creation of electromagnetic waves and the boundary conditions and cannot move arbitrarily in space but is rigidly attached to the earth, strong estimates can be drawn on those terms describing deviations from the usual Maxwell equation. The main point is that if the source emits a unique signal, then this signal usually is distributed over all frequencies and polarizations. In this case we can make two kinds of observations relevant for our analysis: First we compare the time–of–arrival of the signal in one frequency range with the time–of–arrival of the same signal in another frequency range, that is we compare the velocities connected with the two frequencies, $c(\omega_2) - c(\omega_1)$, and second we compare the ar-

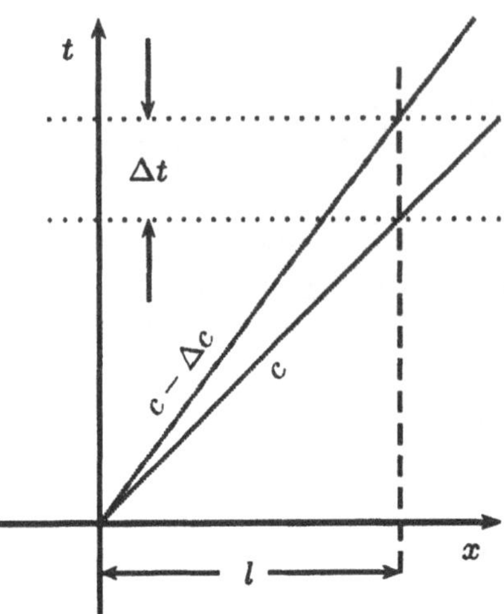

Figure 1 Propagation of signals with different velocities leads to a delay Δt in the arrival of the signals.

rival of the signals analyzed according to the two different polarization states + and −, that is, we compare the two velocities $c_+ - c_+$.

The difference of the time–of–arrival Δt of the same signal is related to the difference of the corresponding velocities according to, see Fig.1,

$$\frac{\Delta c}{c} = \frac{c\,\Delta t}{l}\,. \tag{22}$$

In ordinary Maxwell theory $\Delta c = 0$ so that any measurement of a δt connected with the *same* event is related to a violation of the usual Maxwell equation. In other words, these these measurements search for violations of universality of the freefall of photons.

The advantage of astrophysical observations can be immediately inferred from this formula: The estimate on a violation of the Maxwell equations, namely the estimate on $\Delta c/c$, is better if the event is far away and if the monitored events are very short. The first point is obvious for astrophysics where we have distances of up to 10^{10} light years. In addition, very short events are observable even for very far sources, e.g. for γ–ray bursts which have structures of the scale of microseconds. In addition, astrophysical sources also emit photons with very high energies, up to 10^{20} eV, see below.

The observed objects are quasars, pulsars, γ–ray bursts, active galactic nuclei and supernovae. Any anomalous form of the Maxwell equations leads to dispersion and birefringence. We also will discuss shortly the importance of different velocities of photons and neutrinos.

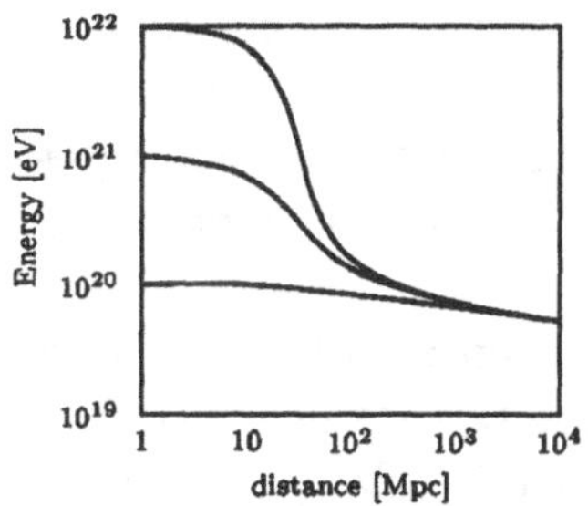

Figure 2 Mean propagation distance of high energy photons due to their interaction with low energy background photons [37].

5.1. DISPERSION

The search for a dispersion in light propagation amounts to a search for a difference in the time–of–arrival for a specific signal when observed in different frequency windows. For the interpretation of such observations one has to be very careful, since matter like interstellar gas may interact with the light and then lead to a dispersive behaviour of the propagation of light. Therefore, one has to know very well the properties of the interstellar gas. A first treatment of this approach with taking interstellar matter into account has been carried through by Feinberg [34]. Fortunately, this is only relevant for low energy photons. High energy photons are not influenced by interstellar matter. The best results have been obtained recently by Schaefer [35] leading to $|(c(\omega) - c(\omega_0))/c(\omega_0)| \leq 6.3 \times 10^{-21}$. In terms of a photon mass this means $m \leq 6.1 \times 10^{-39}$ g.

5.2. HIGH ENERGY COSMIC RAYS

High energy cosmic rays and in particular high energy photons are especially well suited for searches for quantum gravity induced modifications of the dispersion relation (13). Since new aspects enter the discussion, we treat this case in an extra subsection.

First, also here one searches for dispersion effects. An analysis of this kind has been carried through by Biller and coworkers [36]: A short signal of about 280 sec emitted from Mkn 421, which is $3.5c \times 10^8$ ly $\sim$ 112 Mpc away ($z = 0.0031$), arrived, when decomposed into two frequency windows, at earth at the same time. From this analysis one arrives at $\Delta c/c \leq$ and correspondingly, at $E_{QG} \geq 10^{20}$ GeV.

However, another aspect of high energy cosmic rays may be of much more importance: Based on the conventional dispersion relation $p^2 = 0$, it can be shown that the high energy photons react with the infrared background what results in the creation of massive particles. Therefore high energy photons have a finite free path only, which can be calculated to be of the order of 100 Mpc, [38, 39]. This applies to photons with energies larger than 10^{19} eV, see Fig.2. Consequently, it one detects photons with energies larger than 10^{19} eV from sources farer away than 100 Mps, then one reason for that may be a violation of the usual dispersion relation. Beside that, no source is known within 100 Mps which is able to create photons with such high energies.

The HEGRA cooperation measures a certain intensity of TeV γ rays from Mkn 501, which which has a distance of about 120 Mps ($z = 0.033$). If one calculates back the intensity of the emitted TeV γ rays from the observed intensity taking into account the attenuation due to the photon–photon scattering mentioned above, then one obtains a higher intensity of the emitted photons the larger the energy is [40]. This is very strange. In addition, the creation of photons of such high energy always is connected with the creation of neutrinos. However, the AMANDA collaboration didn't found such a large number of neutrinos from Mkn 501 which is necessary for being consistent with the larger number of high energy photons, see [41].

Using now the model of a modified dispersion relation (13) one can show that the GZK cutoff will be shifted to higher energies. This is compatible with the observational data. Therefore, this may be some hint towards a quantum gravity modification of the dispersion relation and thus of Maxwell's equations and finally towards quantum gravity effects.

5.3. BIREFRINGENCE

Another type of observation consists in a polarization analysis. If for two different independent polarizations the signal arrives at the same time, then there is no birefringence observed.

Within our generalized Maxwell equations (17) there are two sources for birefringence, the coefficients $\delta\chi^{abcd}$ and χ^{abc}. For the latter one, a special case has been analyzed in [42], the first one has been discussed in [43] in general.

The first analysis by Carrol, Field and Jackiw employed the model (18) and used data of galaxies which emit polarized light. From their analysis the axial vector has to be smaller than 1.7×10^{-42} GeV or 10^{-26} m^{-1}. This corresponds to a birefringence, that is, a difference in the propagation speed for both polarizations, of $\Delta c/c \leq 10^{-26}$.

Birefringence due to the first term in the generalized Maxwell equation (16) has been studied by Haugan and Kauffmann [43]. The difference in the velocity of the two polarization states is presented in general as function of the various coefficients of $\delta\chi^{abcd}$. From the analysis of light of stars they get an estimate $\Delta c/c \leq 10^{-28}$.

5.4. CHARGE CONSERVATION

If we again associate the time dependence of the fine structure constant α with charge–nonconservation, then astrophysical estimates on $\dot{\alpha}/\alpha$ can be interpreted as test of charge conservation. In a recent paper [44] the cosmological constraint on a time dependence of α is given by $\dot{\alpha}/\alpha \leq 3.5 \times 10^{-15}\ \mathrm{yr}^{-1}$.

5.5. MAGNETIC FIELDS

Since macroscopic objects like stars in general are neutral, there is no way to observe the potential of a charged astrophysical object. However, most astrophysical objects possess a magnetic moment so that there is a magnetic field around. Accordingly, the measurement or observation of the magnetic field of the earth, of our galaxy or of a cluster of galaxies gives very good restrictions on deviations from the usual Maxell equations. The best founded estimate in the framework of a scalar photon mass of this kind is of Goldhaber and Nieto [45, 24] and gives $m_{\mathrm{photon}} \leq 10^{-47}$ g. There are better estimates from the consideration of galactic or intergalactic magnetic fields but these are loaded with more assumptions and uncertainties.

5.6. TESTS FOR SPECIAL RELATIVITY

While it is not possible to test the isotropy of light propagation because we cannot send light in different directions, it is possible to analyze the motion of stars in order to get results about whether the velocity of light depends on the motion of the source or not. Such an analysis has been carried through by K. Brecher [46]. If the velocity of light depends on the velocity of the source, $c = c(v)$, then the light which we observe should possess different velocities according to the velocity the star had during emission. This in general should lead to multiple images of the same star e.g. during evolution around a binary compagnion. This has never been observed. In the case of small velocities we may expand $c(v) = c + \kappa v$ where v is the velocity of the star and c the velocity of light if the star is at rest, the observation of Hercules X1 place a limit $\kappa \leq 10^{-9}$ [46]. If we analyze the short events of γ–ray pulses in the same way, one may reach $\kappa \leq 10^{-20}$ [47].

6. CONCLUSION

Collecting the above listed experiments and observations, one arrives in terms of the parameters defined in the generalized Maxwell equations (17) at the following rough estimates

quantity	astrophysical test	laboratory test
$\delta\chi^{abcd}$	$\lesssim 10^{-28}$	$\lesssim 10^{-15}$
$\chi^{[abc]}$	$\lesssim 10^{-26}\ \mathrm{m}^{-1}$	$\lesssim 1.5 \times 10^{-15}\ \mathrm{m}^{-1}$
χ^{abc}	$\lesssim 10^{-50}\ \mathrm{kg} = 3 \times 10^{-8}\ \mathrm{m}^{-1}$	$\lesssim 10^{-50}\ \mathrm{kg} = 3 \times 10^{-8}\ \mathrm{m}^{-1}$

This table should only provide an indication of the order of estimate for the three classes of parameters; much more analysis on the restrictions of the various components of the involved tensors is necessary.

Until now no confirmed deviation from the usual Maxwell equations has been found. Only the deviations from quantum gravity modifications of the dispersion relation for photons may be somehow promising.

In conclusion we want to remark that if one describes test of Maxwell's equations with the help of spectroscopy or in e.g. Michelson–Morley experiments, then one has to treat the atoms and the solids establishing the interferometer, too. This means first, that the length of the interferometer, for example, is determined by the modified Maxwell equations, too, and second, that one also has to take into account the possibility that also the Dirac equation has to be modified, see. e.g. [48]. However, this makes the interpretation of such experiments much more involved and needs a more detailed theoretical analysis.

Acknowledgments

The author thanks M. Haugan, A. Macias and S. Schiller for useful discussions and A. Macias for his hospitality at the UAM, Mexico City during which part of this work was done. The support of the German–Mexican collaboration by CONACYT–DLR (MEX 98/010) is gratefully acknowledged.

References

[1] W. Ritz, *Ann. Chim. Phys.* **13** (1908) 145.

[2] M. Born and L. Infeld, *Proc. R. Soc. London* **A144** (1934) 425.

[3] W. Heisenberg and H. Euler, *Z. Physik.* **98** (1936) 714.

[4] J. Schwinger, *Phys. Rev.* **82** (1951) 664.

[5] I.T. Drummond and S.J. Hathrell, *Phys. Rev.* **D22** (1980) 343.

[6] P. Podolsky, *Phys. Rev.* **62** (1942) 68.

[7] J. Ellis, N.E. Mavromatos, D.V. Nanopoulos, and G. Volkov, gr–qc/9911055.

[8] R. Gambini and J. Pullin, *Phys. Rev.* **D59** (1999) 124021.

[9] J. Ellis, N.E. Mavromatos, and D.V. Nanopoulos, gr–qc/9909085.

[10] G. Amelino–Camelia et al., *Nature* **393** (1998) 763.

[11] J. Ellis, N.E. Mavromatos, and D.V. Nanopoulos, *Gen. Rel. Grav.* **31** (1999) 1257.

[12] S.L. Dubovsky, V.A. Rubakov, and P.G. Tinyakov, gr–qc/0007179.

[13] A.A. Michelson and E.W. Morley, *Am. J. Sci.* **34** (1887) 333.

[14] A. Brillet and J.L. Hall, *Phys. Rev. Lett.* **42** (1979) 549.

[15] R.J. Kennedy and E.M. Thorndike, *Phys. Rev.* **42** (1932) 400.

[16] D. Hils and J.L. Hall, *Phys. Rev. Lett.* **64** (1990) 1697.

[17] C. Lämmerzahl and M.P. Haugan, *Phys. Lett. A* (2001) to appear.

[18] B.C. Brown, G.E. Masek, T. Maung, E.S. Miller, H. Ruderman, and W. Vernon. *Phys. Rev.* **30** (1973) 763.

[19] R.I. Steinberg, K. Kwiatkowski, W. Maenhaut, and N.S. Wall, *Phys. Rev.* **D12** (1975) 2582.

[20] J.D. Prestage, R.L. Tjoelker, and L. Maleki, *Phys. Rev. Lett.* **74** (1995) 3511.

[21] C. Braxmaier et al., *Proposed test of the time independence of the fundamental constant α and m_e/m_p using monolithic resonators*, preprint, Univ. Düsseldorf.

[22] J.C. Maxwell, *A Treatise on Electricity and Magnetism.* (Dover Publications, New York, 1954. Reprint of the 1891 Edition by Clarendon Press).

[23] E.R. Williams, J.E. Faller, and H.A. Hill, *Phys. Rev. Lett.* bf 26 (1971) 721.

[24] A.S. Goldhaber and M.M. Nieto, *Rev. Mod. Phys.* **43** (1971) 277.

[25] M. Haugan and C. Lämmerzahl. On the mass of the photon, in preparation.

[26] D.F. Bartlett and S. Lögl, *Phys. Rev. Lett.* **61** (1988) 2285.

[27] T. Alväger, F.J.M. Farley, J. Kjellmann, and I. Wallin, *Phys. Lett.* **12** (1964) 260.

[28] H.P. Robertson, *Rev. Mod. Phys.* **21** (1949) 378.

[29] R. Burke et al., *Phys. Rev. Lett.* **79** (1997) 1626.

[30] S.T. Adler, *Ann. Phys. (N.Y.)* **67** (1972) 599.

[31] Wei-Tou Ni., in: *Frontier Tests of QED and Physics of the Vacuum*, E. Zavattini, D. Bakalov, and C. Rizzo, eds. (Heron Press, Sofia, 1998) pp. 83.

[32] V.I. Denisov, *J. Optics A: Pure and Applied Optics* **2** (2000) 372.

[33] V.I. Denisov, *Phys. Rev.* **D61** (2000) 036004.

[34] G. Feinberg, *Science* **166** (1969) 879.

[35] B.E. Schaefer, *Phys. Rev. Lett.* **82** (1999) 4964.

[36] S.D. Biller et al., *Phys. Rev. Lett.* **83** (1999) 2108.

[37] J.W. Cronin, *Rev. Mod. Phys.* **71** (1999) 165.

[38] K. Greisen, *Phys. Rev. Lett.* **16** (1966) 748.

[39] G.T. Zatsepin and V.A. Kuzmin, *JETP Lett.* **4** (1966) 78.

[40] R.J. Protheroe and H. Meyer, *Phys. Lett.* **B493** (2000) 1.

[41] J. Ellis, astro–ph/0010474.

[42] S.M. Carroll, G.B. Field, and R. Jackiw, *Phys. Rev.* **D41** (1990) 1231.

[43] M.P. Haugan and T.F. Kauffmann, *Phys. Rev.* **D52** (1995) 3168.

[44] C.L. Carilli, K.M. Menten, J.T. Stocke, E. Perlman, R. Vermeulen, F. Briggs, A.G. de Bruyn, J. Conway, and C.P. Moore, *Phys. Rev. Lett.* **85** (2000) 5511.

[45] A.S. Goldhaber and M.M. Nieto, *Phys. Rev. Lett.* **21** (1968) 567.

[46] K. Brecher, *Phys. Rev. Lett.* **39** (1977) 1051.

[47] K. Brecher. Talk at the 2000 April Meeting of the APS in Long Beach, California, http://www.aps.org/meet/APR00/baps/abs/S450013.html.

[48] C. Lämmerzahl, *Class. Quantum Grav.* **14** (1998) 13.

LORENTZ FORCE FREE CHARGED FLUIDS IN GENERAL RELATIVITY: THE PHYSICAL INTERPRETATION

Nikolai V. Mitskievich *
Physics Department, CUCEI, University of Guadalajara, Guadalajara, Jalisco, Mexico.

Abstract The Lorentz force free Einstein–Maxwell–Euler self–consistent solutions are interpreted as containing a superposition of the proper electromagnetic field created by the charged perfect fluid and a magnetic–type sourceless field which exactly compensates electric field of the charged fluid in its comoving frame of reference. It is shown that this situation can be easily described also in special relativity, even for essentially non-relativistic charged fluids.

Keywords: Lorentz force, charged fluids, special relativity.

1. INTRODUCTION

In this communication we propose a physical interpretation to the Lorentz force free Einstein–Maxwell–Euler self–consistent solutions (LF-FS : a charged perfect fluid exactly mimicking a neutral fluid; the expression "Lorentz force free" will be denoted as LFF). The absence of Lorentz force is equivalent to vanishing of electric field in the frame comoving with the charged fluid; we conclude that this is due to a superposition of the electromagnetic field generated by its moving charges and a source–free magnetic type field. Such a superposition is influenced by the gravitational interaction between all physical components forming this system, so that, due to nonlinearity introduced by the gravitational interaction, these quite natural physical reasons of seemingly exotic behaviour of a charged fluid are profoundly concealed, and the whole situation gives a false impression of a possibility to find a configuration of moving charged fluid which does not generate any electric field in its proper comoving frame.

*E–mail: nmitskie@udgserv.cencar.udg.mx

First let us recall some properties of the Gödel metric [9] which is of great importance in constructing several types of exact LFFS. For the signature $(+ - - -)$ and after some other transformations, the Gödel metric reads (*cf.* [22, 24])

$$ds^2 = a^2 \left[(dt + \sqrt{2}e^x dy)^2 - dx^2 - e^{2x} dy^2 - dz^2\right], \tag{1}$$

or in "cylindrical" coordinates,

$$ds^2 = a^2 \left[(dt + \sqrt{2}\rho d\phi)^2 - \frac{d\rho^2}{\rho^2} - \rho^2 d\phi^2 - dz^2\right]; \tag{2}$$

the same form in [20] was put as

$$ds^2 = a^2 \left[(dt + \sqrt{2}z dx)^2 - z^2 dx^2 - dy^2 - z^{-2} dz^2\right]. \tag{3}$$

It is interesting that if (1) is rewritten in the form

$$ds^2 = a^2 \left[dt^2 + e^{2x} dy^2 - dx^2 - dz^2 + 2\sqrt{2}e^x dt dy\right], \tag{4}$$

it becomes clear that either of t and y may be interpreted as a time coordinate (obviously related to the existence of closed timelike lines in the Gödel universe). Then, interchanging t and y in this expression, we get

$$ds^2 = a^2 \left[e^{2x}(dt + \sqrt{2}e^{-x} dy)^2 - dx^2 - dy^2 - dz^2\right] \tag{5}$$

(the three-dimensional sector taking here the Cartesian form). Of course, all coordinates used above are, as it is admitted traditionally in the Gödel metric, dimensionless. One can however incorporate a into (at least, some) coordinates variables and redefine t: $t \to t - \sqrt{2}y$, so that, for example, (1) takes the form

$$ds^2 = \left(dt - \sqrt{2}(1 - e^{x/a})dy\right)^2 - dx^2 - e^{2x/a} dy^2 - dz^2; \tag{6}$$

then the limit $a \to \infty$ yields the Minkowski spacetime.

The Gödel spacetime admits five Killing vectors which we give here in the coordinates used in (2), up to the positive constant factor a^2 occurring in the covariant components:

$$\mathbf{I}: \ \xi^\mu \partial_\mu = \partial_t, \quad \xi_\mu dx^\mu = dt + \sqrt{2}\rho d\phi; \tag{7}$$

$$\mathbf{II}: \ \xi^\mu \partial_\mu = \partial_\phi, \quad \xi_\mu dx^\mu = \sqrt{2}\rho dt + \rho^2 d\phi; \tag{8}$$

$$\mathbf{III}: \ \xi^\mu \partial_\mu = \partial_z, \quad \xi_\mu dx^\mu = -dz; \tag{9}$$

$$\mathbf{IV}: \ \xi^\mu \partial_\mu = 2\sqrt{2}\rho^{-1}\partial_t - 2\rho\phi\partial_\rho - (\rho^{-2} - \phi^2)\partial_\phi, \tag{10}$$

$$\xi_\mu dx^\mu = \sqrt{2}\left(\rho\phi^2 + \rho^{-1}\right)dt + 2\rho^{-1}\phi d\rho + \left(3 + \rho^2\phi^2\right)d\phi; \tag{11}$$

$$\mathbf{V}: \ \xi^\mu \partial_\mu = \rho\partial_\rho - \phi\partial_\phi, \ \ \xi_\mu dx^\mu = -\sqrt{2}\rho\phi dt - \rho^{-1}d\rho - \rho^2\phi d\phi. \tag{12}$$

The Killing vectors **I**, **II** and **IV** are everywhere timelike, **III** spacelike, and the Killing vector **V** has a sign-indefinite square, thus in some parts of the space it is timelike and in other parts, spacelike. Due to the commutation properties of the three isometries corresponding to **I**, **II** and **IV**, only two of them (always including **I**) can be simultaneously used to introduce coordinates of which the metric coefficients will not depend (the coefficients of their squared differentials will be therefore both positive – timelike axes). The third Killing coordinate (z) available simultaneously with these two, is clearly spacelike. Of course, it seems strange to use two timelike axes at once, though this is always exactly the case in dealing with the Gödel spacetime. What is not less exotic, there exists the third alternative in introduction of one of this pair of coordinates: it can involve the Killing coordinate corresponding to **V** which naturally excludes the possibility of simultaneous use of the coordinates which are Killingian with respect to **II** and **IV**. The situation really is in its full extent a rare exception in the theory of spacetime.

The real physical signature $(+---)$ of the Gödel spacetime is easily seen from the existence of the orthonormal tetrad (with the normalization properties corresponding to this signature) in this spacetime, for example, the natural tetrads of (1) and (5). The very same consideration saves time in the calculation of the metric tensor determinant: *e.g.*, in the case of (2), $\theta^{(0)} \wedge \theta^{(1)} \wedge \theta^{(2)} \wedge \theta^{(3)} = \sqrt{-g}dt \wedge d\rho \wedge d\phi \wedge dz$ which yields simply $\sqrt{-g} = a^4$.

The fact that there exists only one Killing vector which is spacelike in all spacetime, does not mean, of course, absence of spatial homogeneity in the Gödel universe. There are linear combinations (with constant coefficients) of Killing vectors being as well spacelike, though this property is limited to certain regions of spacetime (similar to the situation with the Killing vector V), the region depending of the choice of constants and never embracing the whole spacetime. Thus the homogeneity property of the Gödel universe is strictly *local*, but it covers all the spacetime piecewise. Similarly, the local essentially stationary property is known for the ergosphere region of the Kerr solution (see [10], p. 331).

Many authors subdivide the multitude of exact solutions into "physical" and "unphysical" subsets. My opinion is that the point of view that in the well established physical theories there could be present unphysical solutions and effects alongside with the "good", physical ones, is completely wrong. All results following from these theories should have proper meanings and interpretations. In any case, the most exotic consequences of fundamental theories, lying at the edge of limits of applicability of these theories (any theory has such limits), ought to correspond to the objective structure of the theory and provide guidelines for its generalization — or permit a more thorough comprehension of its existing generalization, all this being of great physical value too. This is essentially the starting point of the considerations outlined below.

Exact LFFS are now a long-known issue. First the authors who obtained such solutions, had doubts about their physical acceptability; some stressed this point expressing their surprise that Maxwell's theory could lead to these "paradoxical" consequences. It seems that the general relativistic nature of the subject did help to reach an unstable reconciliation with such discoveries: who knows which other wonders could hold this marvellous theory?! Of course, the people soon begun exploring the new LFF grounds applying both spacetime-invariant methods and the theory of reference frames with twist and, sometimes, nongeodesic motion. The corresponding reference frame considerations readily showed a complete self-consistency of the physical processes occurring in spacetimes of these solutions. If they would not show as much as that, this should be only worse for the existing description of relativistic noninertial frames. But this still was not a physical *interpretation* of the phenomenon, the question simply shifted from the mathematical soundness of the solutions *versus* the classical dynamical equations of fields and particles and their decomposition into temporal and spatial parts, to the origin of the vanishing of the Lorentz force: is it due to the physical system (charged fluid) supposed to generate the electromagnetic and gravitational fields, or could it have some alternative physical nature?

2. LFFS IN GENERAL RELATIVITY

First papers, dedicated to LFF charged fluids in general relativity, appeared in 1967-68: [21, 23, 1, 26]. In [2] a comparatively complete and invariant discussion of the properties of LFFS can be found (the paper was focused essentially on rotating dust). In [5] a general LFF solution for axially symmetric rigidly rotating dust was obtained. Several papers on LFFS dedicated to charged differentially rotating dust appeared in 1984: [3, 4, 29, 15]. LFFS with perfect fluid were studied in [17]. In 1985 was published a book by J. N. Islam [14] on rotating configurations including some LFFS. It is an important book, but its structure is somewhat inconvenient when one is looking for concrete form of squared intervals. Later publications were focused on perfect fluid LFFS: [20, 8, 27]. The most recent publications on LFFS, [16, 28], are based on the approach of [20] alongside with that of [5]. In [28] three new LFFS are deduced and thoroughly studied (a good review of the situation in LFFS also being given). In [16], the authors use electric and magnetic potentials *à la* Bonnor; the paper is explicitly focused on Gödel–like solutions, and the authors give cosmological interpretation to their results. Quite a number of previous LFFS represent special cases of two large families of such solutions obtained in [20], so that we shall only use this paper to describe the existing situation.

Our approach in [20] was based on the Horský–Mitskievich (HM) method of constructing Einstein–Maxwell solutions from vacuum metrics [12], later generalized to the use of spacetimes filled with perfect fluids for constructing Einstein–Maxwell–Euler solutions (see [13] where the HM method was applied to the seed metric found in [18]). More general approach can be found in [6], and a discussion of the HM method see in [25]. The method is based on the use of a Killing covector of the seed spacetime as (up to a constant factor) the exact four–potential of the electromagnetic field of the solution to be constructed. Then a metric similar to the seed one, but with still unknown functions instead of those which form the seed metric coefficients, is introduced, and Einstein's equations are written for this metric and the corresponding stress–energy tensors. This system of equations usually is easily solved along the same lines as this was the case for the seed metric.

In [20] the Gödel metric was chosen as the seed one in the form (3) with its Killing vector $a^{-1}\partial_t$. The electromagnetic field tensor then takes form $F_{(\alpha)(\beta)} = F^{(\alpha)(\beta)} = \sqrt{2}ak\exp(-\beta-\delta)\delta^3_{[\alpha}\delta^1_{\beta]}$ in the orthonormal basis

$$\theta^{(0)} = e^{\alpha}(dt + f dx),\ \theta^{(1)} = e^{\beta}dx,\ \theta^{(2)} = e^{\gamma}dy,\ \theta^{(3)} = e^{\delta}dz \tag{13}$$

where α, β, γ, δ and f are functions of z only, and k is an arbitrary constant whose vanishing leads to an Einstein–Euler solution without electromagnetic field (a nontrivial special case is the seed Gödel metric). Thus the electromagnetic field for $k \neq 0$ is of purely magnetic nature, and in the comoving frame of the fluid ($\tau = \theta^{(0)}$) only magnetic field is present. This guarantees vanishing of the Lorentz force.

The dynamical Maxwell equations then are

$$\left(\sqrt{-g}F^{\mu\nu}\right)_{,\nu} = -\sqrt{2}ak\left[e^{\alpha-\beta+\gamma-\delta}\left(\delta^{\mu}_{x} - f\delta^{\mu}_{t}\right)\right]' = -4\pi\sqrt{-g}j^{\mu} \tag{14}$$

with $j = \rho\tau$, $\tau = \partial_t$. Here ρ is the proper (in the comoving frame with the monad τ coinciding with the four-velocity of the fluid) fluid's charge density. Hence Maxwell's condition (only one existing in this case) reads

$$\alpha - \beta + \gamma - \delta = 0 \tag{15}$$

(still without any restrictions on f; one can directly see that f' is always easily integrated from one of Einstein's equations, $R_{(0)(1)} = 0$). Then the proper charge density of the fluid is $\rho = -\frac{\sqrt{2}ak}{4\pi} \exp(-\beta - \gamma - \delta) f'$, while the stress-energy tensors components take a compact form.

The next step of determining exact LFFS is calculation of the curvature tensor from (13) (most immediately with use of the Cartan forms method). The resulting Einstein equations can be readily integrated under Maxwell's condition (15) and specific assumptions concerning one of the five unknown functions encountered in the problem. In [20] two cases were considered: (a) $\alpha =$ const., and (b) $\gamma =$ const. There is no rotating solutions when $\rho = 0$ and $k \neq 0$, so that the HM method automatically generates the complete LFFS without singling out its possible sourceless part which, as we shall conclude, is responsible for vanishing of the Lorentz force when the rotation of fluid is not switched off.

Properties of the obtained solutions are distinct from those of the limiting (and seed) case of the Gödel solution.

While the Petrov type of the Gödel spacetime is D, the family (a) solutions belong to the algebraically general type I everywhere except on a certain well determined hypersurface where they are of type D or type II. In its turn, the family (b) belongs to type D everywhere except on another hypersurface where the spacetime is conformally flat. The both families of solutions admit the obvious three–parameter Abelian isometry group. They also admit closed timelike (nongeodesic) lines, as this was the case for the Gödel spacetime.

In the comoving reference frame of the charged fluid (for the theory describing reference frames see [19]), there exists, as this was mentioned above, only magnetic field (solutions of the both families belong to LFFS), thus the charged fluid's particles do not mutually interact electromagnetically (there is only interaction via their gravitational field). The family (a) charged fluid is moving geodesically, while that of family (b) is non–geodesic. In all cases, expansion and shear are absent. In Maxwell's equations written in the comoving frame of the fluid, for the both families, the charge term is exactly compensated by the kinematic charge which appears due to rotation of the reference frame in the presence of the magnetic field in this comoving system. It is worth mentioning that while in the mechanics of point-like particles in non–inertial frames there appear the so-called forces of inertia ("kinematic forces"), in the field theory appear "kinematic sources". Thus to the usual, dynamical forces and sources, in non–inertial frames additionally appear their kinematic counterparts; moreover, when an electric field is present in a rotating frame (this is not the case for the comoving frame of fluid in our solutions), there appears a kinematic magnetic charge distribution whose dynamic counterpart in classical theory does not exist (the magnetic monopole). More on this subject see in [19]. For earlier studies (in an approximated approach) of the kinematic charges see [7, 11].

3. THE ANSATZ OF ZERO LORENTZ FORCE IN SPECIAL RELATIVISTIC ELECTRODYNAMICS

Since a charged particle in absence of Lorentz force in the Minkowski spacetime should have a straight (geodesic) worldline, this situation is most naturally described in Cartesian coordinates. We denote them by t, x, y and z (the same sub/superscripts, when necessary, will be used as tensor indices). Let all functions depend on z only, with a prime for the corresponding ordinary differentiation. We consider the charged fluid moving with a constant three–velocity v along x axis, so that the four–velocity is

$$u = (1 - v^2)^{-1/2}(dt - vdx); \quad u = (1 - v^2)^{-1/2}(\partial_t + v\partial_x). \tag{16}$$

Choosing the electromagnetic field tensor components as

$$F_{\mu\nu} = -2\delta^z_{[\mu}\left(E\delta^t_{\nu]} + B\delta^x_{\nu]}\right); \quad F^{\mu\nu} = 2\delta^{[\mu}_z\left(E\delta^{\nu]}_t - B\delta^{\nu]}_x\right), \tag{17}$$

where E and B are magnitudes of electric and magnetic field vectors $\mathbf{E}$ and $\mathbf{B}$ in the overall reference frame $K = (t, x, y, z)$, pointing in the positive directions of axes y and z respectively, we may reformulate the ansatz of zero Lorentz force as

$$F_{\mu\nu}u^\nu = -(1 - v^2)^{-1/2}(E + vB)\delta^z_\mu = 0 \tag{18}$$

(this is obviously equivalent to existence of only magnetic field in the local comoving reference frame of the fluid). Since the field tensor (17) represents a simple bivector, the second invariant of the field vanishes identically, $\overset{*}{F}_{\mu\nu}F^{\mu\nu} \equiv 4\mathbf{E}\cdot\mathbf{B} = 0$, while the first invariant is, as usual, $F_{\mu\nu}F^{\mu\nu} = 2(B^2 - E^2)$.

It is obvious that the homogeneous system of Maxwell's equations (structure equations) is satisfied automatically, $\overset{\mu\nu}{F^*}_{,\nu} = 0$, while the inhomogeneous (dynamical) equations read $F^{\mu\nu}{}_{,\nu} = -E'\delta^\mu_t + B'\delta^\mu_x = -4\pi\rho_0 u^\mu$. Here ρ_0 is proper charge density of the fluid. From (18) it follows that $E - BB'/E' = 0$, and the natural condition $v^2 = (B'/E')^2 \leq 1$ leads to $B^2 - E^2 = a^2 = \text{const}$. This means that the electromagnetic field has here to be of purely magnetic type, a having the meaning of the (constant) proper magnetic field intensity. Then it is natural to assume $B = a\cosh\xi$, $E = a\sinh\xi$, so that $v = -\tanh\xi$, and

$$\rho_0 = (a/4\pi)\xi'. \tag{19}$$

There are only two natural approaches to solve the problem under consideration: either (i) one may choose the behaviour of ρ_0 and find $\xi(z)$ via straightforward integration of (19), or (ii) postulate the velocity distribution $v(z)$ and find ρ_0 merely by a differentiation. Thus our task proves to be a trivial one. The general properties of the fluid now are formulated as follows: Its particles are moving inertially (the velocity is constant in the direction of motion x, though it depends on the concrete layer, thus being a function of z only); if the velocity vanishes on two opposite boundaries, the charge density nearby has to take opposite signs; finally, the constant proper magnetic field is directed along y axis.

As illustrations, we just mention four special solutions:

1. $\rho_0 = \text{const.}: \quad \Rightarrow \quad \xi = 4\pi\rho_0 z/a$.
2. $v \to 0$ when $z \to \pm\infty$: *e.g.*, $v = v_0 e^{-b^2z^2}$.

Then, $B = a\left(1 - {v_0}^2 e^{-2b^2z^2}\right)^{-1/2}$, $E = Bv_0e^{-b^2z^2}$, $\quad \rho_0 = b^2 zE/2\pi$.

3. $\rho_0 = \frac{C}{2}(1 + \cos(2\pi z/h))$: now, $\xi = \frac{2\pi C}{a}\left(D + z + \frac{h}{2\pi}\sin\frac{2\pi z}{h}\right)$.
4. The next special case meets the following alternative condition: $v = v_0\cos^2(\pi z/h) \quad \Rightarrow \rho_0 = (av_0/4h)\left(1 - {v_0}^2\cos^4(\pi z/h)\right)^{-1}\sin(2\pi z/h)$.

This solution can be easily joined with a constant sourceless magnetic field $B = a$ (the exterior solution) at one or two (not necessarily consecutive) nodes.

4. CONCLUSIONS

We have shown in this communication that LFFS exist not only in general relativity, but also in flat spacetimes and even in the case when charged fluids are nonrelativistic. These latter cases are of importance since then a superposition of electromagnetic fields is applicable (the Maxwell theory is completely linear, and there is no self–interaction of electromagnetic fields due to their contribution to curvature characteristic for general relativity). In these flat–spacetime solutions the Lorentz force is compensated by influence of a source–free (usually, purely magnetic) field. Our interpretation of the general relativistic LFFS is that they should contain these source–free electromagnetic part which cannot be separated from that created by charged fluid proper in the nonlinear realm of general relativity. Even an overcompensation of the Lorentz force can be foreseen for certain systems when, effectively, the charged particles (of one and the same sign of charge) would experience a mutual "attraction".

Acknowledgments

The author is much indebted to Bryce S. DeWitt for many friendly discussions; in particular, Bryce stressed (in May, 1992) the impossibility for systems of particles possessing electric charges of one and the same sign, to create the field yielding zero Lorentz force acting on these particles. Thus he did implicitly anticipate that vanishing of the Lorentz force should be a result of a superposition of the field mentioned above *and* a source–free electromagnetic field whose presence is obscured by the nonlinear nature of Einstein's equations. My thanks are due to Georgios Tsalakou for the great pleasure of scientific collaboration with him. I gratefully acknowledge both spoken and e-mail discussions with Jan Horský; these contacts always were and continue to be friendly and fruitful.

This work, in its initial stage, was partially supported by the CONACyT (Mexico) Grant 1626P.

References

[1] A. Banerjee and S. Banerji, *J. Phys. A* **1** (1968) 188.

[2] A. Banerjee, N. Chakravarty and S. B. Dutta Choudhury, *Aust. J. Phys.* **29** (1976) 119.

[3] N. van den Bergh and P. Wils, in: *Classical General Relativity*, ed. by W. B. Bonnor, J. N. Islam and M. A. H. MacCallum (Cambridge University Press, Cambridge, U.K., 1984).

[4] N. van den Bergh and P. Wils, *Class. Quantum Grav.* **1** (1984) 199.

[5] W. B. Bonnor, *J. Phys.* **A13** (1980) 3465.

[6] M. Cataldo, K. Kumaradtya and N. Mitskievich, *Gen. Relativ. and Grav.* **26** (1994) 847.

[7] H. Dehnen, H. Hönl and K. Westpfahl, *Zs. f. Physik* **164** (1961) 483.

[8] A. Georgiou, *Nuovo Cim.* **108B** (1993) 69.

[9] K. Gödel, *Revs. Mod. Phys.* **21** (1949) 447.

[10] S. W. Hawking and G. F. R. Ellis, *The Large Scale Structure of Space–time* (Cambridge University Press, Cambridge, U.K., 1973).

[11] H. Hönl and Chr. Soergel–Fabricius *Zs. f. Physik* **163** (1961) 571.

[12] J. Horský and N.V. Mitskievich, *Czech. J. Phys.* **B 39** (1989) 957.

[13] J. Horský and N.V. Mitskievich, *Class. and Quantum Grav.* **7** (1990) 1523.

[14] J. N. Islam, "*Rotating fields in general relativity*" (Cambridge University Press, Cambridge, UK., 1985).

[15] J. N. Islam, N. van den Bergh and P. Wils, *Class. Quantum Grav.* **1** (1984) 705.

[16] P. Klepáč and J. Horský, *Class. Quantum Grav.* **17** (2000) 2547.

[17] S. Kloster and A. Das, *J. Math. Phys.* **18** (1976) 2191.

[18] D. Kramer, *Class. Quantum Grav.* **5** (1988) 393.

[19] N.V. Mitskievich, gr-qc/9606051.

[20] N.V. Mitskievich and G.A. Tsalakou, *Class. Quantum Grav.* **8** (1991) 209.

[21] H. M. Raval and P. C. Vaidya, *Curr. Sci.* **36** (1967) 7.

[22] M. P. Ryan and L. C. Shepley, "*Homogeneous Relativistic Cosmologies*" (Princeton University Press, Princeton, N.J., 1975).

[23] M.M. Som and A. K. Raychaudhuri, *Proc. Roy. Soc. Lond.* **A304** (1968) 81.

[24] H. Stephani, *General Relativity* (Cambridge University Press, Cambridge, UK., 1996), second edition.

[25] H. Stephani, *Ann. Phys.* (Leipzig) **9** (2000) 168.

[26] J. M. Stewart and G.F.R. Ellis, *J. Math. Phys.* **9** (1968) 1072.

[27] G.A. Tsalakou, *Class. Quantum Grav.* **10** (1993) 2191.

[28] G. A. Tsalakou, *Nuovo Cimento* **115 B** (2000) 13.

[29] P. Wils and N. van den Bergh, *Class. Quantum Grav.* **1** (1984) 399.

Index

Prof. Heinz Dehnen: Brief Biography

Heinz Dehnen was born on January 25, 1935 in Essen, Germany. He studied Physics, Mathematics and Chemistry at the University of Freiburg and got his Hauptdiplom in 1959. He obtained the Ph.D. in 1961 under the supervision of Prof. Helmut Hönl at the Institute of Theoretical Physics of the University of Freiburg. Later in 1961 he became Assistant at the same Institut. In 1968 he was substitute Professor at the University of Munich. In 1970 Heinz Dehnen received a call for an Ordinarius Professorship for Theoretical Physics at the University of Konstanz, where he has remained from that time and on. He has been Dean of the Nature Sciences–Mathematics Faculty at the University of Konstanz, and Dean of the Faculty for Physics. Since 1995 up to date he is Studies Dean of the Faculty for Physics also at the University of Konstanz.

Heinz Dehnen began his scientific work in 1958 under the guidance of Prof. Dr. Helmut Hönl at the Institute of Theoretical Physics of the University of Freiburg. The aim of his first investigations was to clarify the role of Mach's principle in the context of General Relativity. Subsequently, he pioneered the research in different topics connected mainly with gravity, which have been his research centre and passion. As examples of them I mention the following: Energy of the gravitational field, relativistic astrophysics and cosmology in the framework of scalar–tensor–theories, quantum field theories on curved space–times, gravitational radiation of thermally excited bodies, cosmic density fluctuations for explaining galaxy formation, unitary spin–gauge theory of gravity, Higgs–field and gravity, bosonic dark matter in the universe, fermionic sector of Kaluza–Klein theories, and induced gravity theories, among others.

In the last times his dominant work consists in the attempt of unification of gravity with the other physical interactions on the basis of unitary gauge groups with respect to quantization. He proposed, starting from a unitary spin gauge theory for gravity, a *spin–Kaluza–theory* where the spin– and isospin–spaces are unified in a higher–dimensional *spin–space*. In this context, it is even possible to begin with totally symmetric fermionic multiplets and to generate the parity violation of the weak interaction subsequently by means of spontaneous symmetry breaking, related with the decoupling of gravity from the unifying theory.

Heinz Dehnen main interests do not concern with single problems of physics but more with the general principles and structures of physical theories.

Heinz Dehnen has been specially significant for Mexico because the long list of his mexican students from the late sixties to the late nineties. Also his scientific collaborations and research projects he has had along the time together with mexicans result important.

For this reason we decided to honour Heinz Dehnen on the occasion of his 65th birthday and to thank him for his contribution to the development of gravity and to the research in Mexico.

ALFREDO MACÍAS

Prof. Dietrich Kramer: Brief Biography

Dietrich Kramer was born in Weida, in a small and nice village in Thuringen, Germany. He studied Physics at the Friedrich Schiller University of Jena and got his Diploma in 1962, at the age of 22. In 1966 he finished his Ph.D. at the same University, working in problems of bispinors in curved spaces.

Since those years curved space–times and General Relativity have been his passions. In 1969 he published together with Gernot Neugabauer a paper in Annalen der Physik (Leipzig), where the definition of the Ernst potentials is given. They discovered and studied the potentials independently at the same time as Fred Ernst did. These potentials are maybe one of the most important results of the last times in General Relativity

The Annalen der Physik Journal was not very well known at the time of the cold war, for this reason only few people were able to read the work of Kramer and Neugebauer. In this work, they proposed a method to generate exact solutions of the axially symmetric exterior fields of the Einstein field equations, by using subgroups and subspaces of the whole isometry group of these space–times. Many other people all over the world developed later this method further. A lot of the well–known exact solutions of the Einstein field equations can be generated using this elegant, powerful and simple method.

Dietrich Kramer got his Habilitation in 1970 with the thesis: *Invariantztransformationen strenger Vakumlösungen in der allgemeine Relativitätstheorie.* In 1980, together with Hans Stephani, Malcolm MacCallum and Eduard Held, he wrote the historical monograph: *Exact Solutions to the Einstein Field Equations*, one of the most important books of General Relativity, indispensable textbook for people working on Differential Geometry, curved spaces or partial differential equations connected with the Einstein equations.

In 1981, together with Gernot Neugabauer, Dietrich Kramer was awarded with the 2nd price of the General Relativity Foundation. Since 1992 he is Professor for Theoretical Physics in Gravitational Theory at Friedrich Schiller University of Jena. Among other positions, Dietrich Kramer has been Dean of the Institute for Theoretical Physics of the same University.

Prof. Kramer has delivered invited talks at many of the most important Universities and international conferences of the world. He maintains good scientific relationships with other groups of relativity, in particular with groups in Madrid and in Mexico at Cinvestav.

We learned a lot from Professor Dietrich Kramer, such as General Relativity, exact solutions, etc. However, his best lecture has been his example as kind and modest person, human being, and physicist. We are honoured for celebrating here his 60th birthday with this conference.

TONATIUH MATOS

Zeitfracht Medien GmbH
Ferdinand-Jühlke-Straße 7
99095 Erfurt, Deutschland
produktsicherheit@kolibri360.de